ELECTRICAL ENGINEERING DEVELOPMENTS

FERROELECTRICS

NEW RESEARCH

Electrical Engineering Developments

Additional books in this series can be found on Nova's website under the Series tab.

Additional E-books in this series can be found on Nova's website under the E-book tab.

Materials Science and Technologies

Additional books in this series can be found on Nova's website under the Series tab.

Additional E-books in this series can be found on Nova's website under the E-book tab.

ELECTRICAL ENGINEERING DEVELOPMENTS

FERROELECTRICS

NEW RESEARCH

XUI LI HUANG
AND
SU LYN MA
EDITORS

Nova Science Publishers, Inc.
New York

For permission to use material from this book please contact us:
Telephone 631-231-7269; Fax 631-231-8175
Web Site: http://www.novapublishers.com

Library of Congress Cataloging-in-Publication Data

ISBN: 978-1-61942-282-7

Published by Nova Science Publishers, Inc. † New York

CONTENTS

PREFACE

This book presents topical research in the study of ferroelectrics. Topics discussed in this compilation include metal ferroelectric insulator semiconductor field effect transistors; resonant and non-resonant microwave absorption studies in multiferroic materials; the nanoscale ferroelectric behavior of GeSbTe films studied by piezoresponse force microscopy; composition design and electrical properties' investigation of silver-containing perovskite lead-free piezoelectric ceramics; low temperature fabrication of high-k and ferroelectric thin films for capacitive applications and coplanar waveguide ferrolectric thin film microwave phase shifters.

Chapter 1 - In this chapter, we introduce metal ferroelectric insulator semiconductor ferroelectric field effect transistor (MFIS FeFET) which is representative research field in device application among the applications of spontaneous polarization of ferroelectric. However, the research flinched from several difficult problems with advanced flash memory and competition with magnetoresistive random access memory (MRAM), phase change random access memory (PRAM), and resistive random access memory (ReRAM) recently. Nevertheless, the research is studied still now due to the unique structure, record in only one transistor and strong application. Our research team describes this chapter as below formation based on our research results.

Chapter 2 - In this work, the effects of temperature and dc magnetic field on resonant and non-resonant microwave absorption measurements at X-band (8.8 9.8 GHz), in the polycrystalline multiferroics $Pb(Fe_{1/2}Nb_{1/2})O_3$ (PFN), $Pb(Fe_{1/2}Ta_{1/2})O_3$ (PFT), $YMnO_3$ (YM) and $YCrO_3$ (YC), are presented. Three techniques are used: electron paramagnetic resonance (EPR), low-field microwave absorption (LFMA) and magnetically modulated microwave absorption spectroscopy (MAMMAS). These studies are performed in the 77-520 K temperature range. Resonant and non-resonant microwave absorption can distinguish between different dissipative dynamics of microwave absorbing centers; these responses showed distinctive features associated with microwave absorption by magnetic moments and/or electric dipoles, to different temperature ranges. Where these measurements are a manifestation of a similar response to electromagnetic absorption, in which similar physical processes take place.

Chapter 3 - For environmental protection reasons, potassium sodium niobate ($(K_{0.5}Na_{0.5})NbO_3$)-based ceramics have been considered as promising alternatives to lead-based materials which are currently used in majority of ultrasonic applications. However, it is very difficult to process the ceramics to full density by ordinary sintering under atmospheric conditions, which deteriorates their properties and restricts their uses in industrial productions. In this chapter, the preparation of KNN-based ceramics with well-densified structure, good piezoelectric and dielectric properties are reported. Novel sintering aids CuO and MnO are used to improve the densification and hence the electrical properties of the ceramics. The additional hardening effects of the sintering aids are also discussed. The piezoelectric properties of the ceramics are further enhanced via the formation of new solid solutions with other ferroelectrics, such as $LiSbO_3$, $LiTaO_3$, $BaTiO_3$ and $Ba(Zr_{0.05}Ti_{0.95})O_3$. After the formation of new solid solutions, the ferroelectric tetragonal-ferroelectric orthorhombic phase transition is shifted to lower temperatures, such that the ceramics contain both the tetragonal and orthorhombic structures at room temperature. Owing to the more possible polarization states arising from the coexistence of the two phases, the piezoelectric and dielectric properties of the ceramics are enhanced significantly.

Chapter 4 - In the past, innovations towards the design of actuators, sensors and ultrasonic transducers were the driving force for the rapid development of piezoelectric ceramics. The high performance piezoelectric activity remains a property confined to lead-based systems, with $Pb(Zr_xTi_{1-x})O_3$ [PZT] as the nominal industry standard. At present, the industry faces the obstacle of environmental regulations, as the high toxicity of lead present in the widely used PZT family (~60%-lead content) of piezoelectrics poses a severe threat to human health. The commercial electronics industry is currently in the process of a transition to lead-free products. Legislations regarding the waste containing lead pollutants have been in force in all member states of European Unions and Japan, as detailed by the directives on waste from electrical and electronic equipment (WEEE), restriction of hazardous substances (RoHS) and end-of life vehicles (ELV). Some states of the United States and China have also followed the trend of lead-free transition and started exploring legislation to reduce the use of lead in electronic industries. It is expected that similar regulations like the EU-RoHS will be established soon in Australia, New Zealand, and Thailand etc.

In order to circumvent the legislated restrictions on the use of toxic lead for civil applications in the near future, bulk lead-free piezoelectric materials have been extensively investigated to explore the potential candidates with electrical properties comparable to their lead-based counterparts. It is no surprise that the search for lead-free piezoelectric ceramics is described as one of the pillars in today's roadmap for ceramics. To date, the reported lead-free piezoelectric ceramics are of perovskite-type, tungsten bronze-type and bismuth layer-structured-type. Perovskite-type lead-free ceramics with an ABO_3 crystal structure seem to be good candidates to replace PZT and have been actively researched in recent years. Two perovskite families of lead-free materials have currently emerged as the most promising candidates to replace PZT: (1) $(Bi_xNa_{1-x})TiO_3$ [BNT] and (2) $(K_xNa_{1-x})NbO_3$ [KNN]. Although no lead-free materials with an overall piezoelectric performance comparable to PZT systems have been found so far, recent studies have shown that both BNT and KNN-based systems do exhibit some advantages over lead-based materials for specific applications.

From the scientific point of view, the intrinsic nature of the high piezoelectric and ferroelectric properties in PZT system are presented in relation to their respective Curie temperatures (T_C) and the existence of a morphotropic phase boundary (MPB). This implication, to some extent, provides a perspective and introduces guidelines on the development of novel lead-free piezoelectrics. In fact, there were very early considerations in the 1970s to replace lead in piezoceramics. Nevertheless, the blossom of searching for lead-free piezoceramics just started in the early 2000's, in particular, the BNT and KNN-based systems, as shown in Figure 1, providing statistics for refereed publications on BNT and KNN -based piezoceramics from 2004 up to now. Current efforts in pursuing and designing lead-free piezoelectric ceramics involve three different approaches: (1) Searching for new lead-free piezoelectric materials empirically, using phenomenological models (structure-property relations) and first principle calculations; (3) Improving the performance of existing lead-free materials by elaborating substitutions and dopants or introducing MPB; (2) Perfecting the existing lead-free materials with textured growth (controlling structural instabilities and extrinsic contributions). Analogous to PZT, enhanced properties are noted for MPB compositions in the BNT binary and ternary systems. The consequence of a ferroelectric to anti-ferroelectric transition well below T_C, however, limits their usefulness. Though comparable with respect to T_C, the high levels of piezoelectricity reported in the KNN family are the result of enhanced polarizability associated with the orthorhombic-tetragonal polymorphic phase transition (PPT) being compositionally shifted downward. Although one can argue that there is room for improvements in bulk, the fact that so far no lead-free ceramics possess similar properties as their lead containing counterparts have stimulated new strategies in the development of novel lead-free piezoelectric materials.

In contrast to the numerous studies on lead-free piezoelectric single crystals and ceramics, literatures tackling with lead-free thin films have been lagging far behind, as demonstrated in Figure 1. The lack of progress in lead-free thin films may be attributed to the challenges associated with preserving volatile A-site cations such as bismuth, sodium and potassium in the films and avoiding significant concentrations of A-site and oxygen vacancies to achieve high-quality thin films. Progress on the growth and characterization of BNT-based and KNN-based thin films, still remains at an infancy stage with rather scattered physical properties. In this sense, the development of high-performance lead-free piezoelectric thin films for MEMS, micro-actuators and transducers applications has proven to be an attractive, albeit challenging, endeavour. A list of BNT and KNN-based thin films in open literatures and their properties is presented in Table 1 and 2, respectively.

Chapter 5 - Ferroelectric and mutiferroic materials have been studied from the viewpoint of pure and applied physics. For the applications, a study from the viewpoint of materials science will be important in the future. Examples of materials science study are substitution of elements, selection of substrate in thin films and in-situ observation. Nano-scale domain configurations of multiferroic $BiFeO_3$ thin films and their dynamic response to the externally applied fields have been investigated by piezoresponse force microscopy. High-resolution piezoresponse images reveal that nano-scale domain walls are approximately with ~2 nm in width. In this article, we focus on $BiFeO_3$ as mutiferroic materials and as ferroelectric materials. We will present our results about electric and magnetic properties, and scanning probe microscopy observations. Local polarization reversal and retention behavior will also be presented and discussed.

Chapter 6 - Ferroelectric materials offer a wide range of useful properties to develop applications for the electronic engineer. The tendency to develop smaller ferroelectric devices implies to have a deep understanding of ferroelectric materials properties at the nanoscale level. Piezoresponse force microscopy (PFM) in a scanning probe microscopy mode (SPM), shows to be an ideal non-destructive technique to study the nanoscale behavior of ferroelectric materials due to its high resolution in the range of nanometers and to its ability to manipulate ferroelectric domains. An improvement of this technique developed to investigate ferroelectric materials with small piezoelectric coefficient is the use of contact resonance

PFM, which allows an amplification of the piezoresponse signal in proportion to the Q-value (10-100) of the cantilever-sample contact resonance. Moreover, switching spectroscopy PFM allows hysteresis loop measurements in a local domain achieving quantitative information about the switching processes at nanoscale resolution, including coercive bias and work of switching. The first part of this chapter will be a brief description about the principles, theoretical background, recent developments and applications of piezoresponse force microscopy. Furthermore, a new development of PFM will be described for the investigation of chalcogenide thin films, GeSbTe alloys, which have application in data storage memory. These alloys are non-typical ferroelectric materials with low electrical resistivity, which make difficult to investigate their ferroelectric properties by other conventional techniques. Different characterization techniques were employed to study GeTe, $Ge_4Sb_1Te_5$ and $Ge_2Sb_2Te_5$ thin films: contact resonance PFM and switching spectroscopy PFM for the investigation of ferroelectric domain structure and impedance spectroscopy technique to obtain Curie temperatures. Because of their ferroelectric properties, these chalcogenide materials can find applications in the super-resolution optical disks where the resolution is far smaller than the diffraction limit, for achieving ultrahigh storage density.

Chapter 7 - Recent theoretical calculations indicate that the situation of Ag in the A site of ABO_3 perovskite is very similar to that of Pb. Good piezoelectrics may be promisingly realized in the Ag-containing perovskite system. In recent years, we have developed some new Ag-containing $(Bi_{0.5}Na_{0.5})TiO_3$ (BNT)- and $(K_{0.5}Na_{0.5})NbO_3$ (KNN)-based perovskite lead-free piezoelectric ceramics, and the piezoelectric and ferroelectric properties of the ceramics were investigated. The design of BNT-based ceramics focuses on the effects of Ag ion substitution and the effects of $Ag_{0.9}Li_{0.1}NbO_3$ (ALN10) addition on BNT–$BaTiO_3$ (BNT–BT) system. The design of KNN-based ceramics focuses on the effects of Ag ion substitution on KNN–$LiTaO_3$ system and the effects of $AgTaO_3$ or $AgSbO_3$ addition on KNN–$LiNbO_3$ system. All the ceramics were prepared by the conventional ceramic technique. Results show that both the substitution of Ag^+ for Na^+ and the addition of ALN10 can improve the piezoelectric and ferroelectric properties of the BNT–BT ceramics. The temperature dependence of electrical properties shows that the depolarization temperature (T_d) of the BNT-based ceramics shifts towards lower temperatures with increasing amount of Ag or ALN10. For the KNN-based ceramics, the orthorhombic–tetragonal polymorphic phase transition (PPT) was modified by the substitution of Ag^+ for A-site ions and the addition of $AgTaO_3$ or $AgSbO_3$. The KNN-based ceramics with a desired PPT near room temperature exhibit the enhanced piezoelectric and ferroelectric properties. In addition, the Curie temperature (T_c) of KNN-based ceramics increases with increasing amount of Ag. Moreover, we also synthesized alkaline elements (Li, Na, K)-modified $AgNbO_3$ (AN)-based ceramics by the conventional ceramic technique under air atmosphere. The results show that AN-based solid solution with pure perovskite structure can be available by optimizing the compositions and using a proper processing, such as low-temperature sintering. A high T_c of 340 °C and a large remnant polarization (P_r) of 14.5 ìC/cm^2 were obtained for the CuO-doped $(Ag_{0.75}Li_{0.1}Na_{0.1}K_{0.05})NbO_3$ ceramics. These results show that Ag-containing perovskite is a promising candidate for lead-free piezoelectric materials.

Chapter 8 - Over the past few years, environmental concerns have stimulated interest in developing lead-free ferroelectric and piezoelectric ceramic compositions as replacements for lead zirconate titanate (PZT). Sodium potassium niobate (NKN) based allied solid solutions have been investigated as one of the most promising alternative candidates for PZT systems.

Given the technological potential of NKN-based compositions, recently, a number of synthesis routes have been developed to synthesize thin films and powders of NKN. Due to its distinct advantages, e.g., lower temperature processing, homogeneous mixing at the molecular level, large area deposition etc, sol-gel and chemical solution deposition methods have found wide applicability in modern advanced technological applications for thin films. The success of these efforts primarily depends on the stability and homogeneity of the NKN precursor solutions, stability of the film substrate with respect to the crystallisation temperatures (interfacial reactions, if any), extent of reactivity of the alkaline earth metal ions (Na^+, K^+) when exposed to high-temperature crystallisation and in ambient conditions. This article reviews the recent developments in NKN thin films via chemical solution-based routes and also sheds light on the fundamental issues involved in the materials chemistry of the NKN system when approached via chemical solution deposition (CSD) routes.

Chapter 9 - A low temperature process can ensure the applicability of high-*k* (high dielectric permittivity) and ferroelectric thin films in above-IC technologies and at the same time benefit from a low thermal budget. Here we report on the growth mechanism and dielectric properties of several important high-*k* and ferroelectric thin films, such as $Bi_{24}Fe_2O_{39}$ (BFO), $Ba_{0.7}Sr_{0.3}TiO_3$ (BST), etc, all of which were successfully prepared by a low temperature process composed of sputtering (ion beam sputtering or rf magnetron sputtering) and postdeposition annealing. The effect of annealing temperature on the structural and dielectric properties of BST thin films was systematically investigated. A sharp transition in their tunable dielectric behaviours was observed in good agreement with the evolution of crystal structure from amorphous to crystalline phase. It was demonstrated that the perovskite phase could crystallize in BST films at very low temperatures (~450 °C), and the low temperature processed BST films showed very good dielectric properties with a high dielectric tunability, typically 66.2% at E (electric field intensity) = 800 kV/cm, and a low dielectric loss, roughly 0.01 at zero field. Similarly, high-*k* BFO thin films were prepared on technologically desirable platinized silicon substrates by the aforementioned low temperature process ($\leq$450 °C). The BFO film not only exhibited high dielectric permittivity (ε_r=113) and relatively low loss tangent ($\tan\delta$=0.012), but also showed a fairly small quadratic voltage coefficient of capacitance (α~800 ppm/V^2) and a small temperature coefficient of capacitance (α_T~790 ppm/°C). The improvement of low temperature crystallization ability in these high-*k* and ferroelectric films is believed to stem from an enhancement of adatom activity by an intensified bombardment of energetic species in the sputtering process. The attractive dielectric and electrical properties obtained make the low temperature processed high-*k* and ferroelectric thin films promising candidates for various capacitive applications in silicon-based integrated circuits and devices.

Chapter 10 - In conventional ferroelectric materials the covalent bond between the anions and cations or the orbital hybridization of electrons on both ions plays a key role in establishing the dipolar arrangement. In contrast, electronic ferroelectricity in which the electric dipole depends on electron correlations rather than the covalency, has been recently proposed and experimentally verified. In this chapter, we report on recent progress in the study of electronic ferroelectricity in a series of charge ordered oxides. In addition to a general description, we focus on the electronic ferroelectricity induced by different forms of charge ordering in two specific systems: (i) strongly charge frustrated system RFe_2O_4 (R=Lu,Yb), and (ii) charge ordered perovskite manganites of the type $(R_{1-x}Ca_x)MnO_3$ (R=La,Pr). The coexistence of electronic ferroelectricity and magnetism in these systems leads to a peculiar multiferroics. Furthermore, the charge ordering can be tuned by applying electric field, laser irradiation, etc., which results in a variety of interesting effects. Especially, the magnetoelectric effects and giant dielectric tunability accompanied with electronic ferroelectricity could be promising for many applications.

Chapter 11 - Ferroelectric thin film, such as barium strontium titanate ($Ba_xSr_{1-x}TiO_3$ or BST), has shown great promise for the fabrication of tunable RF and microwave components. An important tunable microwave device using ferroelectric film is a wideband phase shifter which is the most important component of the phased array antenna. Phase shifters with low insertion loss, low drive power, proper power handling capability, continuous tunability, good impedance match, fast response speed, and low production cost are the key to the development of lightweight phased array antennas. This chapter looks at microwave phase shifter applications of ferroelectric thin films. It starts with the introduction of microwave phase shifter and coplanar waveguide transmission lines, with emphasis laid on the significance and applications of ferroelectric materials. The basic theory of distributed transmission line phase shifters is then discussed. The physical layout design, device fabrication, and microwave characterization of distributed coplanar waveguide transmission line phase shifter using ferroelectric thin film varactors with parallel-plate electrodes are presented in the final two sections.

Chapter 12 - The study in the ferroelectric photovoltaic (PV) effect has been greatly achieved and becomes a key direction in fields of condensed physics and materials sciences. The interest in developing ferroelectric PV devices with specific properties has drastically increased due to the rapidly growing needs in solar energy. In this field, advances in ferroelectric PV cells based on $BiFeO_3$ (BFO) play an important role, and the extended research performed around the world in order to optimize the properties of the ferroelectrics is required. This paper focuses on common and important $BiFeO_3$ ferroelectric materials used for PV effect, and presents a review of the latest developments.

In: Ferroelectrics: New Research
Editors: Xui Li Huang and Su Lyn Ma
ISBN: 978-1-61942-282-7

Chapter 1

APPLICATION OF FERROELECTRICS: METAL FERROELECTRIC INSULATOR SEMICONDUCTOR FIELD EFFECT TRANSISTOR

Hyung-Ho Park* and Hong-Sub Lee
Department of Materials Science and Engineering,
Yonsei University, Seodaemun-Ku, Seoul, Korea

ABSTRACT

In this chapter, we introduce metal ferroelectric insulator semiconductor ferroelectric field effect transistor (MFIS FeFET) which is representative research field in device application among the applications of spontaneous polarization of ferroelectric. However, the research flinched from several difficult problems with advanced flash memory and competition with magnetoresistive random access memory (MRAM), phase change random access memory (PRAM), and resistive random access memory (ReRAM) recently. Nevertheless, the research is studied still now due to the unique structure, record in only one transistor and strong application. Our research team describes this chapter as below formation based on our research results.

1. INTRODUCTION OF FERROELECTRIC FIELD EFFECT TRANSISTOR

A spontaneous polarization of ferroelectric materials is applied to various application fields. One of the applications is ferroelectric random access memory (FeRAM) in next generation nonvolatile memory (NVM). It has been studied by the ideal operation principle with strong candidate memory among the various NVMs, for a substitution of dynamic random access memory (DRAM) and flash memory. But now, the study slowly progress by several difficult problems.

FeRAM can be divided to destructive read out (DRO) type and non-destructive read out (NDRO) type largely. DRO type fabricated by 1transistor (1T)/1capacitor (1C) or 2T/2C and

* Email: hhpark@yonsei.ac.kr.

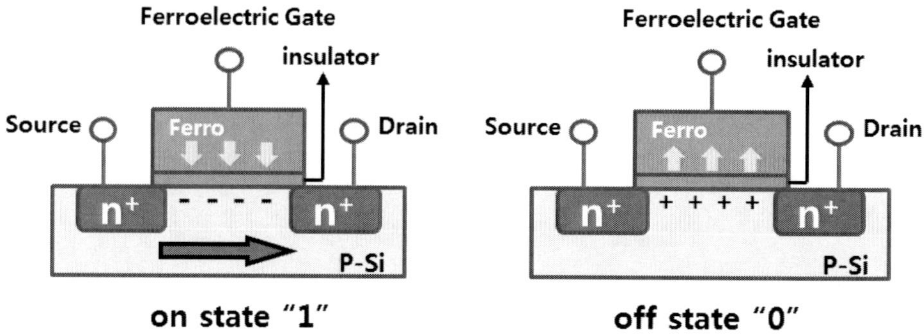

Figure 1. A scheme of the operation mechanism in NDRO FeRAM (FeFET).

the operation method is same with DRAM. Consequently, only a capacitor of DRAM substituted by ferroelectric material. In this DRO type FeRAM, data erase in read process as the name of DRO. In contrast, NDRO type FeRAM fabricated by only one ferroelectric field effect transistor (FeFET) in which a gate of the transistor fabricated by metal/ferroelectric material/insulator/semiconductor (MFIS). Remnant polarization by applied electric field to the ferroelectric gate maintains formed a channel between a source and a drain as Figure 1. And we can confirm the state through a resistivity of the transistor without destructive of recorded data. If the ideal FeFET be realized, it can substitute main memory devices of the day. The FeFET can be applied to a logic device as field-programmable gate array (FPGA). Consequently, it has high potential from nonvolatile memorized transistor [1-4].

In spite of these merits, FeFET stays in a study of one FeFET properties from several technical difficulty and unavoidable structural problems.

So far, ferroelectric materials has been studied such as lead zirconate titanate [$Pb(Zr_{1-x}Ti_x)O_3$] (PZT), strontium bismuth tantalate [$SrBi_2Ta_2O_9$] (SBT), lanthanum-substituted bismuth titanate [$Bi_{4-x}La_xTi_3O_{12}$] (BLT) and pseudo pyrochlore [$(Nd, Ce)_2Ti_2O_7$] (NT, CT) for FeFET [5-9]. Nevertheless, the materials have downside of more than one in properties as fatigue, leakage current, permittivity, phase formation temperature, etc.

Furthermore, because ferroelectric material cannot be formed directly on a Si wafer, an insulator layer is inserted and it reduces ferroelectric properties. Also, there is difficult barrier as high temperature annealing process of ferroelectric material and etching process for patterning. For the application of ferroelectric film to integrated devices, patterning process should be carried out.

Ion etching and ion milling are presently the most widely applied processes for fine patterning of ferroelectric films. However, these processes tend to degrade the properties of the film. During dry etching, changes in the physical and electrical properties of the film due to reactive free radicals and ions in the plasma have been occurred. Furthermore, there is surface grain rounding due to the different etching rate between grain and its boundary during dry etching.

The unavoidable structural problem of FeFET can be divided into retention and disturbance, largely.

Retention

In spite of FeFET purpose the nonvolatile memory, the retention property is not good. That means ferroelectric gate cannot hold the polling state for a long time. As a result, transistor cannot maintain the data.

A cause of the retention problem of a ferroelectric gate is divided into leakage current and depolarization field [10].

Leakage Current

In MFIS structure, charge of leakage current of ferroelectric layer and insulator layer penetrates into inside of ferroelectric material and interface of ferroelectric layer and insulator layer. The penetrated charge compensates ferroelectric polarization. As the result, ferroelectric gate cannot maintain the state and recorded data lose.

Depolarization Field

Ferroelectric layer is sandwiched between metal layer and insulator layer in MFIS structure. The condition induces disproportion of compensated charge in top and bottom of ferroelectric layer. Depolarization field is induced from the disproportion of compensated charge. Therefore, a difference of the capacitance between a ferroelectric material and an insulator material is proportional to the depolarization field.

In MFIS structure, the polarization of polarized ferroelectric layer is decreased by depolarization field at early stage. After the initial decrease, a decrease by leakage current is induced. In other words, after the section of initial decrease, the polarization is offset by penetrating charge per second. For this retention problem, other structures such as metal/ferroelectric/metal/insulator/semiconductor (MFMIS) and metal/insulator/ ferroelectric/ insulator/semiconductor (MIFIS) have been tried, but could not solve the two problems [11, 12].

Disturbance

For use to FeFET memory, the FeFET cells are arrayed on two dimensions. Different from a DRAM, NDRO FeRAM has no cell selection transistor. Therefore, for writing operation to selected cell, we must operate through distribution of writing voltage to top and bottom of ferroelectric gate. For this operation, cells on same word line with the selected cell take half of the writing voltage. In this process, a decrease of polarized domains of ferroelectric is induced [13].

Above mentioned problems as retention and disturbance require a solution of point of materials. First of all, FeFET needs secure retention property and then disturbance should be improved. Therefore, we will show diverse studies of ferroelectric materials and MFIS structure for retention of FeFET in this chapter.

2. Study for FeFET

2.1. Improvement of Ferroelectric Materials for MFIS FeFET

2.1.1. Photochemical Metal-Organic Deposition (PMOD)

A photosensitive precursor ML_n is dissolved in a volatile solvent. A few drops of this solution are dispensed onto a substrate. The substrate is rotated which results in the formation of a film of the complex. The film is exposed by UV light. The exposure of the film results in a photochemical reaction. Upon UV-irradiation, the volatile ligands are ejected from the precursor leaving metal on the substrate (Eq. 1). In the air, the metal reacts with oxygen forming amorphous M_xO_y on the substrate (Eq. 1).

$$ML_n \xrightarrow{h\nu} M + nL \qquad \text{(Eq. 1)}$$

$$M + O_2 \longrightarrow M_xO_y \qquad \text{(Eq. 2)}$$

Compared with sputtering method, there are the following advantages to PMOD. PMOD does not need a high vacuum system. The cost of the equipment used is much lower than sputtering. The PMOD process also enables photoresist-free lithographic processing, and the existing photolithography equipment in the electronic industry can be used in the PMOD process. The PMOD process is also a lot simpler to most other thin film deposition processes in preparing mixed metal oxide thin films. In order to prepare a mixed metal oxide thin film by PMOD, the two metal precursors are dissolved in a solvent in a stoichiometric ratio. A precursor film of the mixed metal precursors is obtained by spin-coating and upon photolysis, a mixed metal oxide thin film with a ratio between the two metals corresponding to the original precursor ratio can be obtained. There is a comparison between conventional etching and photoresist-free etching fabricated by the PMOD process as shown in Figure 2. As you see, many steps including photoresist coating and dry etching steps can be omissible by direct-patterning. Furthermore, we need just solvent washing for removing the unexposed part of the film to obtain a pattern structure.

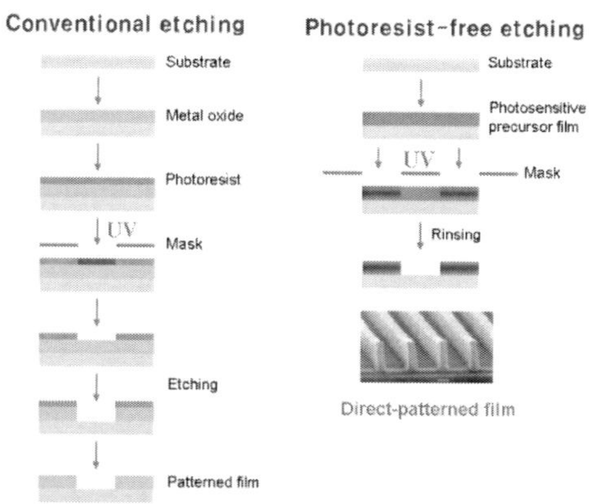

Figure 2. Comparison between conventional etching and photoresist-free etching fabricated by PMOD process.

The PMOD process requires starting precursors with specific qualities. The first requirement for a successful precursor for the PMOD process is that it should be photosensitive since the driving force of the process is via UV-irradiation. The precursor also has to be able to form amorphous films on the substrate. This is because during the photolysis, the precursor decomposes and the ligand or its fragments diffuse out of the film. The precursor also has to be thermally stable in order to be able to perform photolithography with the PMOD process.

A wide range of metal-organic complexes have been studied, including metal carboxylates, metal acetylacetonates, metal carbonyls as well as metal complexes with amines and phosphines [15-20]. Among the classes of metal-organic complexes being studied, widely used metal-organic complexes for photochemistry are metal carboxylates because metal carboxylates are easily able to undergo charge transfer in the UV region upon UV-irradiation [21]. Transitions such as ligand-to-metal charge transfer (LMCT) upon UV-irradiation are able to trigger the photolysis of metal carboxylates, leading to the fragmentation of the precursor molecule [22]. The general structures of various metal carboxylate ligands are illustrated in Figure 3 [21,22]. Among the classes of metal carboxylates, 2-ethylhexanoate ligand is widely used for the photosensitive starting precursor due to photosensitivity and thermal stability. Also, other modes of ligands such as monodentate, bidentate, and bridging are also commonly used in metal carboxylates due to the compatibility with used solvent [21,22]. Due to the above reasons, chosen ligand for this thesis is 2-ethylhexanoate. In this ligand, as shown in Figure 3 (a), the long alkyl chain controls the intermolecular interactions and affects physical properties such as solubility in organic solvents and resistance to crystallization. The low symmetry results in a higher kinetic barrier to crystallization. The 2-ethylhexanoate ligand can be also tailored to be soluble in an organic solvent that is compatible with spin-coating. Thus, amorphous thin films with good optical quality can be deposited. Moreover, the 2-ethylhexanoate ligand has intense charge transfer transitions that control the photochemical reactivity of the precursor films. Bravo-Vasquez et al. reported that the 2-ethylhexanoate ligand has ligand to metal charge transfer (LMCT) transitions in the range from 200 to 400 nm [22]. Hence, in this thesis, 365 nm UV light was exposed to metal-organic precursors containing 2-ethylhexanoate ligand for the photoresist-free direct-patterning of various metal oxides, metal nanostructures, and nano-hybrid metal oxide films.

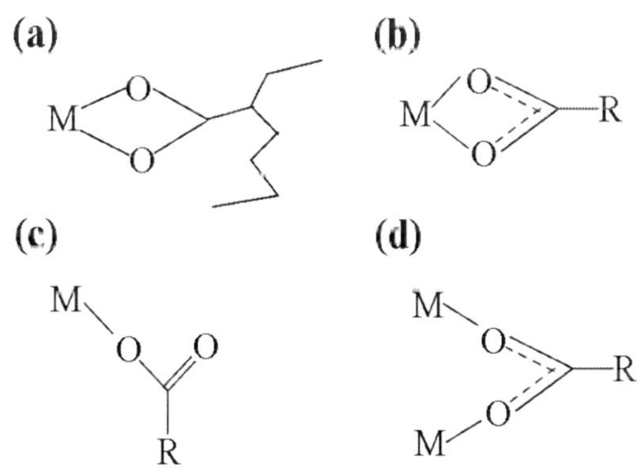

Figure 3. Structures of metal carboxylate illustrating (a) 2-ethylhexanoate, (b) bidentate, (c) monodentate, and (d) bridging ligands (M denotes the metal center while R represents an organic group).

2.1.1.1. Application of PMOD to the Formation of PZT Thin Films

Ferroelectric materials such as PZT have been studied increasingly in recent years, as their suitability for use in non-volatile memory and other microelectromechanical system application has been recognized [23,24]. In this part, a direct patterning method for the lithographic deposition of materials was introduced using PMOD. The properties of self-patterned PZT film with UV-irradiation process were studied.

The precursors used for the photochemical production of PZT film were 2-ethylhexanoates of lead(II) and zirconyl(IV), and titanium-isopropoxides. The composition ratio of Zr/Ti was fixed as 52/48 and 10% of extra lead was added for the compensation of lead-loss during high temperature anneal [25].The PZT precursor dissolved in 67.3 wt.%-solvent was spin-coated at 2000 rpm for 30 s on two types of substrates. One is chemically cleaned Pt(111)/Ti/SiO_2/Si(100) substrate for the analysis of ferroelectric properties of PZT film and the other is p-Si(100) wafer for monitoring the evolution of organic bonds in the film during the photochemical reaction. Spin-coated PZT films were exposed to UV light with 365-nm wavelength for 30 min. The power of UV lamp was 1000 W. Then the film was washed with hexane for removing the unexposed area of the film. The pre-annealing of patterned film was done at 400°C for 10 min by direct insertion to a furnace under O_2 atmosphere. This procedure was repeated when a thick PZT film was required. The final annealing was carried out at 650°C for 30 min by direct insertion to a furnace under O_2 atmosphere. Fourier transform infra-red spectroscopy was used to monitor the evolution of organic groups in PZT precursor during photochemical reaction. Phase formation and crystallization were monitored using X-ray diffraction (XRD).The electrical properties of PZT film, a 370-nm-thick film, was prepared by five times repeated coating. Capacitance voltage (C–V) and current–voltage (I–V) characteristics were obtained using HP4284 and HP4145B (Hewlett Packard), respectively. A standard ferroelectric tester RT66A (Radiant Technology) was used to obtain the ferroelectric properties.

PZT precursor solution for PMOD was spin-coated on p-Si(100) wafer. Figure 4 corresponds to the FT-IR spectra in 400–4000/cm range of spin-coated PZT precursor film according to the variation of UV exposure time.

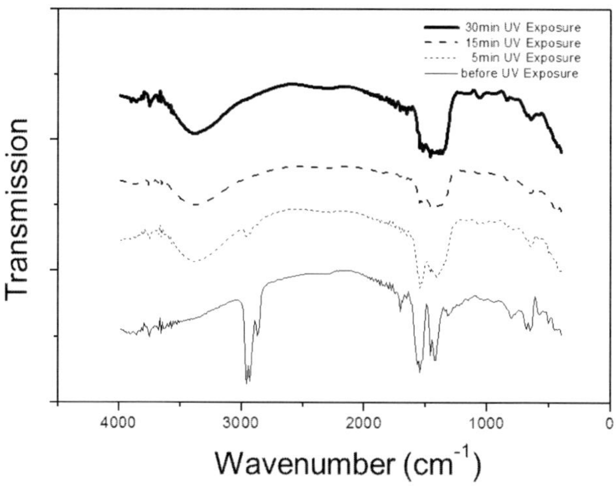

Figure 4. Changes in FT-IR spectra of the spin-coated PZT precursor film upon photolysis for 0, 5, 15 and 30 min with 365-nm UV light.

Figure 5. SEM micrograph of self-patterned PZT film.

The absorption band at 2950 and 2850/cm are assigned to the stretching vibrational modes of alkyl ($–CH_3$) and alkene ($–CH_2$) groups and they decrease with UV exposure. The two bands corresponding to the organic groups disappear after 15 min UV exposure. In the region of 1400 and 1500/cm two broad bands are observed and the band at 1400/cm is believed to be associated with lead oxide, while the other band at 1550/cm is presumably associated with the amorphous metal oxide products.

Therefore, after 15 min UV exposure, the organic groups could be removed completely from PZT precursor film. The possibility of self-patterning by removing the unexposed area to UV using hexane-rinsing was examined and the result is given in Figure 5.

In the SEM micrograph, the bright area corresponds to PZT film and dark area to Si substrate, respectively. 350-nm-thick PZT film could be self-patterned successfully. For the self-patterning of PZT film, the film has to be exposed for 30 min, not less. If film is UV-exposed for less than 30 min, then all the film is removed by hexane.

Even though just after 15 min exposure to UV, C–H organic groups disappeared completely in PZT film as shown in Figure 4, which may signify the completion of photochemical reaction in the film.

The increase of resistance of photochemically reacted film against the dissolution in hexane needed an extra 15 min exposure to UV. However, as shown in Figure 4, an extra 15-min exposure was needed for the completion of micro-structural arrangement including local network bonding.

Surely, exposure time to UV for developing an image should depend on film thickness, UV power, distance between film and UV and so on. With this micrograph, we knew that self-patterning of PZT film could be done nicely.

Figure 6 shows XRD patterns of PZT films coated on $Pt(111)/Ti/SiO_2/Si$ substrate. Figure 6(a) corresponds to UV-exposed PZT film (spin-coated PZT precursor film after the completion of photochemical reaction by UV exposure) after pre-anneal at 400°C for 10 min and Figure 6(b) was obtained after final anneal treatment of the pre-annealed PZT film at 650°C for 30 min under O_2 atmosphere.

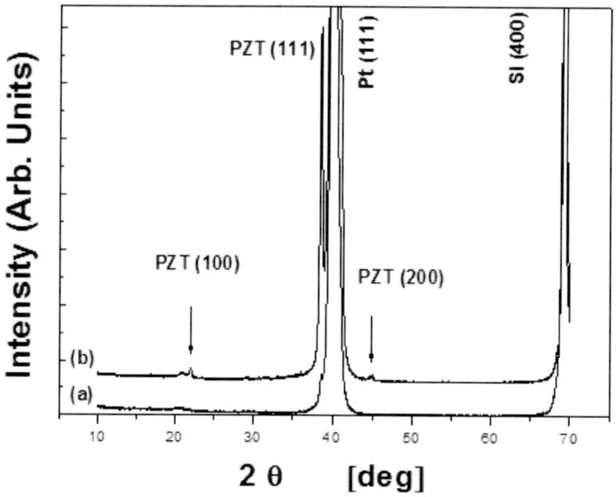

Figure 6. XRD spectra of PZT films after (a) pre-anneal at 400°C and (b) final anneal at 650°C.

For the UV-exposed PZT film, an introduction of pre-anneal treatment was for the complete removal of fragmented organic groups that were originated from the decomposition of precursors during photochemical reaction, however, without the formation of pyrochlore phase for PZT film. From Figure 6(a), it can be clearly seen that neither pyrochlore nor another second phase was formed during the pre-anneal treatment at 400°C. After final anneal of pre-annealed PZT film at 650°C for 30 min, the film was well crystallized with perovskite structure with preferred orientation of (111) as shown in Figure 6(b).This highly preferred orientation meant that the nucleation of PZT was strongly affected from Pt substrate [26]. If PZT film is crystallized on Pt (111) substrate, (111) preferred orientation of PZT film should be dominant due to the lattice matching between Pt and PZT [27]. From this localization of nucleation sites for PZT film, i.e. mainly on Pt substrate, it could be said that through photochemical reaction and pre-anneal treatment the PZT film became free from organic group and compositionally pure. This is why heterogeneous nucleation did not occur entirely for the PZT film. Figure 7 shows polarization-voltage (P–E) hysteresis loops of 370-nm-thick PZT films annealed at 650°C for 30 min under O_2 atmosphere with varying applied voltage (8, 10 and 15 V).The film was prepared by 5 times of repeated coating. The value of remnant polarization (P_r) measured by applying 15 V was approximately 34 μC/cm^2, while the values of coercive field (E_c) were approximately -151 and 108 kV/cm. In the case of conventional 480-nm-thick PZT film prepared by sol–gel procedure, P_r and E_c values were obtained at approximately 30 μC/cm^2 and 47 kV/cm with 15 V of applied voltage [28]. The values of P_r and E_c for self-patterned PZT film were higher than those of conventional PZT film. In our case, due to shift of the loops to the left, P–E hysteresis became asymmetrical. The asymmetry has been attributed to the presence of micro-defects pinning domains at the electrodes. This interaction occurs at two distinct stages. The first is the interaction of the electrode when it transforms from pyrochlore into perovskite. The second is that the one-way evolution of decomposed organic groups from bottom to top during photolysis and pre-anneal treatment and the defect densities at upper and lower electrode-interface should be different. Therefore, the shift in hysteresis loop can be addressed to the changes in defect density and their location in the films. The I–V characteristic of the PZT film is given in Figure 8.

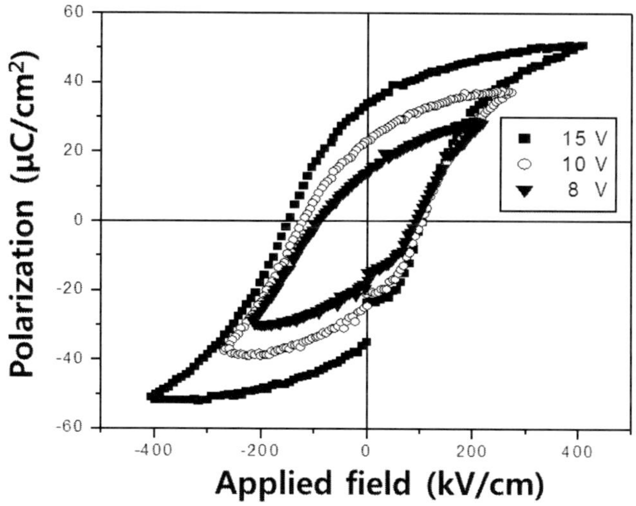

Figure 7. P-E hysteresis loops of PZT film after anneal at 650°C for 30 min with various applied voltages (8, 10 and 15 V).

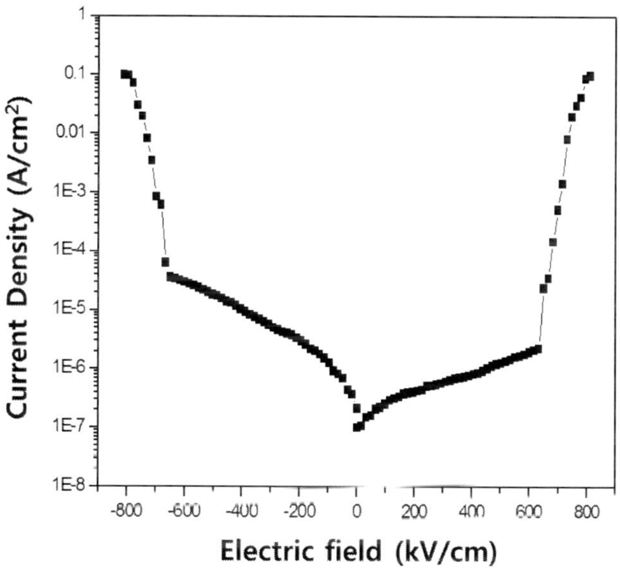

Figure 8. I-V characteristic of the PZT film after anneal at 650°C for 30 min.

The leakage current is measured almost horizontally according to the applied voltage until 640 kV/cm and after dielectric breakdown occurred. In case of conventional PZT film, dielectric breakdown occurred at 400 kV/cm and the leakage current density varied in the range of approximately 10^{-7}–10^{-6} A/cm^2 [29]. Even though self-patterned PZT film was a little more leaky than that of conventional PZT film, the electric field of dielectric breakdown was higher. The I–V characteristic also shows the voltage polarity dependence [30].

When negative bias is applied, the leakage current density varied in the range of approximately 10^{-7}– 10^{-4} A/cm^2. In case of positive bias, it was in the range of approximately

10^{-7}–10^{-5} A/cm^2. At the same magnitude of bias voltage, it was revealed that leakage current density obtained with positive bias is one order lower than that with negative bias. The major leakage current source of crystalline PZT has been known to be Pb or O vacancies created during high temperature anneal for crystallization [31]. However, in this case, micro-defects, such as micro-voids and impurity elements, responsible for the asymmetric behavior of P–E and C–V, might be the major source of the leakage. From the above results, it was revealed that the removal of organic groups during photochemical reaction could induce the generation of micro-defects. The control of photochemical reaction and pre-anneal treatment could bring the amelioration of ferroelectric properties of self-patternable PZT film.

PZT film was successfully formed using photosensitive starting precursors by PMOD and self-patterned without using photoresist and dry etching procedure. Self-patterning of the film could be accomplished after UV exposure and removal of unexposed area of the film by rinsing with hexane. RTA treatment was carried out at 650°C to crystallize 370-nm-thick PZT film with perovskite structure. The crystallization of PZT film on Pt (111) substrate showed highly (111) preferred orientational growth of the film. PZT film with perovskite structure gave a well-developed P–E hysteresis behavior but slightly shifted to a negative bias direction. With leakage current measurement, a more leaky behavior was found with negative applied bias. From the above observations, the electrical properties of PZT film by PMOD seemed to be strongly influenced by the presence of micro-defects that might be formed during the evolution of organic groups under UV exposure.

2.1.1.2. Application of PMOD to the Formation of $Bi_{3.25}La_{0.75}Ti_3O_{12}$ Thin Films

Ferroelectric thin films based on layered perovskite structures have attracted great attention for FeFET due to superior properties, such as high fatigue resistance with a Pt electrode [32,33]. Among related materials of interest, SBT and BLT films have been intensively investigated. Recently, a BLT film has been proposed by Park et al. and Lee et al. [34,35] as a promising ferroelectric material due to its larger polarization compared to SBT and much better fatigue behavior when compared to conventional lead-based materials. Recently, Kim et al. reported that the Ar ion bombardment and chemical reaction have a degrading effect on ferroelectric properties, the destruction of crystalline structure of BLT films and nonvolatile etch by-products such as La-chlorides [35,36]. Here, we study a direct-patterning method for lithographic deposition of BLT films by PMOD, as well as the microstructure and ferroelectric properties of direct-patternable BLT films.

The precursors used for the photochemical production of BLT films were bismuth 2-ethylhexanoate [$Bi(C_7H_{15}COO)_3$], lanthanum 2-ethylhexanoate [$La(C_7H_{15}COO)_3$] and titanium isopropoxide [$Ti[OCH(CH_3)_2]_4$. A 10% excess of bismuth 2-ethylhexanoate was added to compensate for the loss of Bi during high temperature annealing [37]. The BLT precursors dissolved in hexane were spin-coated at 2000 rpm for 30 s on two different kinds of substrate.

One substrate was a chemically cleaned Pt(111)/TiO_2/SiO_2/Si(100) substrate for measuring the microstructure and ferroelectric properties, and the other substrate was p-Si(100) for characterization of the direct-patterning process. A 200-nm thick direct-patterned BLT film by PMOD was obtained after UV exposure.

The precursor films were lithographically exposed for 15 min using the output of a 1000 W mercury arc lamp that was passed through a chrome mask. The exposed film was rinsed with hexane to develop the latent negative pattern. The resultant film was annealed for 10 min

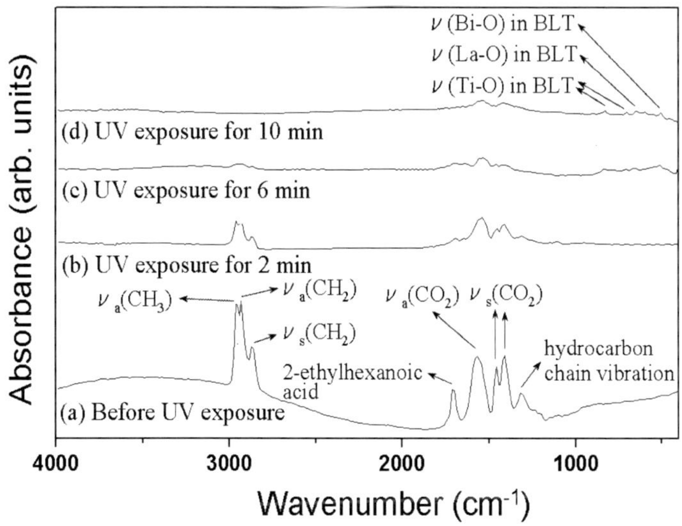

Figure 9. (a)Initiation and (b-d)propagation of M(II) 2-ethylhexanoate during UV exposure(→ stands for the direction of one electron's charge transfer and →stands for the direction of two electrons' charge transfer).

at 400°C. This process was repeated for each layer. The entire structure of multilayer films was annealed at 550, 600, 650 and 700°C for 1 h under ambient O_2.

Fourier transform-infrared (FT-IR) spectroscopy was used to monitor evolution of the organic groups in the photosensitive BLT precursor films during the photochemical reaction. The phase formation and crystallinity of the films were studied with an XRD using Cu Kα radiation. The surface microstructure and direct-patterned film were observed using scanning electron microscopy (SEM) and optical microscopy. A 150-nm thick Pt top electrode with a 100-μm radius was sputter deposited for P-Eand capacitance-voltage (C-V) characterization of the films.

The BLT precursor solution was spin-coated on p-Si(100) wafers. Figure 9 shows the FT-IR spectra (400–4000 cm^{-1}) of spin-coated BLT precursor films prepared using different UV exposure times. The absorption bands at 2953, 2932, and 2861 cm^{-1} were assigned to asymmetric and symmetric stretching vibrations of alkyl (-CH_3) and alkene (-CH_2) groups and their absorption intensities decreased with increasing UV exposure [38]. In the region between 1100 and 1800 cm^{-1}, absorption bands associated with the ethylhexanoate ligand containing CO_2, ethylhexanoic acid, and hydrocarbon chains were observed [38]. However, all the absorption bands from the organic groups nearly disappeared after UV exposure for 10 min, indicating decomposition and removal of organic groups in the spin-coated BLT precursor films. After 6 min of UV exposure, absorption bands appreciably appeared at 707 and 829 cm^{-1}, corresponding to Ti-O vibrations [39].

In addition, absorption bands appeared at 647 and 509 cm^{-1}, which correspond to La-O and Bi-O vibrations, respectively [39, 40]. From these observations, the decomposition and removal of organic groups from the spin-coated BLT precursor films were possible after 10 min of UV exposure in the PMOD process. UV exposure to the photosensitive BLT precursor film results in the cleavage of organic ligands from the metal-organic precursors and converts the film to metal oxide under ambient air and room temperature conditions. The possibility of direct-patterning spin-coated BLT precursor films was examined and the result is given in Figure 10. To obtain the image, a spin-coated BLT film was exposed to UV for 15min and then rinsed in hexane. The dark area corresponds to the direct-patterned BLT film, and the bright area corresponds to the Si substrate. Based on the well-defined width in the direct-patterned BLT film, micron-scale patterning was obtained by lithographicpatterning using PMOD.

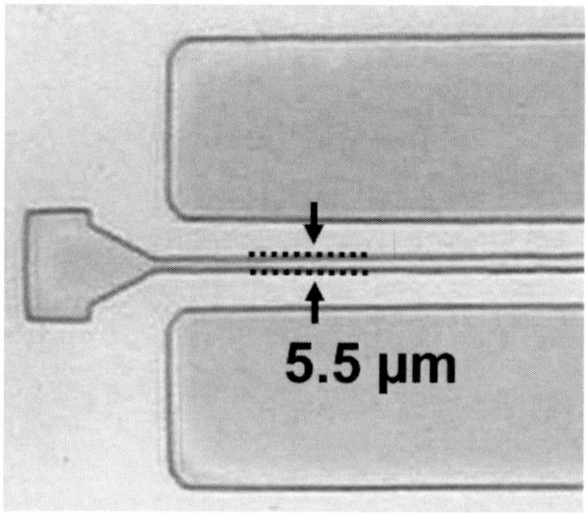

Figure 10. Optical micrograph of a direct-patterned BLT film after rinsing in hexane.

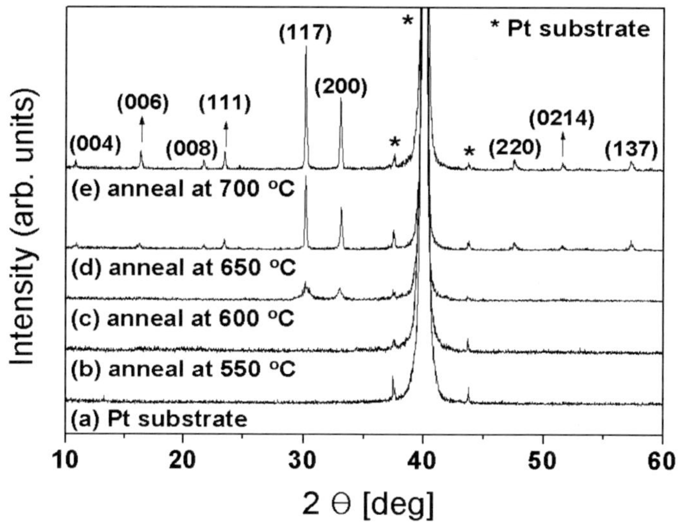

Figure 11. XRD spectra of BLT films on (a) Pt(111)/TiO_2/SiO_2/Si substrate annealed at (b) 550, (c) 600, (d) 650, and (e) 700°C.

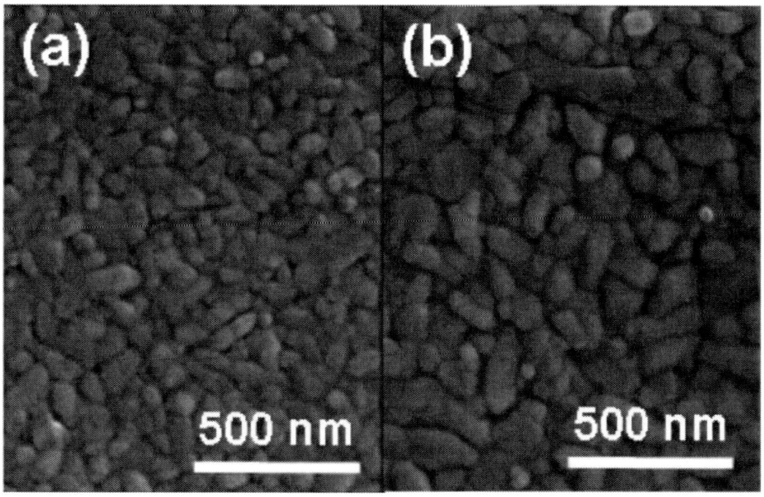

Figure 12. SEM micrographs of BLT films after annealing at (a) 650 and (b) 700°C.

Figure 11 presents XRD spectra of direct-patternable BLT films deposited on Pt(111)/TiO_2/SiO_2/Si substrates and annealed from 550 to 700°C at increments of 50°C. All the diffraction peaks corresponded to BLT and the Pt substrate. The diffraction intensities showed a distribution similar to BLT ceramics [41], and Pt(111) was found to not influence the growth orientation of BLT films [42].

Surface microstructures of 380-nm thick BLT films after annealing at 650 and 700°C were observed using SEM, as shown in Figure 12. The average grain sizes of the BLT films after annealing at 650 and 700°C were approximately 91 and 127 nm, respectively. An increase in annealing temperature resulted in an increase in the grain size, which was obviously related to crystallinity enhancement [43]. Additionally, the surface of the BLT films was closely packed without surface holes. This is because the surface microstructure of BLT films could be achieved by pre-removal of organic groups and pre-drying through UV exposure before annealing processes.

Figure 13 shows the dielectric constant – voltage plots with applied voltage ranging from – 15 V to + 15 V using an AC signal of 1 MHz. At the maximum value of the measured capacitance, the values of dielectric constant of the BLT films annealed at 650 and 750°C were 172 and 276, respectively. These values are comparable to the values obtained from conventional BLT films prepared by metal-organic deposition [44]. There are several factors that affect the dielectric constant of BLT films, including grain size, concentration of space charge and so on [45]. In our case, the C-V curves of the BLT films annealed at 650 and 700° C are symmetric around the zero bias axis, indicating that the films contain few movable ions or charge accumulation at the film-electrode interface [46]. Therefore, the increase in the dielectric constant's values of BLT films with annealing temperature is related to crystallinity enhancement and increase in the grain size of the film.

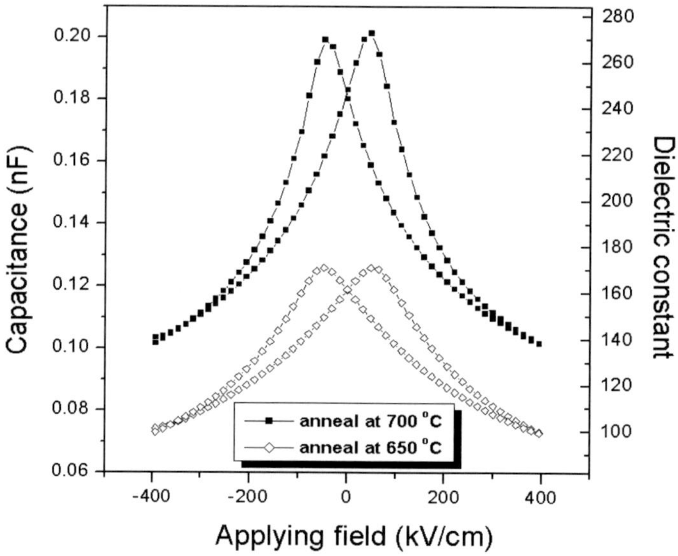

Figure 13. C-V characteristics of BLT films annealed at 650 and 700°C.

Figure 14 illustrates the change in polarization of 380-nm thick BLT films annealed at 650 and 700°C as a function of the number of switching cycles using a bipolar square wave of ± 10 V at a frequency of 100 kHz. The pulse widths (τ_1), switching period (τ) and duty cycle (τ_1/τ) were 8.6 μs, 10 μs and 0.86, respectively. After 10^9 cycles, the remaining remnant polarizations of the BLT films annealed at 650 and 700°C were 96.15 and 94.79 %, respectively. BLT films exhibited good fatigue-free behavior up to 10^9 switching cycles, as shown in the inset of the Figure 14.

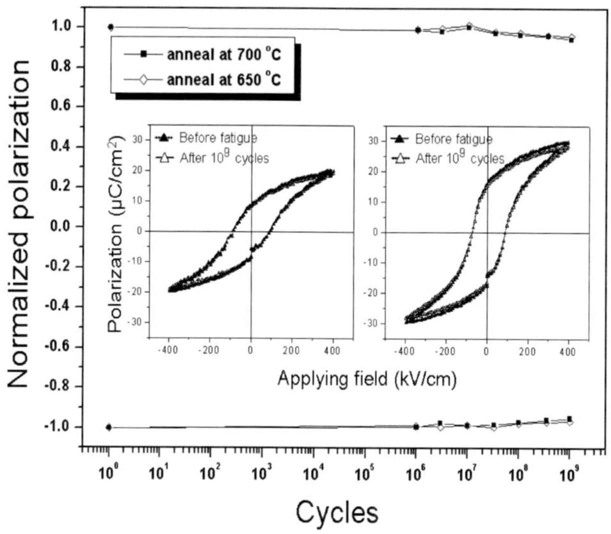

Figure 14. Fatigue characteristics of BLT films annealed at 650 and 700°C.

The measured P_r values for the BLT films annealed at 650 and 700°C before cycling were 8.7 and 16.0 μC/cm^2, respectively. The main reason for this fatigue-free behavior of BLT films could be the charge-compensating role of $(Bi_2O_2)^{2+}$ and the chemical stability of the perovskite layer against oxygen vacancies resulting from substitution of La for Bi atoms [33]. Defect charges can hardly diffuse toward the electrode/film interface, because they need to overcome the barriers formed by Bi_2O_2 layers and/or stable perovskite slabs. From the above results, the direct lithographic patterning of BLT films can be utilized to fabricate a micro-patterned system without using photoresist or a dry etching process.

BLT films were successfully constructed by PMOD using photosensitive precursors. Direct-patterning of BLT films in several micron widths was accomplished after UV exposure for 15 min and removal of the unexposed areas of the film by rinsing with hexane. All the BLT films were polycrystalline and consisted of a single phase in a Bi-layered structure. BLT films annealed at 700°C had well-developed P-E hysteresis, C-V characteristics and fatigue-free characteristics up to 10^9 switching cycles.

From this work, direct-patterning of BLT films was proven to be effectively adopted for fabrication of micro-patterned systems without dry etching.

This time, for the effect of UV-irradiation on the electric and ferroelectric properties, it was investigated of in direct-patternable BLT films. The precursor films were lithographically exposed for 15 min using the output of a 1000 W mercury arc lamp that was passed through a chrome mask. The exposed film was rinsed with hexane to develop the latent negative pattern. The resultant film was annealed for 10 min at 400°C. This process was repeated for each layer. The entire structure of multilayer films was annealed at 650 and 700°C for 1 h under ambient O_2. The UV-irradiation step was omitted when non-irradiated BLT films were prepared. BLT precursor solution for PMOD was spin-coated on p-Si(100) wafers. Figure 15 shows the FT-IR spectra of spin-coated BLT precursor films prepared using different UV exposure times; spectra were obtained for the range of 400–4000 cm^{-1}. The absorption bands at 2957, 2936, and 2865 cm^{-1} were assigned to asymmetric and symmetric stretching vibrations of alkyl ($-CH_3$) and alkene ($-CH_2$) groups and their absorption intensities decreased with increasing amounts of UV exposure. In the region between 1100 and 1800 cm^{-1}, we also found absorption bands associated with the ethylhexanoate ligands containing CO_2, ethylhexanoic acid, hydrocarbon chains, and so on [38].

However, all of the absorption bands from the organic groups almost disappeared after exposure to UV for 15 min, indicating decomposition and removal of the organic groups in the spin-coated BLT precursor films during UV exposure.

After 15 min of exposure to UV, absorption bands appreciably appeared at 707 and 829 cm^{-1}, which corresponds to Ti–O vibrations [39], absorption bands also appeared at 647 and 505 cm^{-1}, which corresponds to La–O and Bi–O vibrations, respectively [39,40]. From these observations we can say that decomposition and removal of the organic groups from spin-coated BLT precursor films was possible after 15 min of exposure to UV in the PMOD process. As a result, amorphous metal oxide films can be obtained from metal-organic precursors at room temperature.

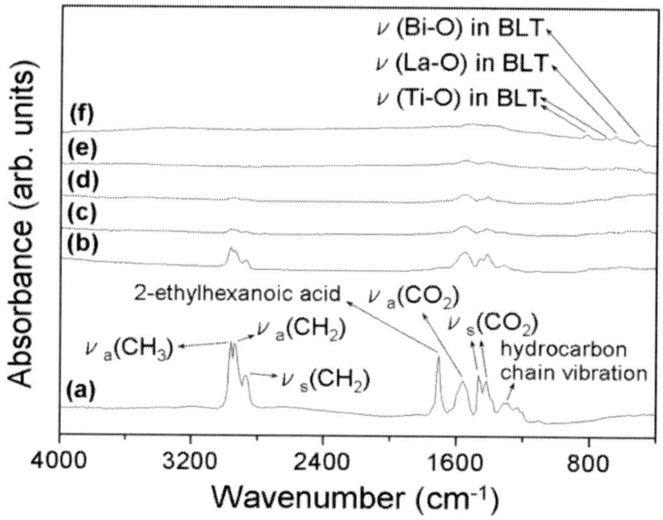

Figure 15. FT-IR spectra of spin-coated BLT precursor films photolyzed for (a) 0, (b) 1, (c) 3, (d) 10, (e) 15, and (f) 60 min with 365-nm UV light.

The possibility of direct-patterning spin-coated BLT precursor films was examined and the result is given in Figure 16. To obtain the image, a spin-coated BLT film was exposed to UV for 15 min and then rinsed in hexane. The dark area corresponds to the direct-patterned BLT film, and the bright area corresponds to the Si substrate. Based on the well-defined width in the direct-patterned BLT film, micron-scale patterning was obtained by lithographic patterning using PMOD.

Figure 17 represents XRD spectra of 380-nm-thick UV-irradiated and non-irradiated BLT films on Pt(111)/TiO_2/SiO_2/Si after annealing at 650 and 700°C. All of the intensities are the total intensity for each BLT film. Pt (111) was observed to not influence the growth orientation of BLT film [40] and the non-irradiated BLT films showed preferred (00*l*) orientations due to interfacial energy minimization between the film and substrate [47].

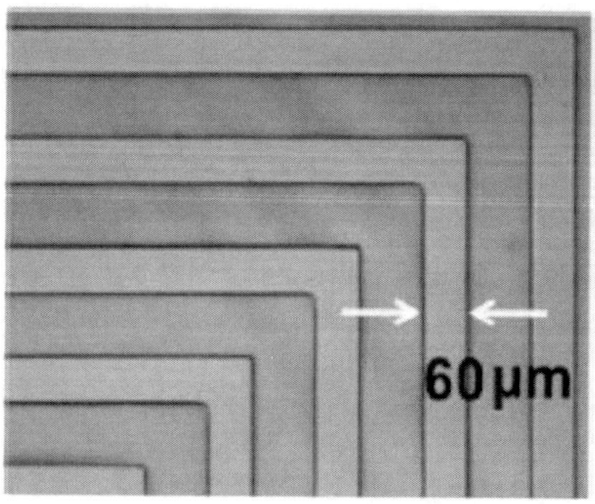

Figure 16. Optical micrograph of a direct-patterned BLT film after rinsing in hexane.

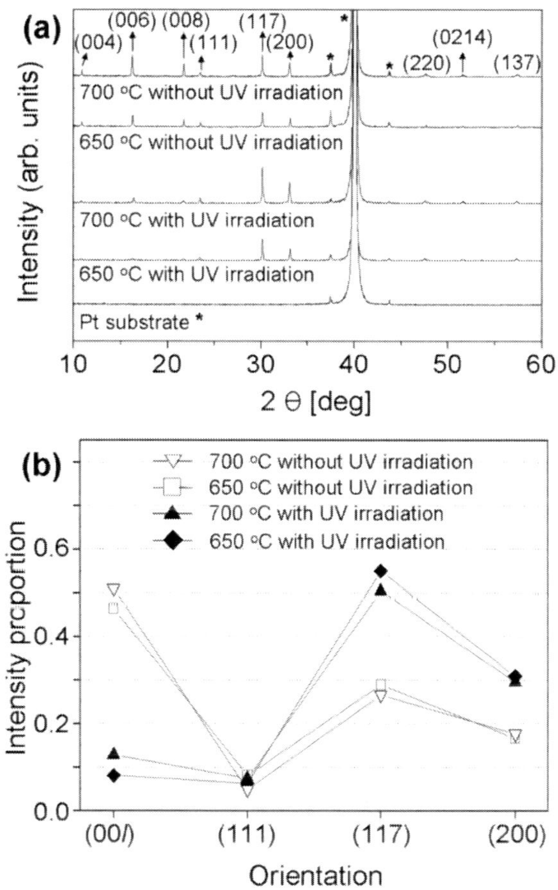

Figure 17. (a) XRD spectra of UV-irradiated and non-irradiated BLT films annealed at 650 and 700°C on Pt(111)/TiO_2/SiO_2/Si and (b) XRD intensity proportion of each diffraction orientation in the BLT films.

However, the UV-irradiated BLT films showed strong diffraction intensities along (117) and (200) orientations similar to BLT ceramics [48]. As shown in the Figure 15, UV-irradiation induced the removal of organic species and the formation of metal–oxygen bonds. The nucleation and growth is possible in the entire of the film and the UV-irradiated BLT films show random orientation of growth escaping out of <00*l*> orientations. Namely, the process of UV-irradiation assisted the growth orientation of BLT film by minimizing the influence on Pt. P–E hysteresis loops for UV-irradiated and non-irradiated BLT films with an applied voltage of 15 V after annealing at 700°C are shown in Figure 18 (a). The measured P_r values of UV-irradiated and non-irradiated BLT films after annealing at 700°C were 16.0 and 10.1 μC/cm^2, respectively. Spontaneous polarization of BLT along the a-axis is known to be appreciably higher than BLT along the c-axis [49]. As shown in the XRD spectra, UV-irradiation promoted (117) and (200) growth orientations of the BLT films. Therefore, UV-irradiation effectively enhanced the ferroelectric properties by crystallization in the non-c-axis orientations in BLT films. Figure 18 (b) illustrates the change in polarization of 380-nm-thick UV-irradiated and non-irradiated BLT films and $Bi_4Ti_3O_{12}$ (BIT) films after annealing at 700°C with switching cycles using a bipolar square wave of ±10 V at a frequency of 100 kHz.

The pulse width (τ_1), switching period (τ), and duty cycle (τ_1/τ) were 8.6 μs, 10 μs, and 0.86, respectively.

After 10^9 cycles, the remaining remnant polarizations of UV-irradiated and non-irradiated BLT films were 94.79 and 93.15 % while those of UV-irradiated and non-irradiated BIT films were 64.78 and 66.29 %, respectively. The main reason for this fatigue-free behavior of BLT films could be the charge-compensating role of $(Bi_2O_2)^{2+}$ and the chemical stability of the perovskite layer against oxygen vacancies from substitution of La for Bi atoms [34]. It is suggested that lanthanum substitution stabilizes the perovskite blocks and hence decreases the concentrations of oxygen vacancies and holes as the conduction carriers.

Direct-patterning of BLT films with width of several microns prepared by PMOD was accomplished by UV exposure for 15 min and removal of the unexposed film area by rinsing with hexane.

UV irradiation effectively enhanced the ferroelectric properties by crystallization in the non-c-axis orientations in BLT films. Regardless of UV irradiation, BLT films exhibited good fatigue-free behavior after 10^9 switching cycles. From this work, direct-patterning of BLT films was proven to be effectively adopted for fabrication of micro-patterned systems without dry etching.

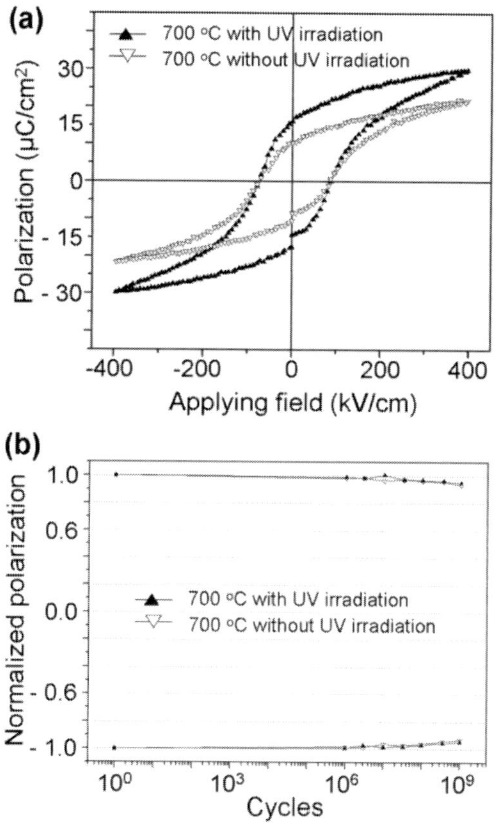

Figure 18. (a) P–E hysteresis loops of UV-irradiated and non-irradiated BLT films and (b) fatigue characteristics of UV-irradiated and non-irradiated BLT films and BIT films annealed at 700°C.

2.1.2. Electron Beam-Induced Metal-Organic Deposition (EMOD)

Due to the scale-down of electronic devices to nanometer range, it needs to develop non-optical lithographic technique because there is a resolution limit in case of the optical lithographic technique as shown in table 1 [18].

For obtaining tens nanometers' resolution, electron beam lithography can be utilized even though it shows some demerits such as high cost and low throughput [50]. Also, in the fabrication of electronic devices, there is a limit of processing temperature due to material compatibility, cost issues and thermal degradation of electronic devices. For example, with a semiconductor circuit, it is important to carry out the annealing for phase transformation below 450°C after metallization in order to avoid thermal degradation to Al-Si interconnect on the underlying circuit [51].

In this part, the EMOD process using SEM or e-beam writer is described in order to realize sub 40-nm patterning of metal oxide films and reduces annealing temperature for phase transformation of metal oxide films.

An electron beam sensitive precursor ML_n is dissolved in a volatile solvent. A few drops of this solution are dispensed onto a substrate. The substrate is rotated which results in the formation of a film of the complex.

The precursor film is exposed by electron beam using SEM or e-beam writer. The result from exposing the electron beam sensitive precursor films is that the molecular linear chain is broken and the average molecular weight of the exposed area is reduced. Namely, upon the irradiation of electron beam, the volatile ligands are ejected from the precursor leaving metal on the substrate (Eq. 3). Due to existence of oxygen in decomposed molecules or in air, the metal reacts with oxygen forming M_xO_y on the substrate (Eq. 4).

The EMOD process also enables photoresist-free lithographic processing with several nanometer dimensions and it is also a lot simpler to most other thin film deposition processes in preparing mixed metal oxide thin films.

$$ML_n \xrightarrow{\textit{electron beam}} M + nL \quad \text{(Eq. 3)}$$

Table 1. Comparison between optical lithography and non-optical lithography

		Merit	Demerit
Optical lithography	Photolithography • g – line: 436 nm • i – line: 365 nm	• low cost • easy process • high throughput	• low resolution (0.5 μm)
	Deep UV lithography • KrF: 248 nm, ArF: 193 nm F$_2$: 157 nm	• high throughput • good resolution (0.13 μm, ArF)	• difficult to make mirror
Non-optical lithography	X-ray lithography	• good resolution (0.1 μm)	• high cost • low throughput
	E-beam lithography	• good resolution (0.01 μm)	• high cost • low throughput

$$M + O_2 \longrightarrow M_xO_y \qquad \text{(Eq. 4)}$$

A precursor film of the mixed metal precursors is obtained by spin-coating and upon irradiation of electron beam. Film development can be performed by rinsing in a solvent to remove the unexposed part of the coated film, leaving an image.

The patterning process by EMOD is similar to PMOD as given in Figure 2. As you see, we need just solvent washing for removing the unexposed part of the film to obtain a pattern structure. Furthermore, comparing with the PMOD process, a mask for lithography is not needed to make a pattern in case of the EMOD process.

2.1.2.1. Ferroelectric Properties of Sub 50-nm Direct-Patterned $Pb(Zr,Ti)O_3$ Thin Films

Electron beam lithography has been considered as a potential alternative to optical lithography for critical level patterning in integrated-circuit manufacturing. For the development of high-density ferroelectric memories (64 Mbit–1 Gbit), ferroelectric capacitor should be patterned in sub-micron size [52]. These findings motivated us to investigate an alternative method for low temperature deposition of direct-patterned PZT films of sub 50-nm width via electron beam sensitive stock solutions using SEM without an electron beam resist or dry etching procedure.

The starting precursors for the electron beam-induced production of PZT films were the lead (2-ethylhexanoates)$_2$, zirconyl (2-ethylhexanoate)$_2$, and titanium(n-butoxide)$_2$(2-ethylhexanoate)$_2$. These were dissolved in a hexane solvent.

The atomic ratio of Zr/Ti was fixed as 52/48 and the molar concentration of the electron beam sensitive stock solution was 0.4 M. The PZT precursor films were spin-coated for 30 s on Pt(111)/Ti/SiO_2/Si(100) substrates with 2x2 cm^2 dimension. Electron beam exposure on spin-coated PZT precursor films was performed by using a Raith e_line e-beam writer with an acceleration voltage of 10 kV for sub 50-nm direct-patterned PZT films. For 500×500-μm^2 direct-patterned PZT films, an FEI Strata 235 SEM with an acceleration voltage of 10 kV was used and the ferroelectric and dielectric properties were measured.

Finally, the electron beam with an acceleration voltage of 5 kV at the 4B1 beam line of Pohang Accelerator Laboratory (PAL) was used for millimeter-scaled films to monitor the compositional evolution with the annealing of the e-beam exposed sample by in-situ photoemission spectroscopy (PES) measurements and to obtain XRD results. Even though the PZT precursor films were exposed to three different electron beams, the exposed dose was identical: 4.5 μC/cm^2.

After exposure of the spin-coated PZT precursor film to the electron beam, the latent image was developed with hexane to yield a negative pattern. The patterned film was annealed at 400°C for 30 min under an O_2 ambient to complete the phase formation with a perovskite structure. The thicknesses of spin-coated, electron beam exposed, and after 400°C annealed PZT films were around 250, 160, and 150 nm, respectively by ellipsometer measurement. A 150-nm-thick Pt top electrode with a 60-μm radius was sputter-deposited to measure the ferroelectric and dielectric properties of the film. Changes in the chemical bonding state of the PZT precursor film were monitored after a 2 h electron beam exposure and an annealing treatment from 100°C to 400°C with an interval of 100°C by in-situ PES measurements. Some of the resultant Pb 4f spectra are given in Figure 19.

The PES spectrum of Pb 4f could be fitted to two spin-orbit doublets, which suggests two chemical bonding states: one at Pb $4f_{7/2}$= 138.1 eV and Pb $4f_{5/2}$= 143.0 eV, and the other at Pb $4f_{7/2}$= 140.2 eV and Pb $4f_{5/2}$= 145.3 eV. The former component was identified as a Pb-O bond [53]. The upward shift of approximately 2.2 eV in the binding energy from one doublet to the other can be ascribed to a Pb–OC bond in the Pb starting precursor, because the Pb precursor was composed only of oxygen and carbon. A similar chemical shift in the Pb $4f_{7/2}$ binding energy was observed with Pb–OS in $PbSO_4$, 2.0 eV from Pb–O [54], due to the similar covalent characteristics of sulfur and carbon with Pb and oxygen.

As shown in Figure 19(b), the doublet peak of Pb–OC was much reduced because the Pb starting precursor was decomposed. Moreover, the doublet peak of Pb-OC had completely vanished after annealing at 400°C (Figure 19(c)). The Pb 4f peak of Pb–O in the PZT film is given in Figure 19(d) after annealing at 650°C under an O_2 ambient as a reference.

The phase formation of the PZT precursor films was found to be completed after electron beam exposure for 2 h and annealing at 400°C. This was also confirmed by XRD measurements. XRD patterns of direct-patterned PZT films deposited on Pt(111)/Ti/SiO_2/Si(100) substrate are given in Figure 20.

It was only with electron beam exposure of the PZT precursor films for 2 h (Figure 20(a)) that the films were partially crystallized with a preferred <111> orientation due to (111) plane matching with Pt substrate [55]. Jeyakumar et al. [56] and Saifullah et al. [57] report that an exposure of electron beam sensitive precursor films to an electron beam results in the cleavage of organic ligands from the central metal atom and converts the metal atom to a metal oxide. In our case, the oxygen content was enough for the metal to convert to a metal

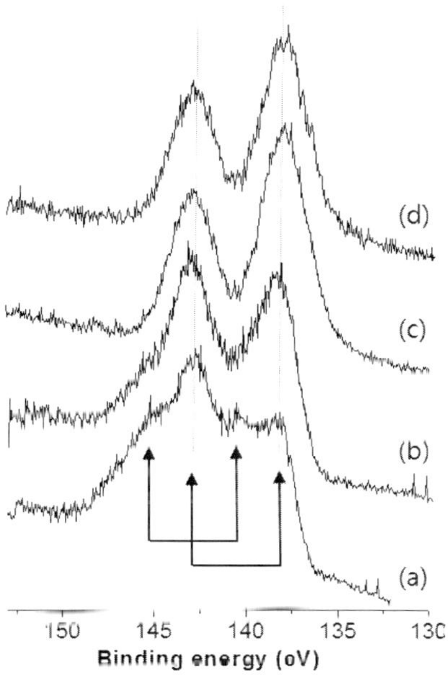

Figure 19. Evolution of Pb 4f photoemission spectra (a) before and (b) after electron beam exposure for 2 h and (c) subsequent in-situ annealing at 400°C or (d) subsequent ex-situ annealing at 650°C under an O_2 ambient as a reference.

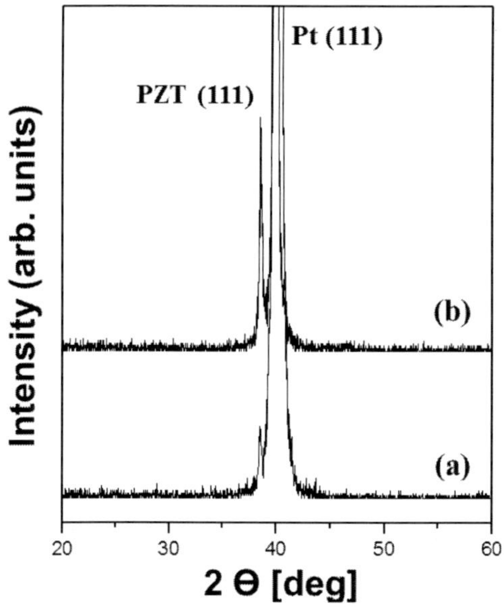

Figure 20. XRD spectra of direct-patterned PZT films after (a) electron beam exposure for 2 h and (b) subsequent annealing at 400°C for 30 min under an O_2 ambient.

oxide during the electron beam exposure because the ratio of metal (Pb, Zr, and Ti) to oxygen was 1 to 5 in the PZT starting precursors. After an electron beam exposure for 2 h and subsequent annealing at 400°C for 30 min under an O_2 ambient, the intensity of the (111) diffraction peak increased as shown in Figure 20(b).

However it has been well known that (111)-oriented $PbPt_x$ alloy could be formed in a low temperature reducing atmosphere and act as a seed for the formation of (111)-oriented PZT [58-60]. To check the formation of interfacial $PbPt_x$ alloy, direct-patterned PZT film of 500×500 μm^2 pattern size by an FEI Strata 235 SEM after anneal at 400°C for 30 min under an O_2 ambient was etched using hydrofluoroboric acid [58]. XRD measurements were carried out before and after hydrofluoroboric acid treatments and energy dispersive spectrometry (EDS) analysis was done on the etched region of the pattern.

Figure 21 represents the EDS analysis result and an inset corresponds to the XRD results. As shown in the figure, the intensity of the PZT(111) diffraction peak was reduced much but still existed even after etching of the PZT with hydrofluoroboric acid and the remained peak should be indexed to (111) diffraction of $PbPt_x$ [58]. The EDS analysis result obtained after acid-etched 500×500 μm^2 pattern showed that some Pb was still observed with major Pt. From these results, it could be concluded that through the 400°C-annealing of e-beam treated PZT precursor film for 30 min under an O_2 ambient, (111)-oriented PZT film was synthesized with the presence of interfacial (111)-oriented $PbPt_x$ alloy in case of Pt(111) substrate.

Figure 22 displays the hysteresis loops of the directpatterned PZT films with pattern size of 500×500 μm^2 under applied voltages of 8, 10, and 15 V using a RT66A ferroelectric tester. The 160 nm thick e-beam exposed PZT precursor film for 2 h did not show a ferroelectric property as shown in Figure 22(a). This paraelectric behavior of the film may indicate that the crystallization of PZT film with a perovskite structure was not complete, as shown in the Figure 20(a). However, after the subsequent annealing at 400°C for 30 min under an O_2 ambient, the 150 nm thick PZT film exhibited a well-defined hysteresis loop, typical of ferroelectric behavior (Figure 22(b)). The P_r and E_c values were measured as 7.0 μC/cm^2 and 97 kV/cm at 10 V, respectively. The P_r value is somewhat small in comparison with that of the PZT films prepared by sol-gel technique [61]. The measured smaller P_r might be associated with low crystalline quality and small grain size due to low annealing temperature, rough interface and possible interfacial mixing between ferroelectric layer and electrode by electron beam bombardment, generated depolarization field, and so on [62–64].

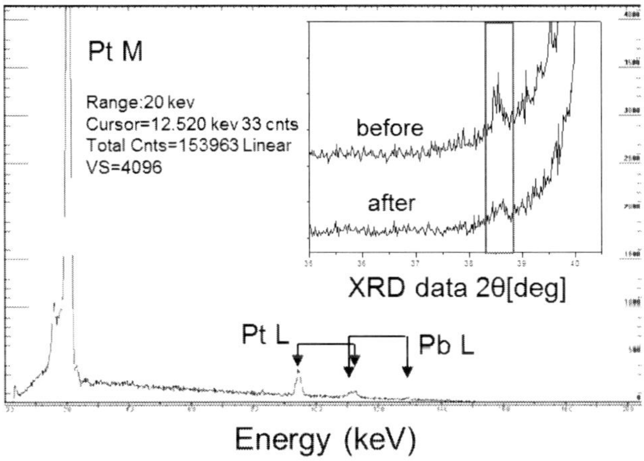

Figure 21. EDS analysis result on the etched region of direct-patterned PZT film of 500×500 μm^2 size with hydrofluoroboric acid (the inset corresponds to XRD spectra of the direct-patterned PZT film before and after the etching treatment).

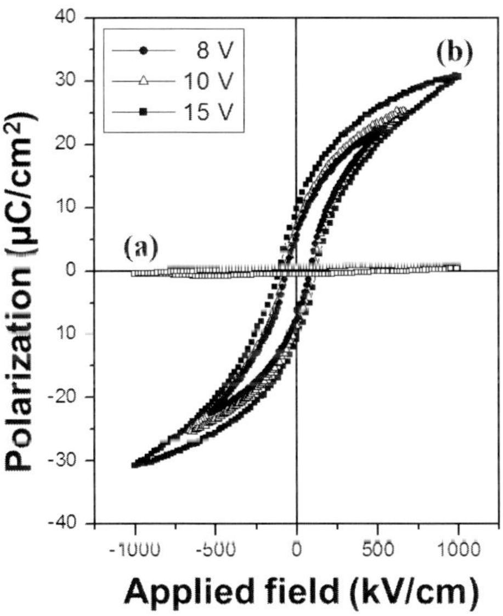

Figure 22. Ferroelectric hysteresis loops of direct-patterned PZT films after (a) electron beam exposure for 2 h and (b) subsequent annealing at 400° C for 30 min under an O_2 ambient.

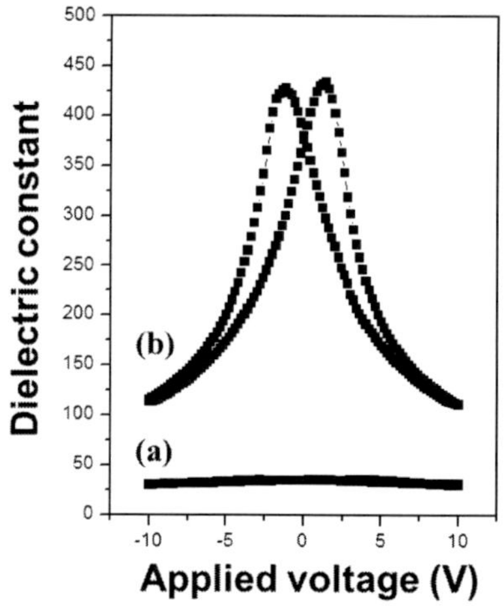

Figure 23. The voltage dependence of dielectric constants for directpatterned PZT films after (a) electron beam exposure for 2 h and (b) subsequent annealing at 400°C for 30 min under an O_2 ambient.

However, the P_r value is somewhat comparable to the value obtained from direct-patterned PZT film formed using a photosensitive stock solution [65].

The capacitance–voltage (C–V) characteristics of the direct-patterned PZT films with pattern size of 500×500 μm^2 with an applied voltage range from −10 to 10 V at 1 MHz are given in Figure 23. As shown in Figure 23(a), the PZT film exposed to an electron beam for 2 h did not show a butterfly loop while a butterfly loop and two maxima of voltage response peaks with a coercive field were observed with the subsequently annealed PZT film at 400°C for 30 min under an O_2 ambient, as shown in Figure 23(b). The value of the dielectric constant was 434 for this PZT film which is comparable with reported data on PZT films [66, 67].

SEM micrographs of sub-micron and sub 50-nm directpatterned PZT films are given in Figure 24. After the PZT precursor films were exposed with a dose of 4.5 $\mu C/cm^2$ using a FEI Strata 235 SEM or Raith e_line e-beam writer, the exposed PZT precursor films were developed by hexane-rinsing to reveal the negative pattern. The relatively bright area corresponds to direct-patterned PZT films and the relatively dark area corresponds to a Pt substrate. Based on Figure 24, we conclude that the electron beam dose (4.5 $\mu C/cm^2$) to which the PZT precursor films were exposed was sufficient for the completion of structural arrangements including local networking. Together, our results indicate that direct-patterned PZT films can be realized using an electron beam sensitive stock solution for NEMS and FeRAMs applications.

In summary, we have demonstrated an electron beam resist-free fabrication technique to deposit sub 50-nm direct-patterned PZT films at low temperature. The PZT precursor film exposed to an electron beam for 2 h did not have ferroelectric properties. However, after subsequent annealing at 400°C for 30 min under an O_2 ambient, PZT film with perovskite structure was synthesized with interfacial (111)-oriented $PbPt_x$ alloy on Pt(111) substrate. The direct-patterned PZT films with pattern size of 500×500 μm^2 exhibited well-defined hysteresis loops and butterfly loops for C-V curves. This electron-beam induced sub 50-nm direct-patterning technology of PZT ferroelectric films can therefore be effectively adopted to fabricate sub 50-nm-patterned systems.

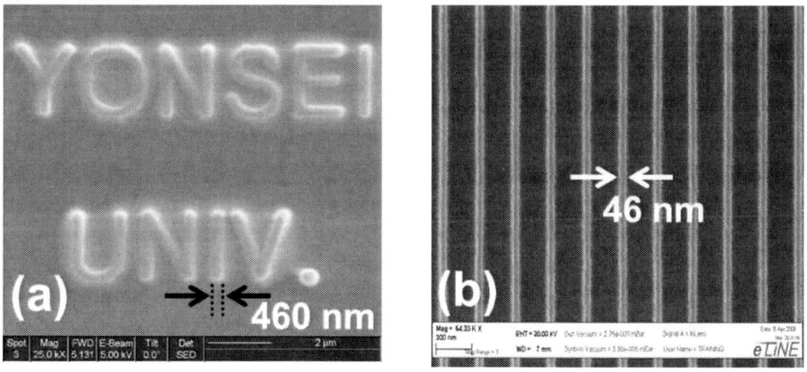

Figure 24. SEM micrographs of direct-patterned PZT films using an electron beam sensitive stock solution by (a) an FEI Strata 235 SEM or (b) a Raith e_line e-beam writer.

2.1.3. Composition Control in the Chemical Solution Process

Fabrication process of MFIS FeFET is very important factor for performance of FeFET. Therefore, this chapter shows that process affect to ferroelectric properties in fabrication of ferroelectric thin film such as crystallization temperature, thickness of thin film, and solvent effect in solution process.

2.1.3.1. The Effects of Solvent on the Properties of Sol-Gel Derived PZT Thin Films

Ferroelectric thin film, PZT has been actively investigated for its wide application. Sol-gel method has been commonly used because of its several advantages such as easiness of composition control, high homogeneity, and low processing temperature [68]. In sol-gel processing, it was reported that the solvent of stock solution strongly effects on the final properties of PZT thin film [69]. 2-Methoxyethanol has been generally used due to its high reactivity. However, to improve the density of film, solvent with light molecular weight seems to be more desirable. But its application has been limited due to the difficulty in the control of reaction kinetic between the solvents and precursor alkoxide [70]. Then the effect of various solvents on PZT film was not widely investigated. In this part, four kinds of solvents were used for the preparation of PZT films. The effects of various solvents on the crystallization and properties of the films were investigated.

The stock solutions were fabricated by using high-purity lead acetate trihydrate, Ti iso-propoxide, and Zr n-buthoxide as stating materials with various kinds of alcohol as solvents. 2-Methoxyethanol, iso-propyl alcohol, ethanol, and methanol were used as solvents. Their molecular weight and boiling point are 76.1, 60.1, 46.07, 32.04 g/mole and 124, 82.5, 78.5, 64.7°C, respectively. Acetylacetone was used as chelating agent. All starting solutions were concentrated to 0.2 M. The Zr/Ti ratio was chosen as 0.4/0.6, and 15% excess lead was added to a precursor to compensate the loss of Pb during the anneal. Films were spin coated at 3500 rpm for 20 sec on $Pt/Ti/SiO_2/Si(100)$ substrate.

The films were pyrolyzed successively at 200 and 400°C on hot plate for 2 min. Then the coated films were annealed at 650°C for 30 min in a tube furnace by direct insertion. Sputter-deposited Pt with a thickness of 150 nm was used as top electrode. After the top electrode deposition, anneal treatment was done at 650°C for 5 min. The weight change and heat exchange as a function of temperature were studied by TG/DTA. The chemical bonding states and stability of solution were characterized by FT-IR. The surface morphology and crystallization quality of films were observed using SEMand XRD, respectively. Especially, the effects of solvent on the temperature of phase formation were observed by in-situ XRD with heating rate of 5°C/min. Ferroelectric properties were observed by using RT66A ferroelectric tester (Radiant Technologies). Figure 25 shows FT-IR spectra of dried PZT sols prepared with various kinds of solvent. The broad peak of 3400-2800 cm^{-1}, corresponding to O-H and C-H stretching was due to residual solvents. Also, peaks for metal-oxygen bond are observed about 450-700 cm^{-1} for all the samples. The occurrence of these peaks might be attributed to the polycondensation reaction.

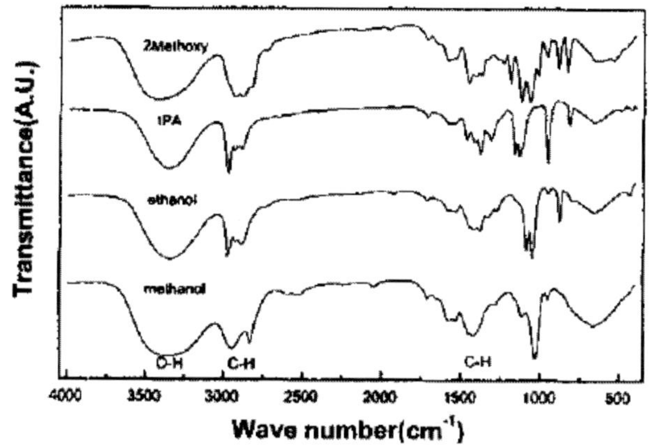

Figure 25. FT-IR spectra of PZT sol as a function of different solvents.

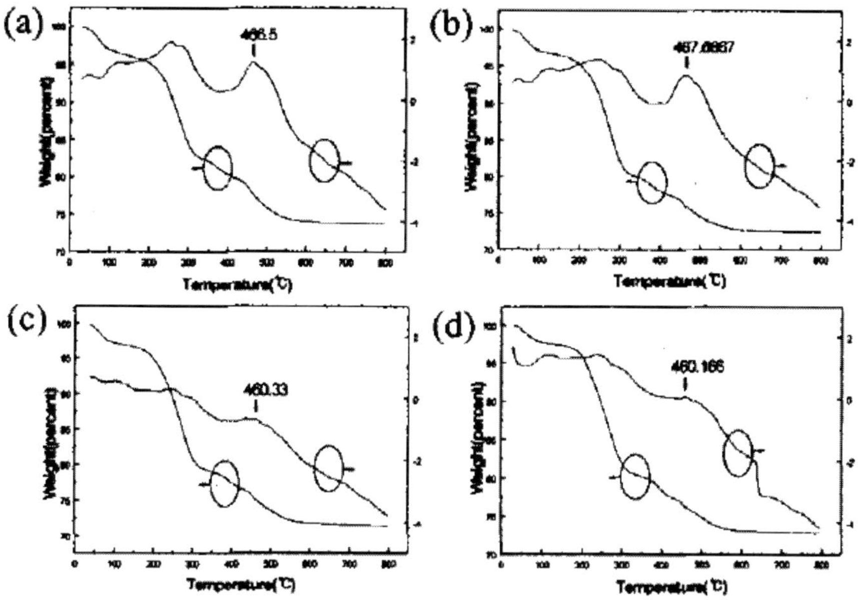

Figure 26. TG-DTA curves of PZT sol powder; (a) 2-Methoxyethanol, (b) IPA, (c) ethanol, and (d) methanol.

The presence of Ti-O stretching at 450 cm^{-1} confirmed the reaction between AcAc and Ti. The peaks around 1600 and 1520 cm^{-1} are due to Zr-O and 650 cm^{-1} is due to Pb-O, respectively. These peaks indicate that AcAc is bound to alkoxide precursor.

Figure 26 corresponds to TGDTA curves of each PZT sol powder. Most of the case, the curves revealed that there were two exothermic peaks around 250 and 460°C. The exothermic peak at 250°C corresponds to the decomposition of organic compounds formed during hydrolysis and condensation. The peak around 460°C can be mainly attributed to the crystallization of PZT sol powder, which seems to be overlapped with the peak arising from the decomposition of reacted AcAc with alkoxide precursor. It was found that the organic components were removed quickly as the molecular weight of solvent decreased. Especially, the temperature of phase formation became low as the molecular weight and boiling point of solvent decreased. These results are clearly confirmed with in-situ XRD results as given in Figure 27. In case of IPA, ethanol, and methanol, PZT (110) peak began to be observed at 500°C. However, in case of 2-methoxyethanol, it was observed at 525°C. It is inferred that the temperature of PZT phase formation is decreased as the molecular weight of solvent decreases.

Figure 28 shows the variation in the diffraction intensity of PZT thin films obtained using different solvents. It was found that the intensity ratio of (111) diffraction, i.e. $I_{(111)}/(I_{(100)}+I_{(110)}+I_{(111)})$ s increased, as the molecular weight of solvent decreased. It means that, nucleation of PZT is strongly affected on Pt(111) substrate as the molecular weight of solvent decreases. We can conclude that as the molecular weight of solvent decreases, PZT phase is formed at low temperature where heterogeneous nucleation is dominant, and then (111) preferred orientation is obtained because Pt(111) substrate acts as a heterogeneous nucleation site.

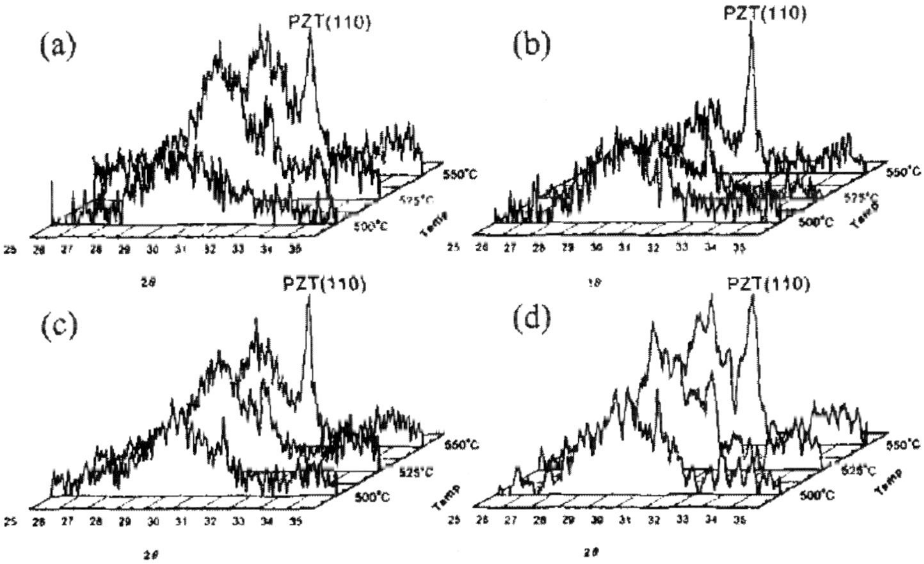

Figure 27. In-situ XRD patterns of PZT sol powder as a function of different solvents; (a) 2-Methoxyethanol, (b) IPA, (c) ethanol, and (d) methanol.

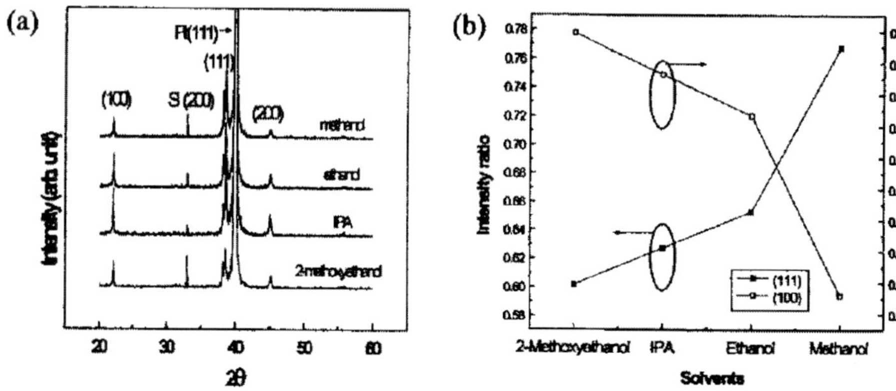

Figure 28. (a) XRD diffraction patterns and (b) intensity ratio of PZT thin films with various solvents of stock solution.

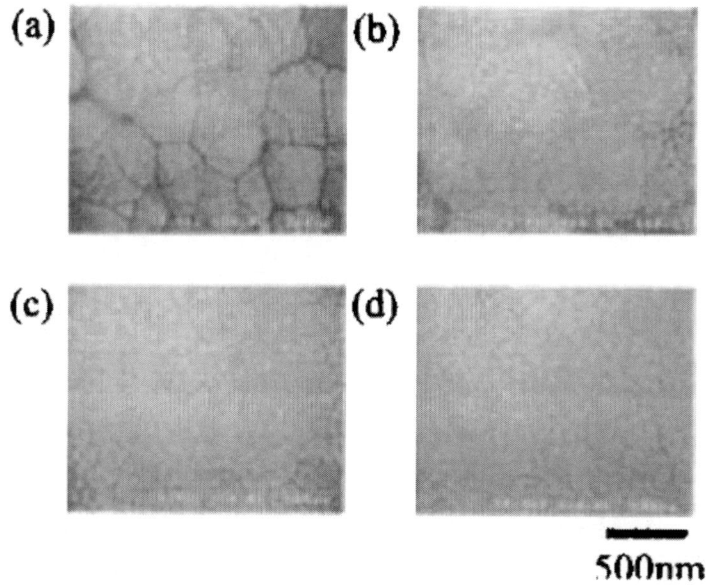

Figure 29. SEM images of PZT films using various solvents for stock solution; (a) 2-Methoxyethanol, (b) IPA, (c) ethanol, and (d) methanol.

The surface morphology of the PZT thin films derived from the different solvents was given in Figure 29. As the molecular weight of solvent decreases, a tendency of decrease in grain size was observed. This agrees well with the above results because at low temperature region, nucleation is dominant than growth. As a result, the observed grain size is small as temperature of phase formation is low. The P-E hysteresis loops of the PZT thin films are given in Figure 30. The maximum and minimum P_r values were obtained as 21.17 and 16.8 $\mu C/cm^2$ with 2-Methoxyethanol and methanol, respectively.

As the molecular weight decreased, the P_r value was decreased but with proper and gentle saturation behavior. To supposed that, domain at the 90 ^{o} field is reduced as the grain size decreases [71].

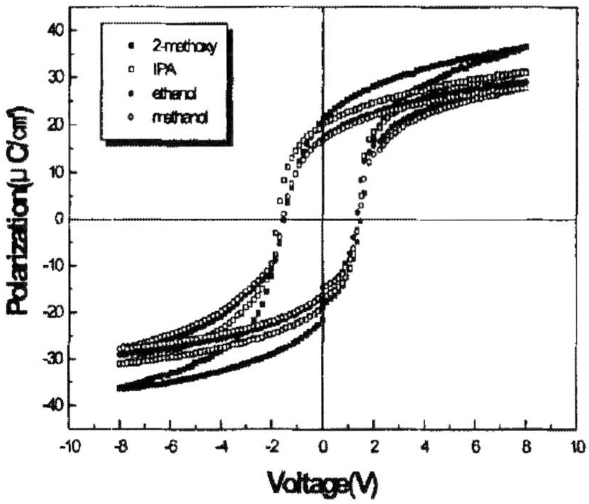

Figure 30. P-E hysteresis loop obtained using various solvents.

PZT thin films were well prepared by using four kinds of solvent (namely, 2-methoxyethanol, IPA, ethanol, and methanol. It was found that the temperature of phase formation was decreased with the decrease of molecular weight of solvent. A tendency of decrease in grain size was observed and XRD intensity ratio of (111) diffraction was increased as the molecular weight of solvent decrease. Also, P_r value decreased. These changes in the material properties according to various solvents could be related with the removal of organic group of alkoxide. Finally, we confined that the different kinds of solvent were effective to crystallization and properties of PZT thin film.

2.1.4. The Effect of Film Thickness and Annealing Temperature

2.1.4.1. The Effects of Film Thickness of PZT on the Crystallization and Ferroelectric Properties

In this part, we investigated the optimum condition for the preparation of self-patterned PZT thin film for minimizing the shrinkage by controlling the molar concentration of PZT stock solution. For applying to memory capacitor in integrated devices, the PZT film with around 100 nm thick was prepared and examined. And over 200 nm thick self-patternable PZT film was also prepared by multiple coating procedure for the effect of film thickness. The material properties of self-patternable PZT films according to film thickness were analyzed.

We prepared the stock solution of PZT with 0.2, 0.4, and 0.6 M to investigate the effect of molar concentration on the behavior of shrinkage. PZT thin films were prepared using mixed solutions of titanium iso-propoxide [$Ti(O_iC_3H_7)_4$], zirconium n-butoxide [$Zr(O_nC_4H_9)_4$], and lead acetate tri-hydrate [$Pb(CH_3COO)_2 3H_2O$] dissolving in 2-methoxyethanol. Negative acting orthonitrobenzaldehyde (NBAL) was used as photosensitizer and HNO_3 as catalyst. PZT thin films were prepared on Pt(111)/Ti/SiO_2/Si substrate by spin coating, followed drying. UV light with 365 nm wavelength was irradiated on dried PZT photo-sensitive gel film for 2 min. The non-irradiated area was etched off using 2-methoxyethanol as developer. After patterning of PZT films, the obtained negative patterns of the films were baked. Finally the films were annealed at 650°C by direct insertion to a tube furnace under O_2 ambient. After then, platinum was deposited on the PZT films as top electrode by RF magnetron sputtering. The etching morphology and film thickness were observed by SEM. The orientation of the films was determined by XRD method. Standard ferroelectric tester RT66A was used for obtaining ferroelectric properties such as polarization-voltage (P-V).

Self-patterning process of PZT films prepared with stock solutions of 0.2, 0.4, and 0.6 M was performed, respectively. PZT film prepared with stock solution of 0.6 M shows better etching shape and lower shrinkage than the other films. The shrinkage was determined from the patterned size difference between etched film and annealed film at 650°C. 3, 1.4, and 0.8 % of shrinkage was found for the films with stock solutions of 0.2, 0.4, and 0.6 M, respectively. The higher shrinkage of the lower concentrated stock solution was due to the smaller thickness of the film, i.e. relatively large contraction under film stress during anneal. For realizing the self-patterned PZT film of appropriate thickness, multi-coating of PZT sol on the self-patterned and dried film is inevitable. This means that if a relatively thick PZT film is desired, a precious alignment of micro-pattern and shrinkage of pre-formed film will be very important. With PZT stock solutions of 0.2 and 0.4 M, the final PZT film was found to be very thin with large shrinkage. But PZT stock solution with 0.6 M, the final thickness of PZT film was around 120 nm with acceptable shrinkage. However mole-concentration of the stock solution exceeds 0.6 M, a formation of precipitate is usually happened. Then it can be said that 0.6 M-stock solution is appropriate for the preparation of self-patterned PZT film for applying to memory devices.

Figure 31 shows the SEM image of self-patterned PZT thin film prepared with 0.6 M stock solution. Bright area corresponds to the PZT films and dark area to Pt substrate, respectively. As discussed earlier, for applying PZT film to piezoelectric micro-mechanical system, we need several hundreds nm thick film. That means we have to carry out multiple coating for obtaining an appropriate thickness.

Figure 32 (a) and (b) correspond to the SEM planar images of self-patterned PZT films after one and two times of coating, respectively. The thickness of two times-coated PZT film was found to be about 240 nm after anneal at 650°C. In the images, it was easily observed that a finer grain microstructure was obtained with one time-coated PZT film.

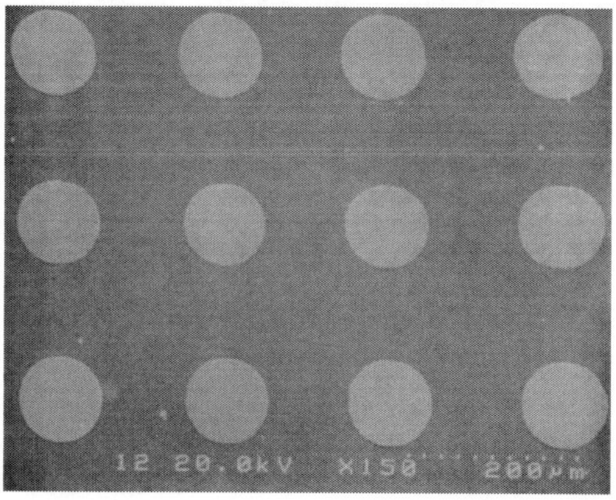

Figure 31. SEM image of self-patterned PZT film.

Figure 33 corresponds to their XRD patterns that show (111) preferred orientation is more important in one time-coated PZT film than the other film. The above results mean that the nucleation and growth of one time coated PZT film is more influenced by Pt substrate than that of multiple-coated PZT film. If the nucleation of PZT phase is localized on Pt substrate, (111) preferred orientation should be dominant due to the lattice matching between Pt and PZT [27]. Furthermore the grain should be small because the nucleation on Pt substrate is heterogeneous one at which nucleation is superior than grain growth.

Figure 34 represents hysteresis loops of one and two times-coated PZT films. Their P_r and E_c values are measured 27, 36 $\mu C/cm^2$ and 100, 80 kV/cm, respectively. Even though smaller P_r and larger E_c values for one time-coated film are obtained than two time-coated film due to its small film-thickness, we can find that one time-coated PZT film with about 120 nm thick shows a well saturated - typical ferroelectric behavior. From the above results, it can be said that 0.6 M-PZT stock solution is an optimum molar concentration for applying it to self-patternable PZT thin film.

Figure 32. SEM planar images of self-patterned PZT films after coating (a) one time and (b) two times.

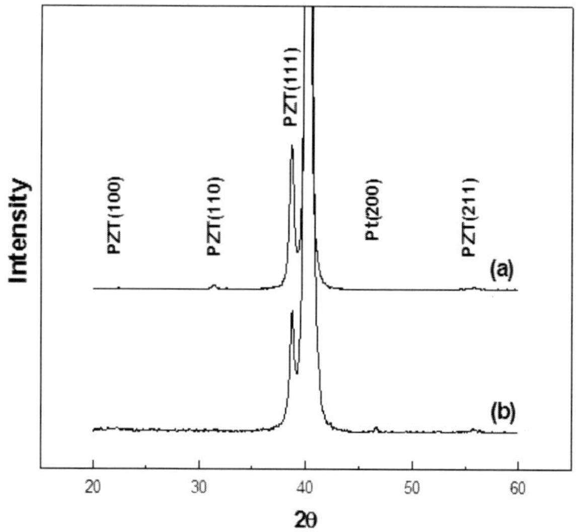

Figure 33. X-ray diffraction spectra of PZT films after coating (a) one time and (b) two times.

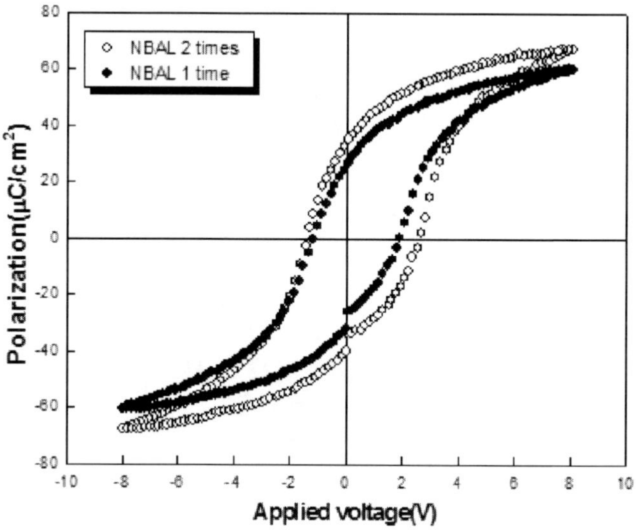

Figure 34. P-E hysteresis loop of PZT films.

Self-patterned PZT films were successfully fabricated with 0.6 M-PZT stock solution using photosensitizer and UV-irradiation process. Etching of the film is accomplished from photo-reaction of photosensitive solutions. The finer grain microstructure was obtained, and highly (111) oriented PZT film was observed with one time coated PZT film with 120 nm thick. Remnant polarization of the one time-coated film was relatively lower than the two times-coated film with 240 nm thick due to its small film thickness. However self-patterned PZT film with 120 nm thick showed a well saturated-typical ferroelectric behavior.

2.1.4.2. Substrate Modification for the Direct Formation of PZT Films with Perovskite Structure

For applying ferroelectric PZT to integrated devices, low temperature anneal for PZT crystallization is desirable. However, it is inevitable to adopt a high temperature anneal as above 600°C in conventional deposition process due to the presence of intermediate phase, i.e. pyrochlore. Recently, in-situ anneal of substrate during the sputter deposition has been investigated as a promising technique to lower the crystallization temperature [31,72]. However, it was also reported that serious Pb deficient problem was raised during initial stage of deposition due to a short residence time of Pb oxide molecules on Pt substrate [73]. To overcome this problem, Pb source has been supplied in excess. In this part, we investigated the effect of excess supplement of Pb on the crystallinity and ferroelectric properties of the PZT film.

Furthermore we established the relationship between interfacial states and ferroelectric properties based on the 'space charge layer model' [74] and provided an effective sputtering method to lower the crystallization temperature of perovskite - PZT using a target containing moderate content of PbO.

PZT thin films were sputter-deposited from a sintered ceramic target ($PbZr_{0.52}Ti_{0.48}O_3$ with 10, 30, 50 and 75 % excess PbO) in an off-axis RF magnetron sputtering system. The sputtering chamber was initially pumped down to a base pressure of $3x10^6$ Torr, then argon and oxygen gases with a mixed ratio of 0.9:0.1 were introduced up to a working pressure of 1 x 10^{-2} Torr. The substrates used were Pt (150 nm)/Ti (10 nm)/SiO_2/Si (hereafter called Pt/Ti), Pt (150 nm)/SiO_2/Si (Ti/Pt) and e-beam deposited Ti (3nm)/ Pt (150 nm)/Ti (10 nm)/SiO_2/Si (Ti/Pt). During the deposition of PZT films, the substrate was maintained at room temperature and 550°C respectively. The thickness of the deposited PZT films was about 3 and 250 nm. The film deposited at room temperature was annealed for 2 min in air using rapid thermal anneal (RTA) system. Phase and crystalline state analyses of the PZT films were performed using XRD. The composition and chemical state of PZT films were investigated using X-ray photoelectron spectroscopy (XPS). Hysteresis loops and fatigue behavior analyses were done using RT66A. High voltage source unit on HP4155A was used to measure leakage current density. The positive and negative curves are taken after the application of a negative or positive voltage on the top electrode (TE) and bottom electrode (BE), respectively.

Figure 35 shows XRD pattern of in-situ deposited PZT films on the Pt/Ti substrate heated at 550°C. The perovskite phase without any intermediate phase directly formed and the film was preferentially (111)-oriented.

Figure 36 shows fatigue test results of PZT films deposited at 550°C. Fatigue behavior was measured at ±5 V of square pulse mode. 5 V of applied voltage was sufficient to saturate the hysteresis loops. The PZT film shows an asymmetric degradation of polarization. The positive polarization loss (58.5 %) is larger than negative loss (48 %) of the PZT capacitor after 10^{10} cycles. This fatigue behavior may be explained using a 'space charge layer model' where during switching cycles, mobile ions and oxygen vacancies diffuse to the interface between PZT film and bottom electrode. They are locked up at the interface and produce positive space charge layer. This space charge layer builds up internal field with an external switching field and results in the reduction of the external positive field [74,75].

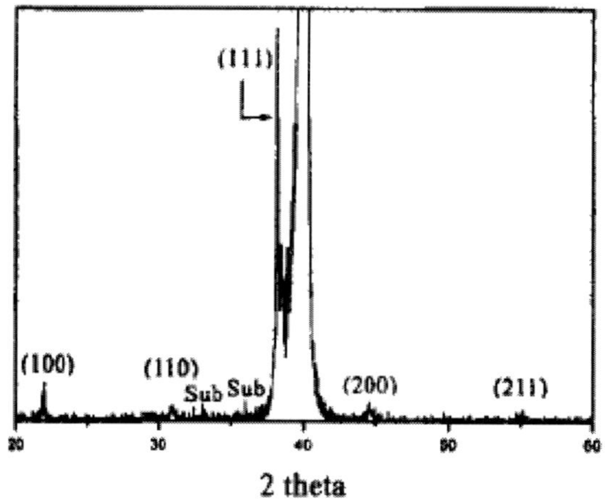

Figure 35. XRD patterns of PZT film deposited on Pt/Ti at 550°C.

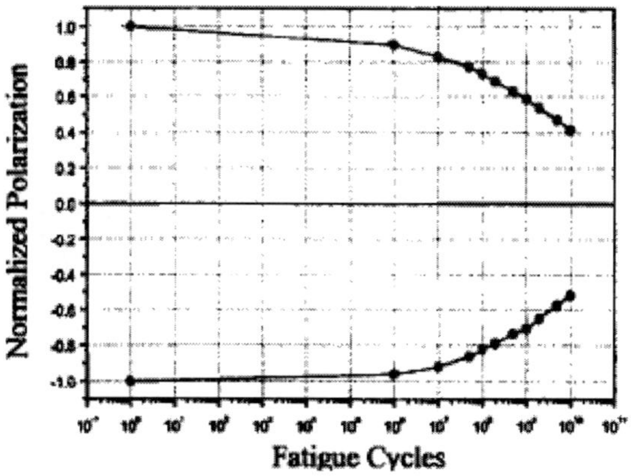

Figure 36. Fatigue behavior of PZT film deposited on Pt/Ti at 550°C.

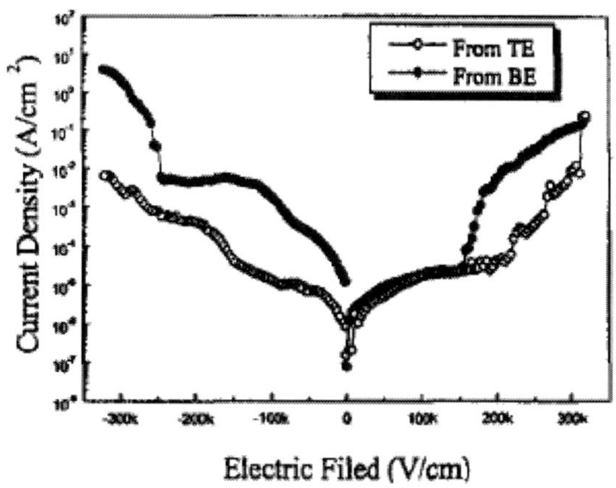

Figure 37. Change of leakage current of PZT film with applied field on TE and BE.

Figure 37 shows change of leakage current with applied field on the TE and BE. For negative bias on BE, the curve shows high leakage current density and unstable behavior, while the curve for negative bias on TE shows stable one. This higher leakage current for negative bias on BE might be caused by electron trap site such as oxygen vacancy at the BE interface [76]. The leakage current curve for positive bias on BE shows higher take-off point than the curve for positive bias on TE. It should also be noted that the leakage current behavior on BE shows strong polarity dependence, in spite of symmetric leakage current behavior on TE. These suppose that the interface state between PZT film and BE is different from that of TE. To get a better understanding on the bottom interfacial state, we deposited 3 nm PZT thin film on Pt/Ti and Pt substrates heated at 550°C using targets containing 30, 50 and 75 % of excess PbO.

Figure 38 shows the variation in the composition ratio of Pb/(Zr+Ti) and Zr/(Zr+Ti) in three different cases. Pb/(Zr+Ti) ratio increased while Zr/(Zr+Ti) ratio decreased. Pb/(Zr+Ti) ratio shows that initial-deposited PZT films had less Pb than the stoichiometric composition even using a target containing 75 % excess PbO. From these results, it can be concluded that the electrical anomalies in PZT capacitor actually come from space charges due to a deviation from the stoichiometry near the bottom interface. The PZT film is known to have a tendency to grow with Pb-deficient pyrochlore phase due to the high volatility of Pb oxide on pt substrate [77]. However, in case of Pt/Ti substrate, underlying Ti diffuses through Pt grain boundaries to the surface during in-situ anneal of the substrate. This influenced the lowering of Zr/(Zr+Ti) ratio with Pt/Ti substrate. The out-diffused Ti enhances the incorporation with sputtered Pb component so that it suppresses the volatility of Pb oxide and serves as nucleation seed for the perovskite phase. To verify the effects of out-diffused Ti at Pt surface as nucleation seed, 250 nm thick PZT films was deposited on Ti(3 nm)/Pt and Pt, at room temperature using a target containing 10 % excess PbO and RTA-treated at 550°C.

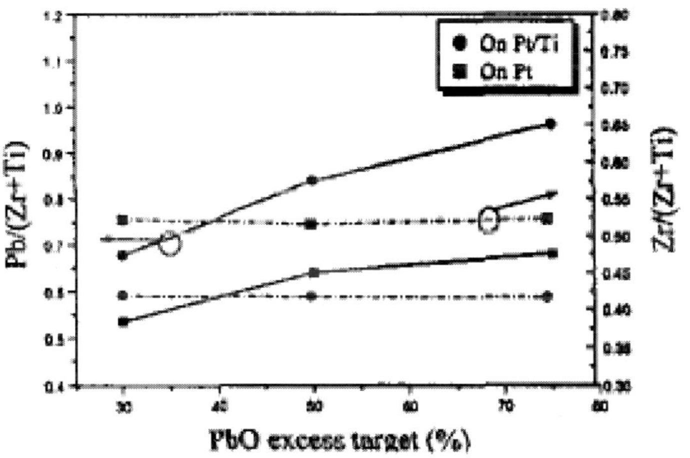

Figure 38. The variation of Pb/(Zr+Ti) and Zr/(Zr+Ti) ratio as a function of target composition.

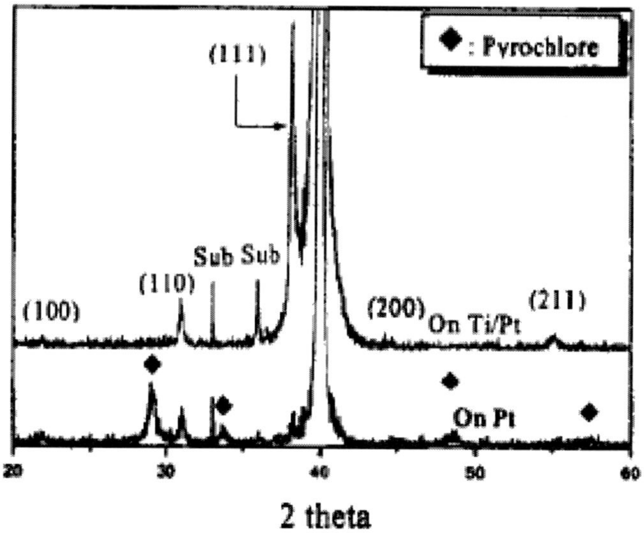

Figure 39. XRD patterns of PZT films deposited at room temperature (bottom) and RTA-treated at 550 °C (top)..

The XRD spectra of the films are shown in Figure 39. A direct formation of PZT film with perovskite phase was observed only with Ti/Pt substrate. This structure was preferentially (111)-oriented. These results confirmed that Ti formed nucleation seed for PZT crystallization. From the above results, it could be concluded that PZT film with perovskite structure was successfully formed at low temperature post-anneal (550°C) with minimizing excess PbO contents, i.e. 10 %, by inserting 3 nm thick Ti layer on Pt as seed layer for the nucleation of PZT. PZT films with the perovskite phase were obtained through post-anneal of room temperature sputter - deposited PZT at 550°C without forming a pyrochlore phase. The lowering of crystallization temperature and excess Pb content in perovskite - PZT film was accomplished by depositing 3 nm thick Ti seed layer on Pt electrode and using a target containing 10% excess content of PbO.

2.1.5. Element Substitution in Ferroelectric Materials

This part shows an improvement of fatigue, remnant polarization, and leakage current in ferroelectric thin films by element substitution.

2.1.5.1. Ferroelectric Properties of La Substituted $Bi_4Ti_3O_{12}$ Thin Films

To apply for our PMOD technique to BLT materials, comparison of the microstructure and ferroelectric properties between BIT and BLT is essential to clarify the effect of La substitution. This part report a direct-patterning method for lithographic deposition of BLT films by PMOD was introduced. Effects of La substitution on the microstructure and ferroelectric properties were investigated in direct-patternable BLT films.

The precursors used for the photochemical production of BLT films were bismuth 2-ethylhexaoate [$Bi(C_7H_{15}COO)_3$], lanthanum 2-ethylhexanoate [$La(C_7H_{15}COO)_3$], and titanium isopropoxide [$Ti[OCH(CH_3)_2]_4$]. The composition of $Bi_{4-x}La_xTi_3O_{12}$ precursors was varied with x = 0 [$Bi_4Ti_3O_{12}$], 0.25 [$Bi_{3.75}La_{0.25}Ti_3O_{12}$], 0.50 [$Bi_{3.50}La_{0.50}Ti_3O_{12}$], 0.75 [$Bi_{3.25}La_{0.75}Ti_3O_{12}$], and 1.00 [$Bi_{3.00}La_{1.00}Ti_3O_{12}$]. The chosen La^{3+} contents were limited because the Curie temperature (T_c) of $Bi_{4-x}La_xTi_3O_{12}$ at x > 1.25 decreases below 0°C, which makes these compositions paraelectric at room temperature [78]. A 10 % excess of bismuth 2-ethylhexanoate was added to compensate for the loss of Bi during high temperature anneal [37]. The BLT precursors dissolved in hexane were spin-coated at 2000 rpm for 30 s on two different kinds of substrate. One substrate was a chemically cleaned $Pt(111)/TiO_2/SiO_2/Si(100)$ substrate for measuring the ferroelectric properties, and the other substrate was p-Si(100) substrate for characterization of direct-patterning process. A 200-nm thick direct-patterned BLT film by PMOD was obtained after UV exposure.

The precursor films were lithographically exposed for 15 min using the output of a 1000 W mercury arc lamp that was passed through a chrome mask. The exposed film was rinsed with hexane to develop the latent negative pattern. The resultant film was annealed for 10 min at 400°C. This process was repeated for each layer. The entire structure of multilayer films was annealed at 700°C for 1 h under ambient O_2.

The images of the surface microstructure and direct-patterned film were observed using SEMand optical microscopy. The phase formation and crystallinity of the films were studied with an XRD using Cu k_α radiation. A 150-nm thick Pt top electrode with a 100 μm radius was sputter deposited for P-Eand current-voltage (I-V) characterization of the films.

Surface microstructures of BLT films annealed at 700°C as a function of La contents were observed using SEM, as shown in Figure 40. The average grain sizes of $Bi_{4-x}La_xTi_3O_{12}$ films with x = 0, 0.25, 0.50, 0.75, and 1.00 were 214, 178, 140, 119, and 54 nm, respectively. An increase in La contents resulted in a decrease in the grain size because the substituted La acts as a grain-growth inhibitor in the BLT films [79]. Additionally, surface holes were observed on the BLT films with x = 0, 0.25, and 0.50 while the surface of the BLT films with x = 0.75 and 1.00 was closely packed without surface holes. It was observed that La substitution (x = 0.75 and 1.00) to BIT film was effective to induce an improved microstructure of the BLT films without the formation of surface holes.

Figure 41 (a) represents XRD spectra of 380-nm thick BLT films as a function of La contents on $Pt(111)/TiO_2/SiO_2/Si$ substrate. All the diffraction peaks corresponded to BLT and the Pt substrate. The diffraction intensities showed a distribution similar to BLT ceramics [41], and Pt(111) was found to not influence the growth orientation of all BLT films [42]. As shown in Figure 41 (b), XRD intensity proportion of each diffraction orientation in the all BLT films was calculated.

The intensity proportion was calculated as the sum of (00*l*), (111), (117), and (200) diffraction intensities is the total intensity for each BLT film. According to the increase with La contents, BLT films tended to increase (117) and (200) orientations and decrease (00*l*) orientations. This tendency can be explained in terms of the lattice spacing of the (117), (200), and (00*l*) planes [80].

As the La contents increase from x = 0 to 1.00, calculated lattice spacing of the (117) and (200) planes were increased from 2.961 to 2.975 Å and from 2.708 to 2.713 Å, respectively, while that of the (00*l*) plane was decreased from 5.438 to 5.405 Å. The observed changes can be attributed to the incorporation of La into the BIT lattice. For higher La contents, the observed diffraction intensities became less intense probably due to a degradation of the crystallinity of BLT films. P-E hysteresis loops for BLT films annealed at 700°C with an applied voltage of 15 V are given in Figure 42 (a). The measured P_r and E_c values of $Bi_{4-x}La_xTi_3O_{12}$ films with x = 0, 0.25, 0.50, 0.75, and 1.00 were 3.7, 7.1, 11.3, 16.0, and 8.1 μC/cm^2 and 120, 113, 111, 90, and 73 kV/cm, respectively. As shown in Figure 42 (b), the value of P_r increased gradually with increasing the La contents below the critical level, x = 0.75 and decreased thereafter [81,82]. Spontaneous polarization of BLT along the a-axis is known to be appreciably higher than BLT along the c-axis [49].

Figure 40. SEM micrographs of $Bi_{4-x}La_xTi_3O_{12}$ films annealed at 700°C as a function of La contents.

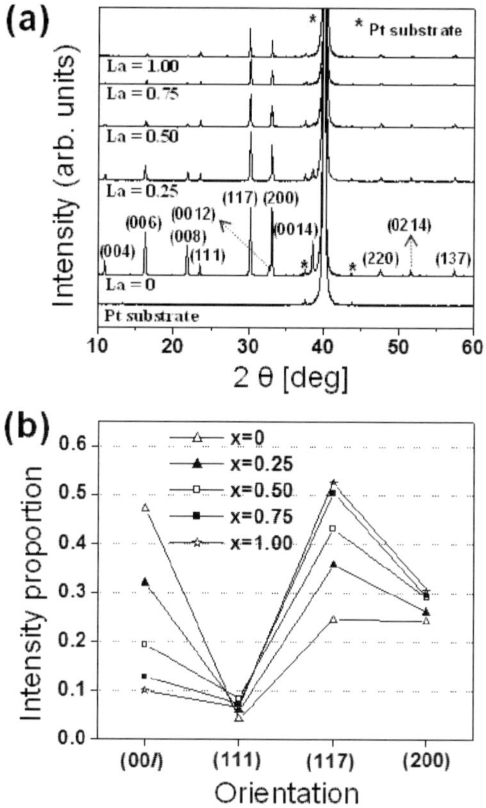

Figure 41. (a) XRD spectra of BLT films as a function of La contents and (b) proportion of XRD intensities in BLT films with various orientations.

As shown in the XRD spectra, according to the increase with La contents, BLT films tended to grow toward the non-c-axis orientations. However, BLT film with x = 1.00 showed slim hysteresis loop due to the reduced ferroelectric distortion represented by their low T_c [81]. The value of E_c decreased gradually with increasing the La contents. Sugita et al. reported that T_c and crystallization temperature of the BLT films decreases with increasing La contents [83].

For higher La contents, since the phase transition temperature reduces with increasing La contents, the residual stress decreases during cooling. As a result, the value of E_c was decreased with increasing La contents.

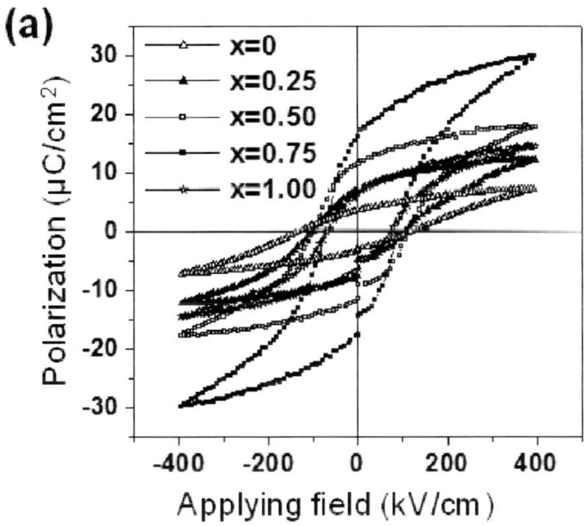

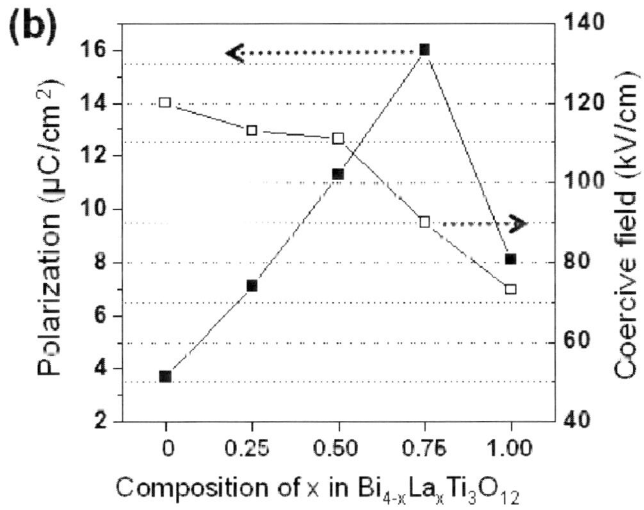

Figure 42. (a) P-E hysteresis loops and (b) P_r and E_c values of $Bi_{4-x}La_xTi_3O_{12}$ films annealed at 700°C as a function of La contents.

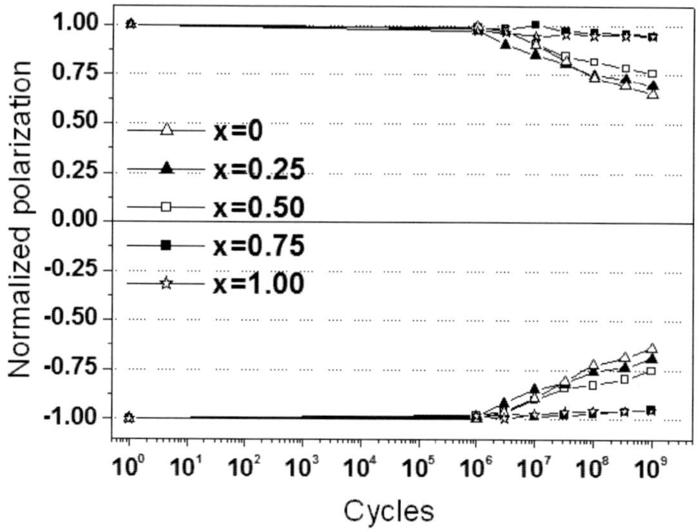

Figure 43. Fatigue characteristics of $Bi_{4-x}La_xTi_3O_{12}$ films annealed at 700°C as a function of La contents.

Figure 43 illustrates the change in polarization of 380-nm thick BLT films as a function of the number of switching cycles using a bipolar square wave of ± 10 V at a frequency of 100 kHz. The pulse widths (τ_1), switching period (τ), and duty cycle (τ_1/τ) were 8.6 μs, 10 μs, and 0.86, respectively. After 10^9 cycles, the remaining remnant polarizations of $Bi_{4-x}La_xTi_3O_{12}$ films with x = 0, 0.25, 0.50, 0.75, and 1.00 were 64.9, 69.5, 75.9, 94.8, and 95.0 %, respectively. According to increasing La contents, fatigue resistance of BLT films was improved. Because of the volatility of Bi and the possible reduction of Ti^{4+} ions to Ti^{3+}, oxygen vacancies may occur in BLT films [81]. It is suggested that lanthanum substitution stabilizes the perovskite blocks and hence decreases the concentrations of oxygen vacancies and holes as the conduction carriers. In addition, as shown in the surface microstructure of Figure 40, surface holes were observed on the BLT films with x = 0, 0.25, and 0.50 while the surface of the BLT films with x = 0.75 and 1.00 was closely-packed without surface holes. This closely-packed smooth surface microstructure might be one of important factors for the fatigue-free behavior because charged defect accumulation which induces a degradation of mobility of domain boundaries could be minimized [84,85].

The possibility of direct-patterning spin-coated BLT precursor films with x = 0.75 was examined and the result is given in Figure 44. To obtain the image, a spin-coated BLT film was exposed to UV for 15 min and then rinsed in hexane. As can be seen in Figure 44, the dark area corresponds to the direct-patterned BLT film and the bright area corresponds to the Si substrate. Based on the well-defined width in the direct-patterned BLT film, micron-scale patterning was successfully obtained by lithographic patterning using PMOD. From the above results, direct lithographic patterning of BLT film can be accomplished by PMOD for ferroelectric device applications.

BLT films as a function of La contents were successfully prepared using photosensitive precursors. Direct-patterning of BLT films by PMOD was accomplished after UV exposure for 15 min and removal of the unexposed area of the film by rinsing with hexane.

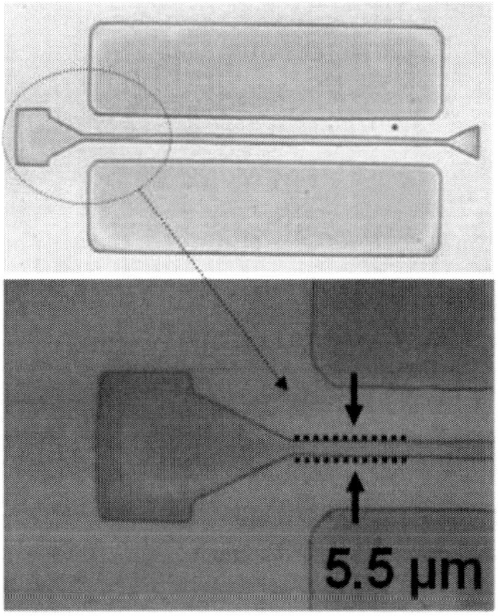

Figure 44. Optical micrograph of direct-patterned $Bi_{3.25}La_{0.75}Ti_3O_{12}$ film after rinsing in hexane.

According to the increase with La contents, BLT films tended to increase (117) and (200) orientations and decrease (00*l*) orientations. The value of P_r increased gradually with increasing the La contents below x = 0.75 due to the growth of BLT films toward non-c-axis orientations and decreased thereafter due to the reduced ferroelectric distortion represented by their low T_c. Also, the value of E_c decreased gradually with increasing the La contents due to the reduced residual stress.

The fatigue resistance of BLT films with x = 0.75 and 1.00 showed better than that of BLT films with x = 0, 0.25, and 0.50 due to the reduced concentrations of oxygen vacancies and holes and the enhanced surface microstructures.

2.1.5.2. Structural and Electrical Properties of $Bi_{4-x}Nd_xTi_3O_{12}$ Thin Films

Several studies have investigated the properties that make ferroelectric thin films suitable for application to nonvolatile memory; these properties include large polarization, low leakage current, and high fatigue resistance. Representative ferroelectric materials include materials with a layered perovskite and perovskite structures. The Aurivillius family, generally represented by $M_2A_{n-1}B_nO_{3n+3}$, contains fluorite-like $[M_2O_2]$ units and perovskite-like $[A_{n-1}B_nO_{3n+1}]$ blocks. In the Aurivillius phase, the perovskite A site (A being 12-coordinated) is normally occupied by large cations such as Ca^{2+}, Sr^{2+}, Ba^{2+}, Bi^{3+}, and Ln^{3+}, and the B site by d^0 cations such as Ti^{4+}, Nb^{5+}, and W^{6+} according to the formula $[M_2O_2][A_{n-1}B_nO_{3n+1}]$. 'n' is the number of perovskite-like layers and can be 2, 3, 4, or 5. $SrBi_2Ta_2O_9$ and $Bi_4Ti_3O_{12}$ are representative Aurivillius family ferroelectric materials [34,86,87]. In particular, $Bi_{4-x}Nd_xTi_3O_{12}$ (BNT) of the Aurivillius family has been intensively investigated as a material suitable for nonvolatile memory due to its relatively low phase formation temperature, large polarization, and good fatigue properties [88,89]. We fabricated BNT films using a direct-patternable PMOD method without a conventional dry etching process in order to study the changes in the structural and electrical properties of the films according to the amount of Nd^{3+} substituted for Bi^{3+}. During the phase formation of BNT at high temperatures,

Bi^{3+} evaporates in the form of Bi_2O_3 and induces oxygen vacancies [91]. These defects result in fatigue and leakage, limiting the application of the resulting films to ferroelectric devices [34,95].

Nd^{3+} substitution for Bi^{3+} in $Bi_4Ti_3O_{12}$ ferroelectric films has been investigated with respect to oxygen defects and growth orientation, stress, tilt of the titanium octahedral, lattice distortion, and polarization. However, conflicting results for oxygen defects, polarization, and electrical properties have been reported [89,91,96-99]. In this part, variations in the structure, Ti-O octahedrals, and electrical properties of $Bi_{4-x}Nd_xTi_3O_{12}$ ferroelectric thin films substituted with various amounts of Nd^{3+} of small ionic radius [r(Nd^{3+})= 1.27 Å] and subjected to strong ionization were investigated.

The precursors for the photochemical production of BNT films were bismuth(III) 2-ethylhexanoate, titanium(III) isopropoxide, and neodymium(III) acetylacetonate-hydrate. A 10% excess of bismuth 2-ethylhexanoate was added to compensate for the loss of Bi during high temperature annealing. The BNT precursors dissolved in hexane were spin-coated at 3000 rpm for 30 s on two different substrates. One substrate was a chemically cleaned Pt(111)/TiO_2/SiO_2/Si(100) used to measure the electrical properties of the resulting films, and the other substrate was p-Si(100) used for direct-patterning. For PMOD, the precursor films were lithographically exposed for 15 min using the output of a 1000 W mercury arc lamp that was passed through a chromium mask. The exposed film was rinsed with hexane to develop the latent negative pattern. The resultant film was annealed for 10 min at 400°C. This process was repeated for each layer. The entire multilayer film structure was annealed at 700°C for 1 hr under ambient O_2 in a tube furnace. Images of the direct-patterned BNT films were obtained using optical microscopy.

The phase formation and crystallinity of the films were studied with XRD θ-2θ mode scans using Cu K_α radiation. A 150-nm-thick Pt top electrode with a 100-μm radius was sputter-deposited for P-Eand I-Vcharacterization using RT66A and HP4145 semiconductor parameter analyzers, respectively. The experiments using a synchrotron radiation source were performed at the Pohang Light Source. To acquire the Ti $2p_{3/2}$ core level and the O 1s absorption spectra, photoemission spectroscopy (PES) and X-ray absorption spectroscopy (XAS) were employed at the 4B1 beam line. The XRD spectra of $Bi_{4-x}Nd_xTi_3O_{12}$ films formed on the Pt(111) substrate are shown in Figure 45. All films were well-crystallized with a (117) preferred orientation and no secondary phase. The diffraction data revealed a small shift in the main (117) peak with variations in x. Variations in the lattice parameters according to x were calculated from all of the diffraction peaks, and the results are presented in Figure 46. A decrease in the a and b axis parameters and an increase in the c-axis parameter were observed as x increased. $Bi_{4-x}Nd_xTi_3O_{12}$ also showed a tendency to form tetragonal structures with reduced orthorhombicity due to a decrease in the difference between the a-axis and b-axis parameters as x increased. The changes in lattice parameters up to x=0.65 indicate that substitution of Bi^{3+} with Nd^{3+} in $Bi_{4-x}Nd_xTi_3O_{12}$ reduces lattice distortion, decreases lattice volume, and results in a more tetragonal crystal structure [91,92]. However, the BNT films showed a significant decrease in the c-axis parameter and an increase in the a-axis parameter at x= 0.85. An abrupt change in the lattice parameter could be explained by the triggered phase transition reported by Makoto Iwata et al. [93,94]. The energy states of the titanium $2p_{3/2}$ electrons and the X-ray absorption of the oxygen 1s electrons were obtained to analyze the electronic structure modification according to the substitution with Nd.

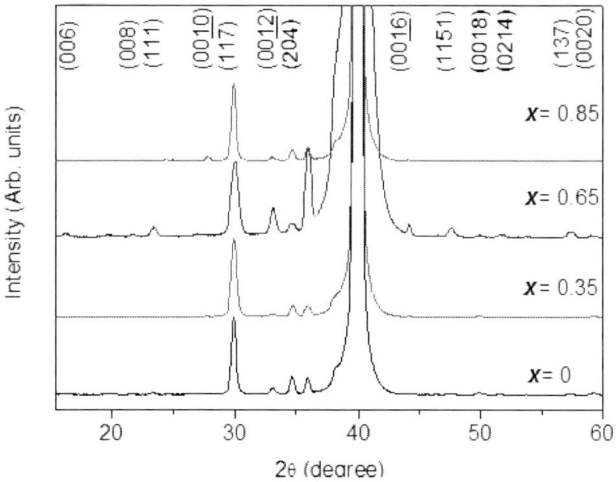

Figure 45. X-ray diffraction patterns of $Bi_{4-x}Nd_xTi_3O_{12}$ films.

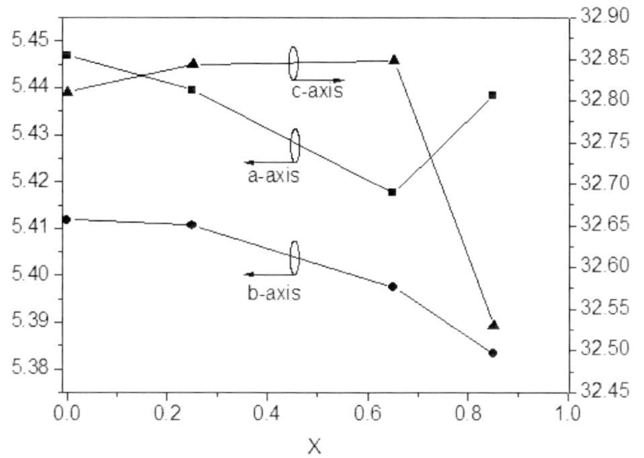

Figure 46. Lattice parameter variations in $Bi_{4-x}Nd_xTi_3O_{12}$ films according to x.

Figure 47 shows the measured binding energies of the Ti $2p_{3/2}$ core electrons in the $Bi_{4-x}Nd_xTi_3O_{12}$ ferroelectric films with various amounts of Nd^{3+}. The binding energy of the $2p_{3/2}$ photoelectron of Ti^{3+} was 1.5 eV lower than that of Ti^{4+} at x= 0.85. The trivalent titanium $2p_{3/2}$ electron peak was shifted due to a negative charge effect, and this corresponds to the occurrence of oxygen vacancies. This indicates that oxygen vacancies did not form until x= 0.65. However, as the Nd^{3+} content increased, a shift in the binding energy of the Ti $2p_{3/2}$ photoelectron peaks was observed, for example, 457.96, 457.83, 457.66, and 457.56 eV for x-values of 0, 0.35, 0.65, and 0.85, respectively. Furthermore, the values of full width at half maximum (FWHM) for the Ti $2p_{3/2}$ peaks varied with increasing x, for example, 1.25, 1.13, 1.08, and 1.83 eV for x-values of 0, 0.35, 0.65, and 0.85, respectively. Through Nd^{3+} substitution for Bi in the perovskite layer, more stable orbital hybridization between the Nd 5d and the O 2p states occurred because ionization of Nd is easier than that of Bi.

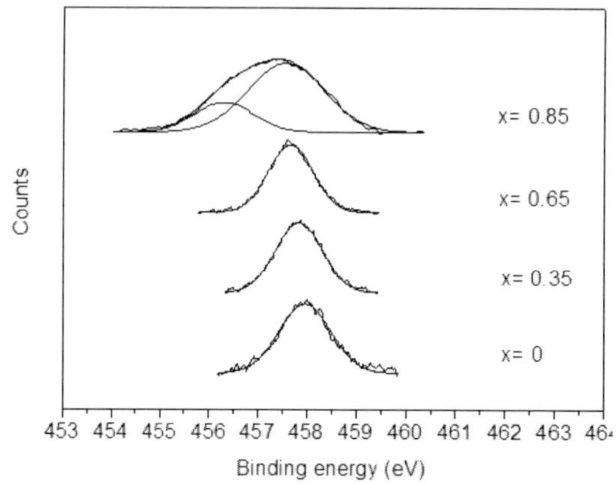

Figure 47. Binding energies of Ti $2p_{3/2}$ photoelectrons in $Bi_{4-x}Nd_xTi_3O_{12}$ films according to x.

The increased stabilization of Ti in the Ti-O octahedron and the reduced distortion in the perovskite layer with increasing values of x could be responsible for the shift in the binding energy of Ti 2p to a lower value and the decreased FWHM.

However, an abrupt increase in FWHM at x=0.85 was observed due to lattice distortion and oxygen defects [96]. Concerning with this lattice distortion at a high substitution rate, Raman spectral analyses of the resulting films at x= 0.85 could be suggested that this phenomenon is closely related to the substitution of Bi in the $Bi_2O_2^+$ layer with Nd^{3+} [100,101]. The decrease in Ti $2p_{3/2}$ photoelectron binding energy was confirmed by the increase in covalent mixing as shown in the XAS results in Figure 48 [102].

Figure 48 shows the X-ray absorption patterns of the O 1s electrons that show orbital hybridization with Bi 6p and Ti 3d. The XAS spectra reveal splitting of the t_{2g} and e_g states, and these results show good agreement with those predicted by the octahedral crystal field theory. An increase in the I_{t2g}/I_{eg} ratio was observed with increasing Nd^{3+}, but a significant decrease in this ratio was observed at x= 0.85. The increase in the I_{t2g}/I_{eg} ratio corresponded to an increase in covalent mixing [103].

This is consistent with a previous report that demonstrated an increase in covalent mixing and stabilization as Bi^{3+} was substituted with La^{3+} [104].

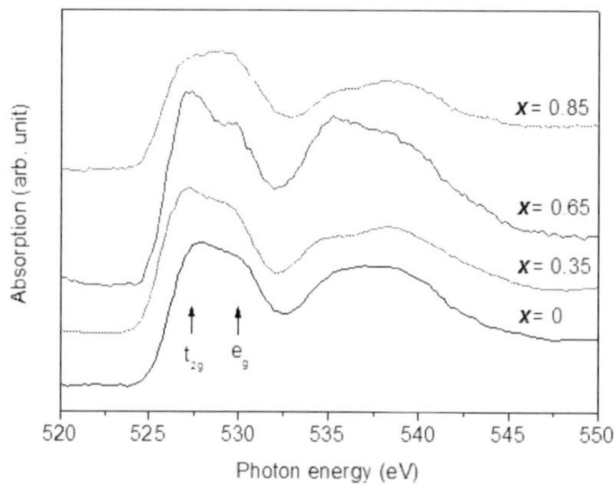

Figure 48. X-ray absorption spectra of the oxygen K-edge in $Bi_{4-x}Nd_xTi_3O_{12}$ films according to x.

The decreases in t_{2g} and e_g transition energies of oxygen 1s clearly demonstrate an increase in covalent mixing. The XAS of oxygen 1s also agrees with the shift in binding energy of Ti $2p_{3/2}$ shown in Figure 47. The covalent character of the Ti-O bond increased due to an increase in the amount of Nd-O bonds because of the greater ionic character of Nd^{3+} than Bi^{3+}. Furthermore, the sharpness of the t_{2g} and e_g peaks were enhanced with a decrease in the spreads of the t_{2g} and e_g energy states.

These decreases can be understood in terms of a reduction in distortion and an increase in the centro-symmetry of the films [105]. A significant broadening and decrease of the I_{t2g}/I_{eg} ratio were observed at x= 0.85 due to the generation of oxygen vacancies; the existence of oxygen vacancies was confirmed by the PES results shown in Figure 47. Non-bonding electrons generated due to the oxygen vacancies occupied the relatively stable d_{xy} orbital in the Ti t_{2g} state, reducing the transition probability of the O 1s electron to that of the Ti t_{2g} state. A significant decrease in the I_{t2g}/I_{eg} ratio was therefore observed. Jahn-Teller distortion and a decrease in Ti octahedral symmetry were induced by the Ti d^1 electron configuration, both of which could affect the electrical properties of the BNT films. Figure 49 shows the hysteresis loops recorded from $Bi_{4-x}Nd_xTi_3O_{12}$ films by pulsing in virtual ground mode at 15 V. The films showed a small decrease in polarization with increasing Nd^{3+} until x= 0.65, but an abrupt increase in the remnant polarization by about 5 μC/cm^2 was observed at x= 0.85. This increase in remnant polarization was caused by Jahn-Teller distortion, which is responsible for TiO_6 octahedral distortion as discussed in Figs. 47 and 48 [96].

The typical current-voltage dependence of Pt/BNT/Pt capacitors is shown in Figure 50. A decrease in leakage current and an increase in the break-down voltage were observed with increasing Nd^{3+} due to an increase in structural stability.

However, an abnormal increase in leakage current and a transition of the conduction mechanism from Schottky to Poole-Frenkel were observed around 1.6 V at x= 0.85 due to the generation of oxygen vacancies. Figure 51 shows a rectangular patterned image of a $Bi_{3.35}Nd_{0.65}Ti_3O_{12}$ film formed on a silicon substrate using a chromium mask and exposure to UV radiation. BNT films of size 75 μm x 50 μm were found to direct-pattern well with inscription of a picture.

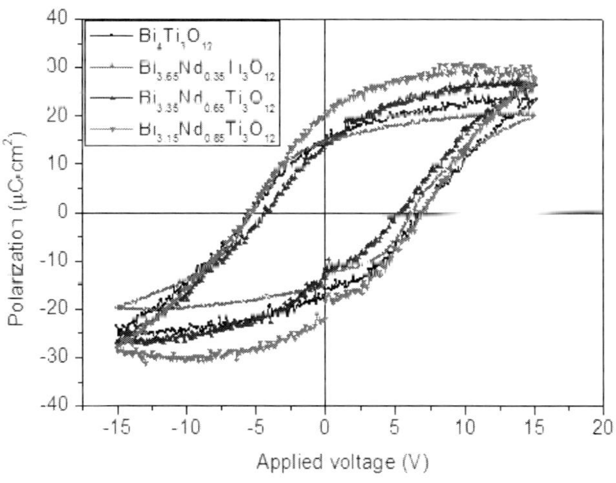

Figure 49. P-E hysteresis loops of $Bi_{4-x}Nd_xTi_3O_{12}$ films according to x.

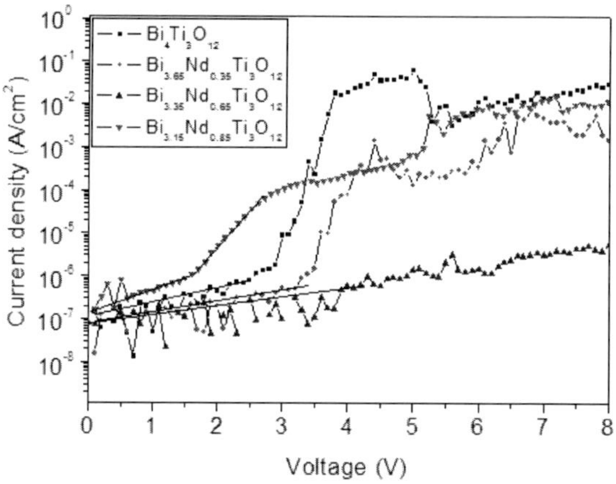

Figure 50. Current-voltage characteristics of $Bi_{4-x}Nd_xTi_3O_{12}$ films according to x.

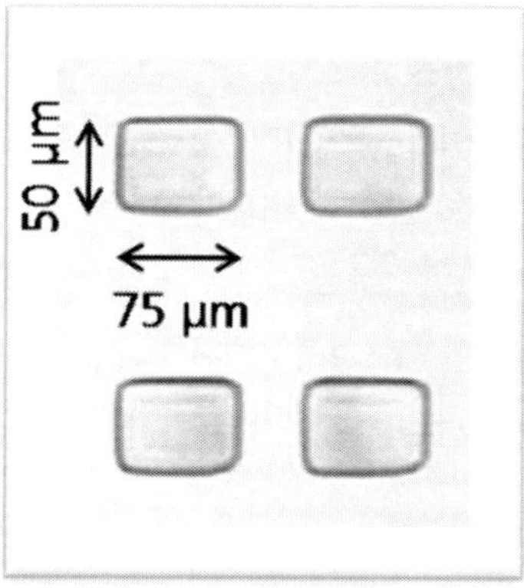

Figure 51. A direct-patterned image of $Bi_{3.35}Nd_{0.65}Ti_3O_{12}$ film.

In this part, we investigated changes in the electrical properties, crystal structures, and chemical bonds of $Bi_4Ti_3O_{12}$ ferroelectric films according to substitution of Nd^{3+} for Bi^{3+}. Variations in the lattice parameters were measured by XRD. Increases in covalent mixing, centro-symmetry, and Ti-O octahedron stabilization upon substitution of Nd^{3+} for Bi^{3+} were observed by X-ray absorption analysis of the oxygen K-edge. A shift to a lower binding energy in the Ti $2p_{3/2}$ core level was also observed with the substitution of Nd^{3+} ions for Bi^{3+} ions. Stress release between the perovskite layer and the fluorite-like layer was observed with Nd substitution, but distortion, oxygen defects, and abnormal lattice parameter change from the triggered phase transition were observed at x= 0.85 as Bi was substituted with Nd in the $[Bi_2O_2]^{2+}$ layer. The $Bi_{4-x}Nd_xTi_3O_{12}$ ferroelectric films showed controllable ferroelectric properties when Nd was substituted for Bi up to x=0.65.

2.1.5.3. Development of Sol-Gel Precursor System for PZT Thin Films Containing Various La Contents

In this part, the improvement of ferroelectric properties could be obtained by doping with A (Pb)-site donors as La [106].

The influence of donor dopants on the films as well as their ferroelectric and other properties were measured. Ferroelectric $Pb_{1-x}La_x\ (Zr_{0.52}Ti_{0.48})_{1-x/4}O_3$(x = 0, 2, 4, and 6 mol%) thin films were prepared by sol-gel procedure. Zirconium n-butoxide [$Zr(O_nC_4H_9)_4$, ZNB], titanium iso-propoxide [$Ti(O_iC_3H_7)_4$], lead acetate tri-hydrate [$Pb(CH_3CO_2)\cdot 3H_2O$] and lanthanum iso-propoxide {$La[OCH(CH_3)_2]_3$} were used as starting materials. Di-ethanolamine and NBAL were dissolved in ethanol solvent and were used as stabilizer and photosensitive agent, respectively. 0.5M concentration of PLZT solutions were prepared with Zr:Ti ratio of 52:48 at morphotropic phase boundary where a meta stable coexistence of tetragonal and rhombohedral phases results in the maximum poling efficiency and electromechanical activity and induces high dielectric constant and piezoelectric coefficient [107].

To prevent Pb loss during high temperature anneal, an excess 15 mol% Pb was added. PLZT precursor solution was then obtained after mixing and stirring for 24 hr. The solution was deposited on Pt(111)/Ti/SiO_2/Si(100) for PLZT film with spinning speed of 3500 rpm for 30 s. After the spin coating, the films were dried at 70°C for 3 min using hot plate and successively irradiation to UV for 3 min using a mask for negative patterning of film.

The developed substrates were pre-fired at 400°C for 20 min in oxygen atmosphere to eliminate organic components completely. Before final anneal, this step is repeated three times to obtain the desired thickness of film. The films were finally annealed at 650°C by direct insertion to furnace under O_2 ambient for 45 min and 500 nm thick PLZT films could be obtained. In order to measure the electrical properties, platinum was sputter-deposited for top electrode. The phase formation and orientation of film were investigated using XRD with $Cu_{K\alpha}$ radiation in the angle range from 20° to 60°. Polarization-electric field, current-voltage, and capacitance-voltage properties of films were measured using RT66A standard ferroelectric test system, HP4145B semiconductor parameter analyzer, and HP4248A precision LCR meter, respectively.

XRD patterns for various La doping concentrations (La = 0, 2, 4, and 6 mol%) coated on the Pt(111)/Ti/SiO_2/Si substrate after final anneal treatment at 650°C are shown in Figure 52. The diffraction patterns of PLZT films represent that only perovskite phase was formed for the crystal structure. The pyrochlore phase was not detected. The PLZT films are strongly textured having a <111> orientation from the influence of Pt (111) substrate. However the relatively different distribution of XRD intensity was observed with the La concentration.

The increase of relative intensity of (100) and (200) diffraction peaks with increasing La concentration implies that La promoted the nucleation in PLZT films and the growth orientation of PLZT films was less influenced from Pt substrate [70,108].

Figure 53 shows the dielectric constant-voltage plots with applied voltage range from −8 to 8 V. The calculated dielectric constant of the PZT film with 0 mol% of La was 1236 at a maximum capacitance voltage.

However, with the addition of La, the dielectric constant increased. The doping with La reduces the domain pinning which is induced by space charge originated from the accumulation of oxygen vacancies at PZT/PT electrode interface. La^{3+} ion acts as a donor defect in PZT perovskite crystal structure, which can compensate oxygen vacancies.

As a consequence, an enhanced domain mobility as well as higher dielectric constant than PZT can be obtained by the reduction of oxygen vacancies [108-110].

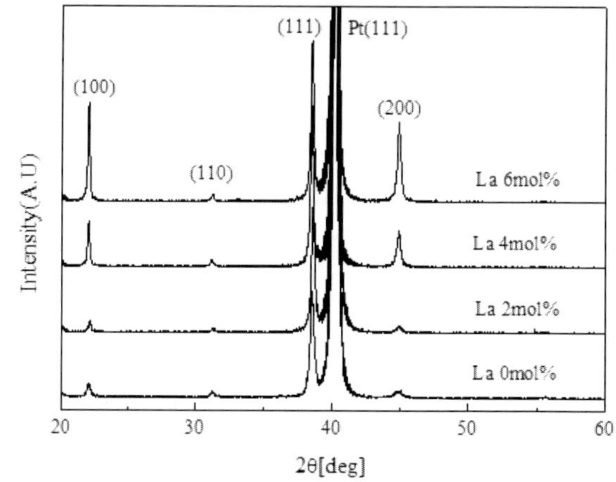

Figure 52. X-ray diffraction spectra of PLZT films annealed at 650˚C with various La concentrations.

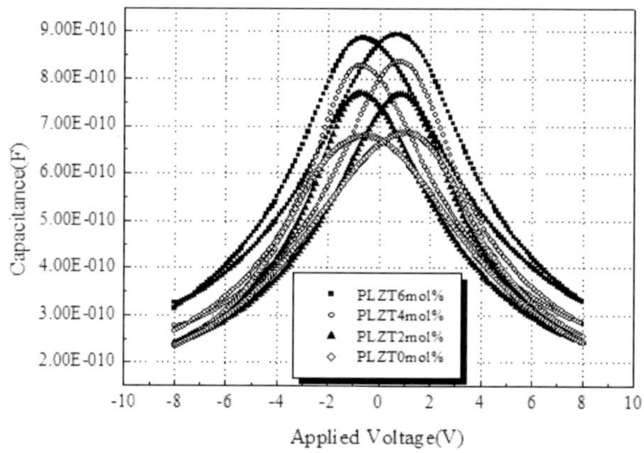

Figure 53. C-V characteristics of PLZT films annealed at 650˚C with various La concentrations.

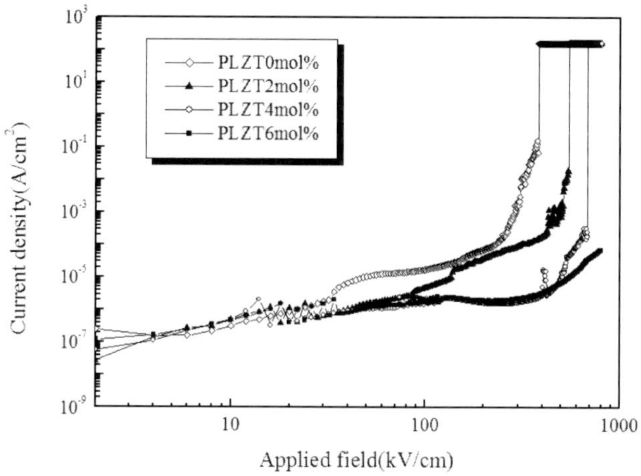

Figure 54. I-V characteristics of PLZT films annealed at 650˚C with various La concentrations.

Figure 54 shows the I-V characteristics of 500 nm thick PLZT films annealed at 650°C for 45 min under O_2 atmosphere. Leakage current density of the PZT (La 0 mol%) thin film was on the order of 10–5 A/cm^2 at the voltage less than 40 V. However the leakage current density characteristic was improved with increasing La concentration.

The breakdown voltage value of PLZT (6 mol%) is more higher than the value of PLZT (4 mol%) even if the leakage current density value is similar to that of the PLZT (4 mol%).

The major leakage current source of crystalline PZT related compounds has been known as Pb and O vacancies created during high temperature anneal for crystallization [29]. From this, it can be said that doping with La reduced the defect dipole concentration [111].

P-E hysteresis loops of diect-patternable PLZT films under applied voltage of 3, 5, 8, and 10 V using RT66A ferroelectric tester are shown in Figure 55. It exhibited well-defined hysteresis loops and saturated behavior. With 0 mol% of La, the P_r and E_c values were measured as 25.8 μC/cm^2 and 32.2 kV/cm, respectively. It can be seen that P_r and E_c values decrease with increasing La concentration. It is expected that the introduction of La reduces the tetragonality of PLZT films on the basis of relative ionic radii, 1.32 Å for Pb^{2+} and 1.22 Å for La^{3+}. With increasing La concentration, PLZT film transformed from ferroelectric state to paraelectric state [112,113].

Figure 56 illustrates the degradation of polarization of PLZT film as increasing the number of switching cycle through bipolar square wave of ±10 V at 500 kHz. The remained remnant polarization value with increasing La concentration increased after 10^9 cycles. As

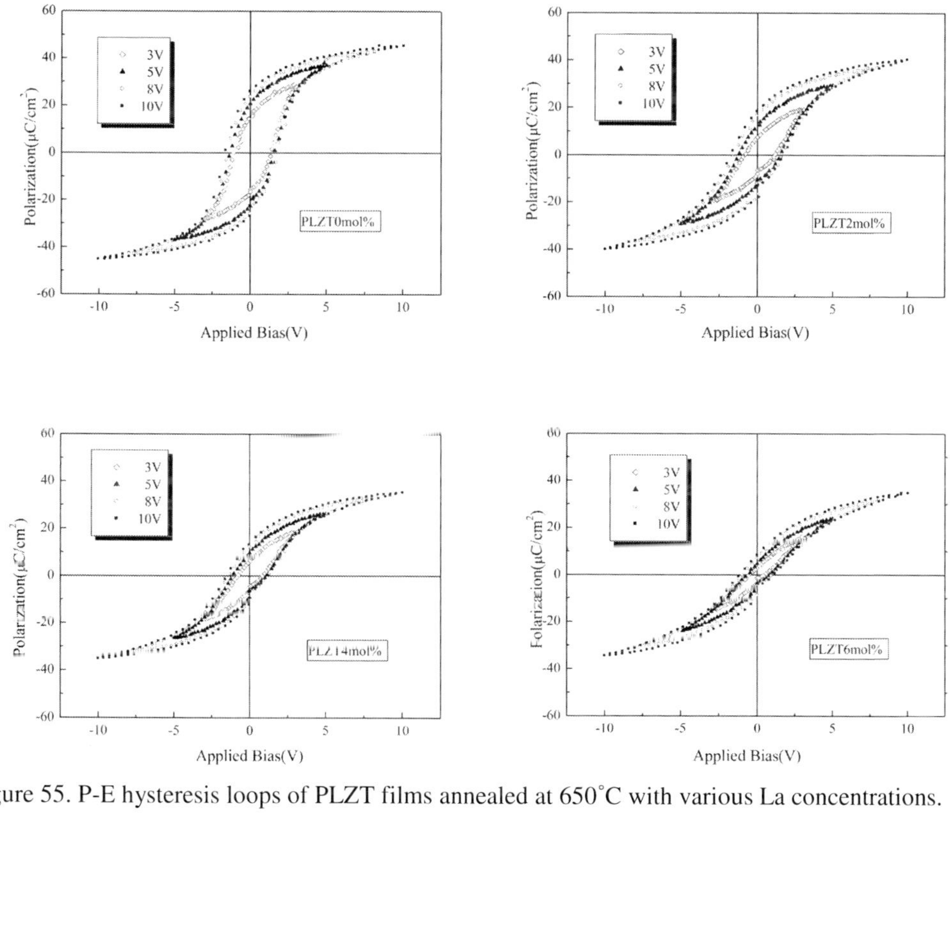

Figure 55. P-E hysteresis loops of PLZT films annealed at 650°C with various La concentrations.

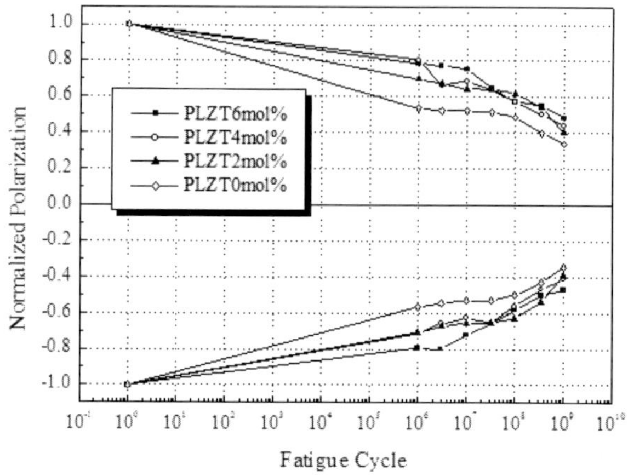

Figure 56. Fatigue characteristics of PLZT films annealed at 650˚C with various La concentrations.

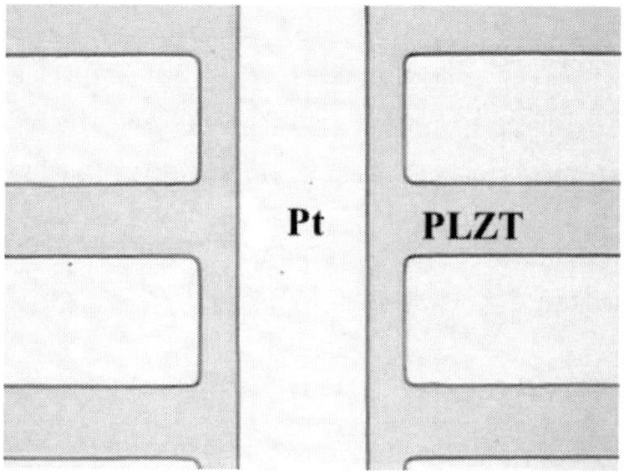

Figure 57. Optical micrograph of direct-patterned PLZT film with 4 mol% of La.

mentioned above in the C-V and I-V measurement results, with increasing La concentration, the reduction of defect dipole such as Pb and O vacancies near the interface induces an amelioration of the electrode charge trapping and domain wall pinning [34]. An optical micrograph of direct-patterned PLZT with 4 mol% of La was given in Figure 57. Dark area corresponds to the PLZT films and bright area to Pt substrate. Direct-patterned PLZT films were clearly formed by UV irradiation with complete removal of unexposed region. From this result, it can be said that direct-patternable PLZT film is suitable for micro electromechanical system. The improvement of leakage behavior and fatigue characteristic was obtained through the doping with La. La^{3+} would tend to reduce the trapping charge such as Pb and O vacancies near the film/electrode interface. The tendency was also confirmed by the result of C-V measurement. However remnant polarization and coercive field values decreased due to the reduction of tetragonality.

2.1.5.4. Electrical Properties of PLZT Thin Films with Various Zr/Ti Ratios

A number of papers have reported on the effects of doping on the electrical properties of PZT films [114]. One of alternatives to improve the electrical properties is doping with A (Pb)-site donors like La^{3+}. However, a structure composition relation of PLZT film at morphotropic phase boundary (MPB) [115] composition is modified due to the doping with La. Therefore, it is important for PLZT film to find the MPB composition exhibiting the maximum ferroelectric properties. The electrical properties of PLZT films by PMOD with various Zr/Ti ratios are still unknown near the MPB composition.

In this part, a direct patterning method for the lithographic deposition of PLZT films with various Zr/Ti ratios was introduced by using PMOD. The ferroelectric and electrical properties of direct-patternable PLZT films with UV-irradiation process were investigated near MPB composition through the variation of Zr/Ti ratio. The precursors used for the photochemical production of PZT or PLZT films were lead(II) 2-ethylhexanoate, lanthanum(III) 2-ethylhexanoate, zirconyl(IV) 2-ethylhexanoate, and titanium(IV) isopropoxide. The composition ratio of Zr/Ti was varied from 56/44 to 44/56 with fixed La concentration of 3 mol%. Also, 10% of extra Pb was added for the compensation of Pb-loss during high temperature anneal. PZT or PLZT precursors dissolved in hexane were spin-coated at 2000 rpm for 30 s on two different kinds of substrate. One was chemically cleaned Pt(111)/Ti/SiO_2/Si(100) for monitoring of ferroelectric properties and the other was a p-Si(100) wafer, to show a generality of direct-patterning process. 350 nm thick PLZT film by PMOD was prepared after both a spin-coating of PLZT precursor film and UV exposure once in order to demonstrate a possibility of directpatterning.

The precursor films for lithography were exposed for 30 min with 1000W of mercury arc lamp passed through a chromium mask. The latent image was developed by rinsing with hexane to reveal a negative pattern. The patterned film was annealed for 10 min at 400°C. This process was repeated for each layer. Following the formation of direct-patterned structure, the patterned films were annealed at 650°C for 30 min. A 150 nm thick Pt top electrode with a 200 μm diameter was sputter-deposited and then annealed at 650°C for 10 min under O_2 ambient. Phase formation and crystallization were monitored using XRD. For the measurement of electrical properties, 300 nm thick film was prepared by repeated coating. The I-Vcharacteristics were obtained using HP4145B (Hewlett Packard). The ferroelectric hysteresis loop and polarization fatigue were measured using a RT66A (Radiant Technology) ferroelectric tester. XRD patterns of PLZT films coated on Pt(111)/Ti/SiO_2/Si substrate are given in Figure 58. PLZT films with various Zr/Ti ratios were fully crystallized after anneal treatment at 650°C for 30 min under O_2 ambient. The diffraction peaks were indexed only with perovskite structure with preferred orientation of (111). The highly preferred orientation means that the nucleation of PLZT films was strongly affected from Pt substrate due to the lattice matching between Pt and PLZT [46]. According to the decrease with Zr/Ti ratio, relative intensities of (111) diffraction increased. Due to smaller size of La compared with Pb, average size of A site cation decreases (ionic radii of Pb^{2+} and La^{3+} are 1.21 Å and 1.15 Å, respectively [116]) and in this case, PLZT film with ABO_3 structure shows higher lattice matching with Pt (111) substrate when B site cation size decreases. For example, because ionic radius of Ti^{4+} is smaller than that of Zr^{4+}, PLZT film with Zr/Ti ratio of 44/56 showed the most intensified (111) diffraction in Figure 58.

P-E hysteresis loops of PZT and PLZT films as a function of Zr/Ti ratio with 15 voltages are given in Figure 59. The measured P_r values of PZT films with Zr/Ti ratio of 60/40, 56/44, 52/48, 48/52, and 44/56 were 26, 32, 29, 27, and 24μC/cm^2, respectively. Maximum P_r value was obtained with Zr/Ti ratio of 56/44, which relates to MPB composition. However, PZT bulk ceramics show maximum P_r value of the vicinity of Zr/Ti ratio of 52/48 [117]. The MPB

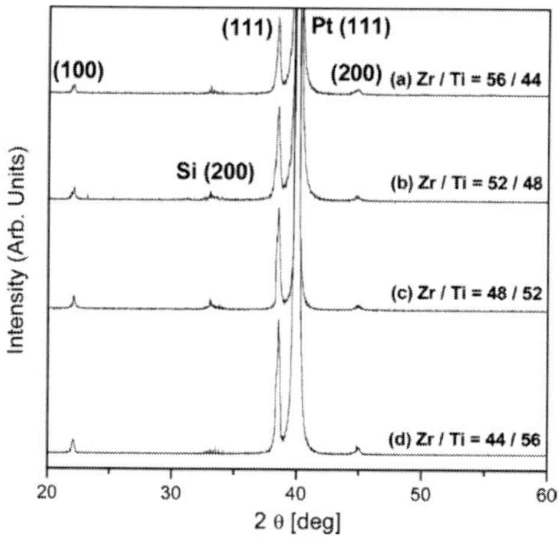

Figure 58. XRD patterns of PLZT films annealed at 650°C as a function of Zr/Ti ratio.

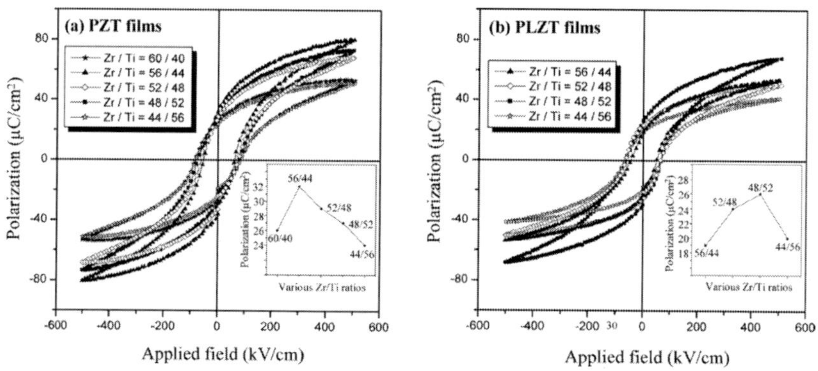

Figure 59. P-E hysteresis loops of (a) PZT and (b) PLZT films annealed at 650°C with various Zr/Ti ratios.

composition of PZT films tends to move toward rhombohedral phase field, Zr-rich side, under the influence of in-plane film stress, regardless of whether it is tensile or compressive [118]. In case of PLZT films, the measured P_r values with Zr/Ti ratio of 56/44, 52/48, 48/52, and 44/56 were 19, 24, 26, and 20 μC/cm^2, respectively. MPB composition of PLZT thin film could be assigned between 52/48 and 48/52 of Zr/Ti ratio when considering a symmetric slope change in P_r values with Zr/Ti ratio. This result is contrary to the observation of G.H. Haertling that the MPB compositions of PLZT ceramics shift to more Zr-rich side than that of PZT ceramic [119].

However Y. Honda found that the MPB composition of PLZT films was expanded in Zr/Ti ratio (for example, in case of 5% of La/(Pb + La) ratio, coexistence region of tetragonal and rhombohedral phases was found as 50/50–65/35 of Zr/Ti ratio), different with that of PLZT bulk ceramics [120]. This might be due to the variation in the residual stress, internal microstructure, growth orientation affected by substrate, and so on. Furthermore in case of PSZT ceramics, PZT-modified by Sr (smaller ionic size than Pb, same as La), MPB composition is changed from Zr-rich side to Ti-rich side according to incorporated amount of Sr [121]. As a result, the shift in the MPB composition toward Ti-rich region in our PLZT film comparing with PLZT bulk ceramics could be explained by the reduction in residual stress due to the incorporation of La and high {111} growth orientation relation with Pt substrate.

Figure 60 shows capacitance-voltage plots of PZT and PLZT films with applied voltage ranging from −10 to 10 V as a function of a Zr/Ti ratio. A dielectric constant variation was given as the inset of the figure. In case of PZT film, maximum dielectric constant was obtained with the film of Zr/Ti ratio of 56/44. The composition is identical with that obtained from the P-E measurement. The dielectric constants of PLZT films with Zr/Ti ratios of 56/44, 52/48, 48/52, and 44/56 were 904, 1078, 1167, and 1051 as given at the inset of Figure 60(b). Measured maximum dielectric constant, 1167 was obtained with PLZT film of Zr/Ti ratio of 48/52. However, considering the positive and negative slopes of dielectric constant values according to Zr/Ti ratio, the Zr/Ti ratio for real maximum dielectric constant could be assigned between 52/48 and 48/52. This is well agreed with the observation through P-E measurements.

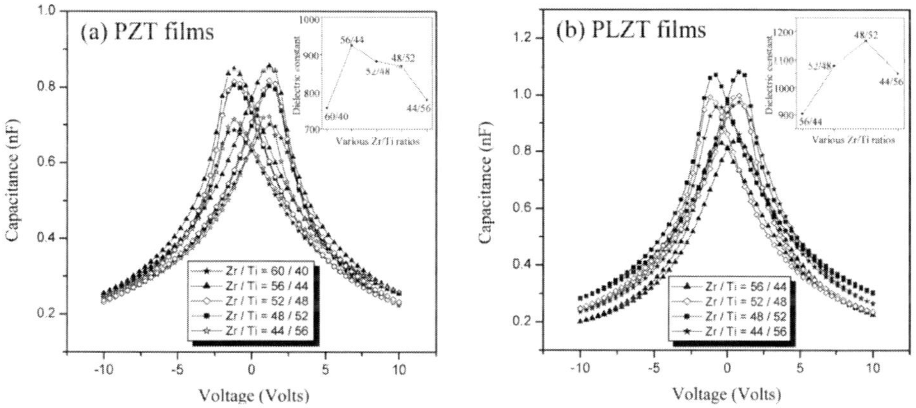

Figure 60. C-V characteristics of (a) PZT and (b) PLZT films annealed at 650°C as a function of Zr/Ti ratio.

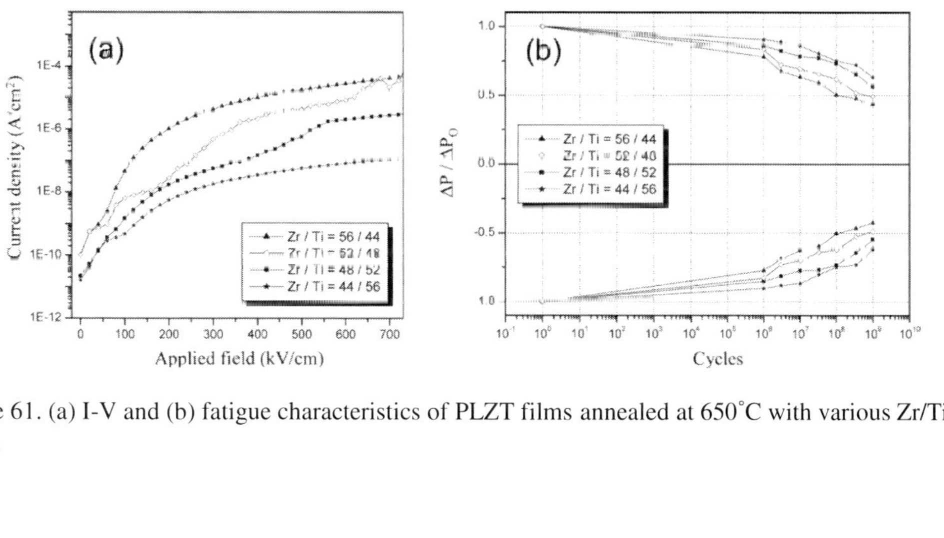

Figure 61. (a) I-V and (b) fatigue characteristics of PLZT films annealed at 650°C with various Zr/Ti ratios.

I–V characteristics of 300 nm thick PLZT films annealed at 650°C for 30 min are shown in Figure 61(a). According to the decrease with Zr/Ti ratio, I–V characteristics of PLZT films were enhanced. The major leakage current source of crystalline PLZT has been known to be Pb or O vacancies created during high temperature anneal for crystallization [122]. With the decrease of Zr/Ti ratio in PLZT, perovskite phase formation temperature is lowered [123]. The defect concentration of Pb and O vacancies was decreased due to lowered temperature of perovskite phase formation, namely, Zr/Ti ratio of 44/56 showed the lowest leakage. Furthermore, the reduction of grain size with decreased Zr/Ti ratio was found from the changes in the diffraction peak width by using Scherrer relation [124]. The lower leakage current density of small grained film might be due to the inter-grain depletion of grain boundary limited conduction [125-128]. Namely, local space charges near grain boundaries inhibit current flow and the overlapped depletion region of neighboring grain boundaries are substantial cause of low leakage current. Therefore, the leakage current density of PLZT film was improved due to the decrease with Zr/Ti ratio.

The leakage current of PLZT film was enhanced by La-doping through the reduction of oxygen vacancies with excess electrons. Figure 61(b) illustrates the change of polarization of PLZT films as a function of Zr/Ti ratio with a number of switching cycle through a bipolar square wave of ±10 V at a frequency of 500 kHz. After 10^9 cycles, the remaining remnant polarizations of Zr/Ti ratios of 56/44, 52/48, 48/52, and 44/56 were about 44, 50, 57, and 63 %, respectively. The PLZT film with Zr/Ti ratio of 44/56 showed not-excellent but a somewhat improved fatigue resistance. This behavior is consistent with I-V characteristic. Namely, according to the decrease Zr/Ti ratio, I-V and fatigue characteristics were improved. The possibility of direct-patterning of a PLZT film by removing the area unexposed to UV using hexane was examined and the result is given in Figure 62. In the SEM micrograph, the dark-gray area corresponds to a PLZT film with Zr/Ti ratio of 48/52 and the gray area to the Si substrate. The direct-patterned lines are clearly visible with 3μm × 180μm in size. This micrograph indicates that direct lithographic patterning of PLZT films could be accomplished by PMOD.

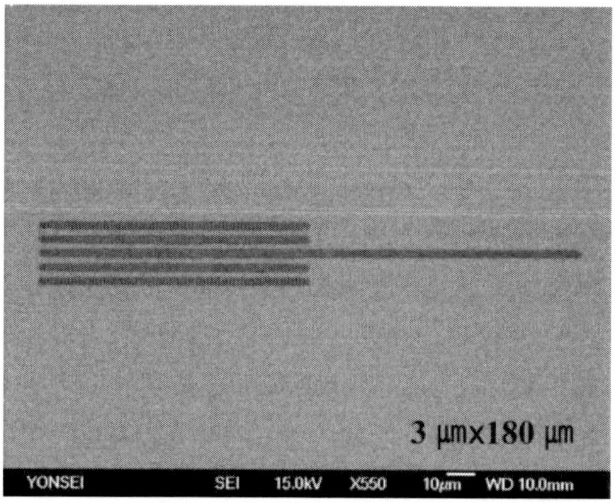

Figure 62. SEM micrograph of UV-direct-patterned PLZT film with the Zr/Ti ratio of 48/52.

PZT or PLZT(3 mol% La) films as a function of Zr/Ti ratio were successfully characterized by using photosensitive starting precursors. Direct-patterning of PLZT films in several microns width by PMOD was accomplished by UV exposure and removal of the unexposed area of the film by rinsing with hexane. Due to high lattice matching between Ti-rich phase and Pt substrate, PLZT film with the lowest Zr/Ti ratio(in this experiment, 44/56) relatively showed the most intensified (111) diffraction. From P-E and C-V measurements, MPB composition of PLZT thin film could be assigned between 52/48 and 48/52 of Zr/Ti ratio. In leakage current and fatigue measurements of PLZT films, a lowering of Zr/Ti ratio induced an improvement of leakage current and fatigue properties due to the lowered perovskite phase formation temperature.

2.1.5.5. Synthesis and Characterization of Ferroelectric Properties of $Ce_2Ti_2O_7$ Thin Films with Ce^{3+}

There are some problems that remain unsolved in FeFET, such as the ineffective distribution of applied voltage inside the gate stack structure due to the presence of low dielectric constant interfacial SiO_2 ($\varepsilon = 3.9$) [129]. In an effort to address these problems, Ma reported that a ferroelectric material with low permittivity and remnant polarization is advantageous towards minimizing side effects from depolarization fields, such as applied-voltage distribution [10]. In addition, the problem of leakage currents can be significantly improved through effective applied-voltage distribution.

Ferroelectric materials with a pseudo-pyrochlore structure of the general formula $A_2B_2O_7$, such as $Sr_2Nb_2O_7$, $La_2Ti_2O_7$ (LT), and $Nd_2Ti_2O_7$ (NT), exhibit a high coercive field, good thermal stability, and a low dielectric constant [130,131]. In addition, compared with conventional ferroelectric materials, such as lead zirconium titanate or strontium bismuth tantalate, these materials do not contain volatile element which may cause leakage problem due to a possible increase of defect density. In this paper, we report the deposition of pseudo-pyrochlore phase ferroelectric $Ce_2Ti_2O_7$ (CT). The film formation of ferroelectric CT with pseudo-pyrochlore structure could be expected from the ionic radii relation of Ce with La, Nd, and Ti. Ce^{3+} is required to form a pseudo-pyrochlore phase with Ti^{4+} ions; thus, we obtained Ce^{3+} from more stable Ce^{4+} by annealing the sample in a vacuum.

CT thin films were deposited by a chemical solution deposition process. High-purity titanium iso-propoxide and cerium acetylacetonate were used as starting materials, and acetic acid was used as a solvent. Acetylacetone was used as a chelating agent for stabilizing the coating solution. The molar ratio of cerium acetylacetonate:titanium iso-propoxide:ethanol:acetic acid:acetylacetone for the CT precursor solution was 1:1:6:82:1. CT films were formed by a spin-coating method on $Pt/Ti/SiO_2/Si$ and Y_2O_3/Si substrates, and subsequent annealing was performed under low-vacuum (6.664 Pa) at 800°C for 30 min in a tube furnace by direct insertion. For low vacuum anneal, a vacuum pump was installed directly at the position of the sealed tube of the furnace and 6.664 Pa was obtained within 1 min. This low-vacuum annealing process is necessary for the formation of the pseudo-pyrochlore phase in CT with Ce^{3+} ions and it also helps to minimize the formation of interfacial silicon oxides during the CT anneal. Y_2O_3 films were deposited by direct current sputtering of the yttrium target in an Ar/O_2 ambient environment and annealing at 800°C for 20 min. Platinum top electrodes were sputter-deposited.

The thicknesses of the Y_2O_3, CT, and platinum films were measured to be approximately 20, 240, and 150 nm, respectively by using spectroscopic ellipsometry and field emission scanning electron microscopy. The crystallinity of CT and the oxidation state of cerium were characterized by synchrotron XRD and X-ray absorption-near-edge-structure (XANES) measurement using the Pohang Light Source facility in Korea. The C-V and ferroelectric characteristics were measured using an HP4284A LCR meter at 1 MHz with a bias sweep rate of 0.18 V/s and a RT66A ferroelectric tester, respectively. The I-Vcharacteristics were measured with HP 4145B semiconductor parameter analyzer. To eliminate the contribution of leakage currents to ferroelectric properties, Radiant Technology's analysis version 2.4 (1995) was used.

The diffraction pattern of CT in powdered form, prepared by sintering at 1200°C for 7 h at 6.664 Pa, was used to confirm the crystal structure of CT and to analyze the diffraction pattern of the CT films. Although Brixner et al. previously reported the synthesis of CT [132], a detailed structural description has not yet been published. A phase-recognition method was used to analyze the crystal structure of CT using the XRD data of LT, which is expected to have the same crystal structure as CT.

As shown in Figure 63 (a), the diffraction pattern of CT is almost identical to the theoretical diffraction result of LT [133], suggesting that CT and LT have the same crystal

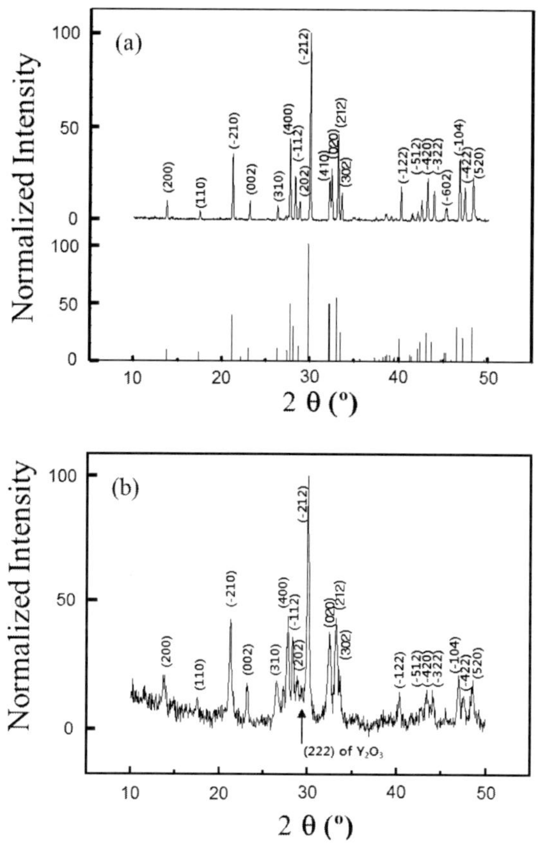

Figure 63. XRD diffraction results of powdered and filmed form of $Ce_2Ti_2O_7$; (a) Theoretical XRD data of $La_2Ti_2O_7$ and experimental data of $Ce_2Ti_2O_7$ in powdered form and (b) Experimental XRD data of $Ce_2Ti_2O_7$ film deposited on an Y_2O_3/Si substrate.

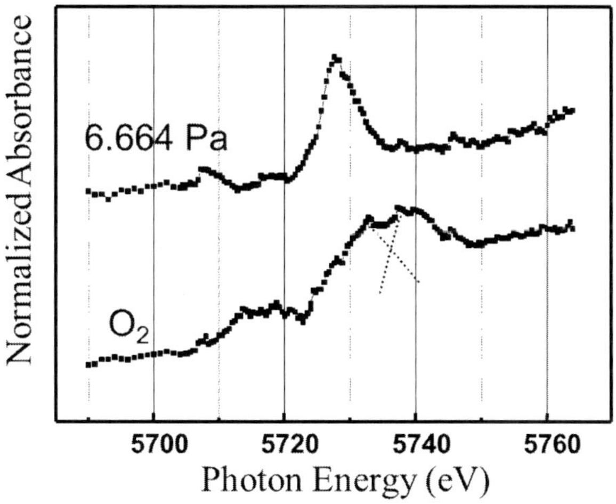

Figure 64. Cerium L_3-edge XANES spectra of $Ce_2Ti_2O_7$ films on Y_2O_3/Si after annealing under two different atmospheric conditions: O_2 and low vacuum (6.664 Pa).

structure. The diffraction peaks of CT could be indexed in a monoclinic system with a lattice parameter of a = 13.018 Å, b = 5.523 Å, and c = 7.749 Å, with β = 98.61°.

Figure 63 (b) represents the XRD pattern of the CT film that was deposited on Y_2O_3/Si. By comparing the LT diffraction pattern with that of the CT powder pattern, we confirmed that the CT thin films are purely of the pseudo-pyrochlore structure and that there is no evidence of a second phase. The pseudo-pyrochlore structure of CT can be also deduced by examining the radius ratio of Ce^{3+}/Ti^{4+}, calculated to be 2.21. Generally, for compounds of the $A_2B_2O_7$ family, a pyrochlore structure is expected for an A^{3+}/B^{4+} radius ratio between 1.46 and 1.80; a pseudo-pyrochlore structure is expected for a ratio between 1.97 and 2.24 [134]. To verify the oxidation state of cerium, we obtained L_{III}-edge XANES spectra of the CT films after annealing under two different atmospheric conditions: O_2 and low vacuum (6.664 Pa). As shown in Figure 64, the low-vacuum sample showed a strong absorption peak at 5728 eV. This peak corresponds to the electron transition from 2p to 5d in Ce^{3+}, where electron transition is symbolized as $2p(4f^1)5d^*$, 2p denotes a hole in the 2p shell with J = 3/2, 5d* refers to the excited electron in the 5d state, and $(4f^1)$ corresponds to the one valence electron of Ce^{3+}.

However, with the O_2-annealed CT, two strong absorption peaks were observed at 5732 and 5739 eV. The peak at 5732 eV corresponds to the electron transition from 2p to 4f of cerium or from 2p of the bonding oxygen to 5d of Ce, symbolized as $2p4f^15d^*L$, where L denotes a 2p electron hole in the bonding oxygen to cerium; the peak at 5739 eV corresponds to the $2p(4f^0)5d$ transition.

These two absorption peaks are typical of Ce^{4+} [135-138]. From these observations, the oxidation state of cerium is suggested to be controlled by the appropriate choice of annealing atmosphere, that is, Ce^{3+} can be obtained by low-vacuum anneal and Ce^{4+} by O_2 anneal. Figure 65 represents the electrical properties of vacuum-annealed CT in a Pt/CT/Pt/Ti/SiO_2/Si structure. The dielectric constant of the CT films was 48.7 at 1 MHz and the dielectric loss was 0.013. These values are comparable with the well-known values of NT and LT [139,140].

Figure 65(b) shows a P-E (polarization-applied field) hysteresis loop obtained for samples with a Pt/CT/Pt/Ti/SiO_2/Si structure. We are certain that the ferroelectric properties of relatively large coercive field (126.8 kV/cm) and small remnant polarization (4.19 μC/cm^2) have an advantage with the metal ferroelectric insulator semiconductor (MFIS) stacking structure.

However the round shape of P-E loop might correspond to the leakage property of CT film and then the leakage current behaviors of Pt/CT/Pt metal ferroelectric metal (MFM) stacking structure and Pt/CT/Y_2O_3/Si (MFIS) were monitored and given in Figure 66.

Figure 66(a) corresponds to the I-V characteristic of $Ce_2Ti_2O_7$ in MFM and the leakage current value is in the nearly the same range with $Sr_2(Ta_xNb_{1-x})_2O_7$ [141]. Then it can be said that this kind of the round shape of P-E loop seems due to the relatively small polarization value of pseudo-pyrochlore ferroelectric $A_2B_2O_7$ phase. Because even the leakage current value is not important, the effect of leakage could be strongly reflected on the polarization behavior when the polarization is small. Figure 66(b) represents the I-V result obtained from CT in MFIS structure.

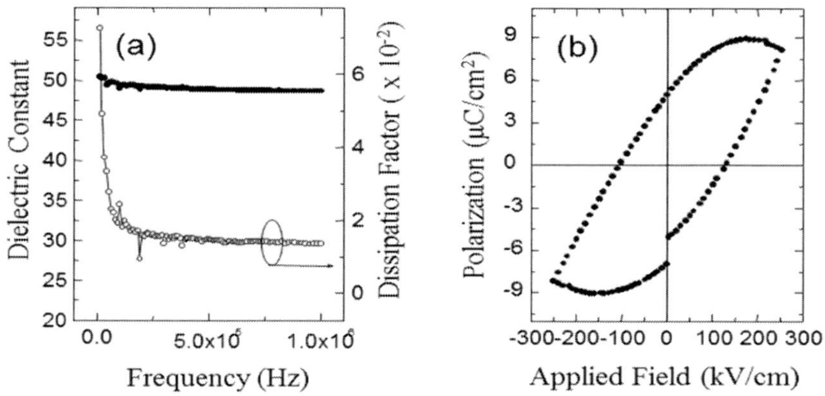

Figure 65. Dielectric properties of $Ce_2Ti_2O_7$; (a) Dielectric constant and loss and (b) P-E hysteresis loop of the Pt/$Ce_2Ti_2O_7$/Pt structure.

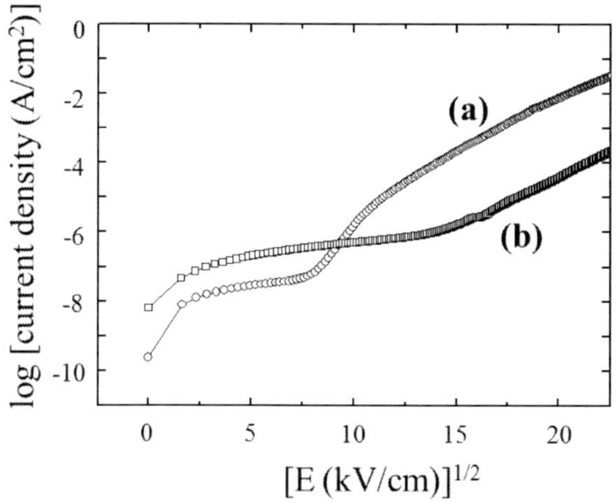

Figure 66. I-V characteristics of $Ce_2Ti_2O_7$ in (a) Pt/$Ce_2Ti_2O_7$/Pt and (b) Pt/$Ce_2Ti_2O_7$/Y_2O_3/Si structures.

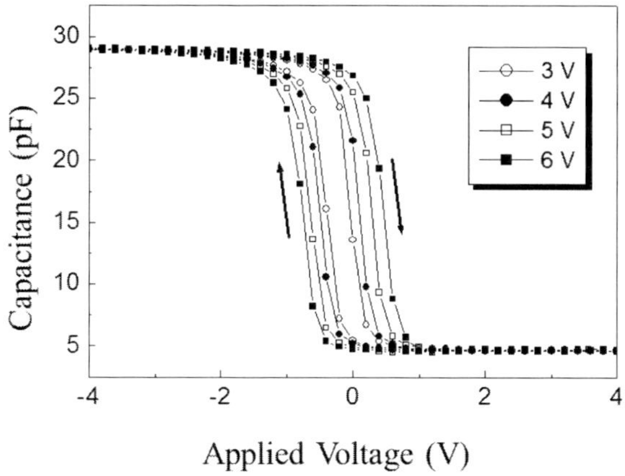

Figure 67. C-V characteristics of the $Pt/Ce_2Ti_2O_7/Y_2O_3/Si$ capacitor.

Normally an improved I-V characteristic was expected due to the presence of Y_2O_3 insulator layer than the case of MFM, however at small applied voltage region; MFIS shows an almost one order higher leakage value than MFM. This might be due to the vacuum anneal effect of Y_2O_3 insulator; a possible generation of defect from the absence of O_2 atmosphere during the annealing at high temperature. Anyway at high applied voltage region, MFIS shows an improved I-V characteristic due to a positive role of Y_2O_3 on the interfacial formation.

The memory window values were obtained with $Pt/CT/Y_2O_3/Si$ structure and the results were given in Figure 67. This structure shows a very large memory window value of 1.24 V at an applied voltage of 6 V without a shift of the C-V curve. We conclude, therefore, that low-vacuum annealing is effective to control the oxidation state of cerium. Furthermore it may be helpful to depress the generation of interfacial defects, including the formation of silicon oxides, because a positive inclination is frequently observed in MFIS-stacked or high-k metal oxide semiconductor structure that results from the internal negative charge by oxygen thermal treatment [142,143].

In summary, we identified CT as a ferroelectric material for use as a ferroelectric gate material for MFIS-stacked structure. We showed the possibility of fabricating CT thin films with a pseudo-pyrochlore structure. A vacuum-annealing process was used to synthesize CT to obtain the Ce^{3+} oxidation state in CT. C-V measurements confirmed the absence of both silicon oxide and interfacial defect formation that result from the fabrication of MFIS-stacked structures. The permittivity, remnant polarization, and coercive field of the CT thin film were measured to be 48.7, 4.19 $\mu C/cm^2$, and 126.8 kV/cm, respectively. In addition, the memory windows of the Pt/CT (240 nm)/ Y_2O_3 (20 nm)/Si structure were in the range of 0.36–1.24 V, which correspond to gate voltages between 3–6 V.

2.1.6. Ferroelectric Multilayer Structure

This part shows the multilayer structure of ferroelectric materials for overcoming limitation of improvement of a material property.

2.1.6.1. Stacking Effect on the Ferroelectric Properties of PZT/PLZT Multilayer Thin Films

A number of papers have reported the ferroelectric properties of PZT and PLZT films of interest for use in non-volatile memory applications [144]. La doping of PZT films is known to lower both the leakage current and the P_r [145]. The remnant polarization decreases due to a reduction in the tetragonality of the film. Therefore PZT/PLZT multilayer films are of interest in that they may provide improved ferroelectric properties. In this part, direct-patterning by PMOD using UV exposure was used to prepare PZT/PLZT multilayer films and the effect of stacking order on the ferroelectric properties of the resultant PZT/PLZT multilayer films was investigated.

The precursors used for the photochemical production of PZT and PLZT films were the 2-ethylhexanoates of lead(II), zirconyl(IV), and lanthanum(III) and titanium(IV) isopropoxide. The atomic ratio of Zr/Ti was fixed as 52/48. Then 10% of extra Pb was added for the compensation of Pb-loss during the high temperature anneal [146].

In PLZT a 3 mol% reduction of the lead was compensated for by the addition of 3 mol% lanthanum. The PZT and PLZT precursors were dissolved in hexane and spin-coated at 2000 rpm for 30 s on two substrates. One substrate was chemically cleaned Pt(111)/Ti/SiO_2/Si(100) ideal for monitoring of ferroelectric properties and the other substrate was a p-Si(100) wafer, to show process generality and demonstrate the patterning of 350 nm thick PZT and PLZT films by PMOD. The precursor films for lithography were exposed for 30 min with the output of a 1000 W mercury arc lamp passed through a chromium mask. The latent image was developed with a hexane rinse to reveal the negative pattern.

The precursor films for multilayer structures were exposed for 30 min with the output of the 1000 W lamp. The resultant material was annealed for 10 min at 400°C. This process was repeated for each layer. Following the deposition of the entire structure the multilayer films were annealed at 650°C for 30 min. A 150 nm thick Pt top electrode with a 100 μm radius was sputter-deposited and then annealed at 650°C for 10 min under O_2.

The image of the surface microstructure and patterned film was observed using SEM. Phase formation and crystallization were monitored with XRD. For the measurement of electrical properties, a 300 nm thick ferroelectric multilayer film was prepared by repeated coating. The final thickness of the ferroelectric layers was almost uniform regardless of the order of stacking or the composition of the layers. The I–Vcharacteristics were obtained using a HP4145B (Hewlett Packard). The ferroelectric hysteresis loop and polarization fatigue were measured using a RT66A (Radiant Technology) ferroelectric tester.

The surface microstructures of the PZT/PLZT multilayer films as observed using SEM are shown in Figure 68. The grain size of the 4-PLZT film is smaller than that of the 4-PZT film. This is consistent with enhanced nucleation in the PLZT film dominating grain growth.

A similar microstructrue is observed with the PLZT/2-PZT/PLZT and 4-PLZT films (Figure 68(c) and (d)). This result implies that the bottom ferroelectric layer, the first deposited layer on the Pt substrate, has a significant influence on the surface microstructure of film. This is also evidenced by the observation of a similar microstructure in Figure 68(a) and (b).

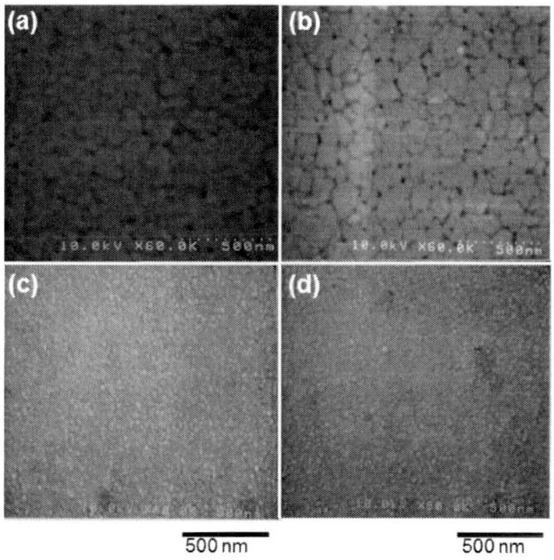

Figure 68. SEM micrographs of the PZT/PLZT multilayer films annealed at 650°C: (a) 4-PZT, (b) PZT/2-PLZT/PZT, (c) PLZT/2-PZT/PLZT and (d) 4-PLZT.

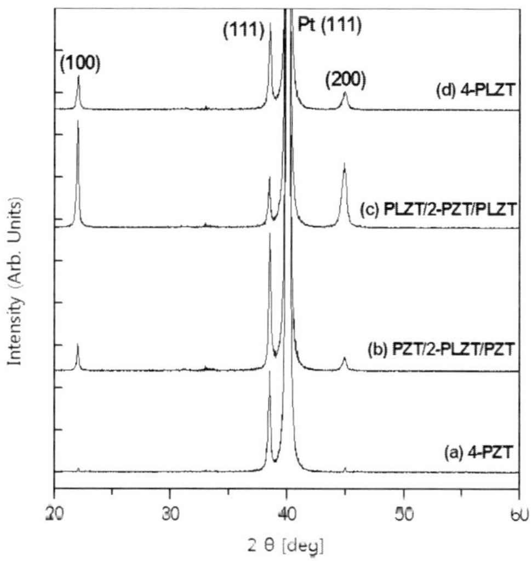

Figure 69. XRD patterns of the PZT/PLZT multilayer films annealed at 650°C: (a) 4-PZT, (b) PZT/2-PLZT/PZT, (c) PLZT/2-PZT/PLZT and (d) 4-PLZT.

XRD patterns of the various PZT/PLZT multilayer films coated on the Pt(111)/Ti/SiO_2/Si substrate are given in Figure 69. The diffraction patterns of all of the PZT/PLZT multilayer films indicate the presence of only the perovskite phase. In any case the pyrochlore or other phase is not observed. The 4-PZT film crystallized with a preferred (111) orientation due to (111) plane matching between Pt and PZT as shown in Figure 69(a) [27]. This highly preferred orientation indicates that the nucleation and growth of PZT is strongly affected by the Pt substrate. In contrast the diffraction of the 4-PLZT film, with 3 mol% La, showed increased relative intensities of the (100) and (200) reflections and reduced grain size. Our results are consistent with the observation of Q. Zou et al. [108] showing the reduction of grain size and increasing intensity of the (100) reflection with La-doping. This is explained by the fact that (100) oriented perovskite lattice has the lowest growth activation energy [109].

In Figure 69(b), the presence of the intermediate PLZT film prevents the Pt substrate from affecting the growth orientation of the ferroelectric film. In case of the PLZT film deposited on the Pt substrate, as shown in Figure 69(c), the relative intensity of the (100) and (111) reflections are reversed in comparison with Figure 69(b). This implies that the first layer on the Pt substrate in the PZT/PLZT multilayer films is an important influence on the growth orientation of film.

However in case of the PLZT/2-PZT/PLZT film (Figure 69(c)), (100) growth is enhanced compared to the 4-PLZT (Figure 69(d)). This abnormal behavior could be explained by a layer discontinuity effect. The growth orientation of the PLZT is also affected by the Pt substrate, although less than PZT, as expected by considering the similarity in the lattice parameters of PZT and PLZT. However, the continuity through the layers in the 4-PZT film is larger than the PLZT/2-PZT/PLZT film, hence, the substrate more strongly affects the growth of the 4-PLZT film.

P–E hysteresis loops of the various PZT/PLZT multilayer films with varying applied voltage (8, 10 and 15 V) are shown in Figure 70.

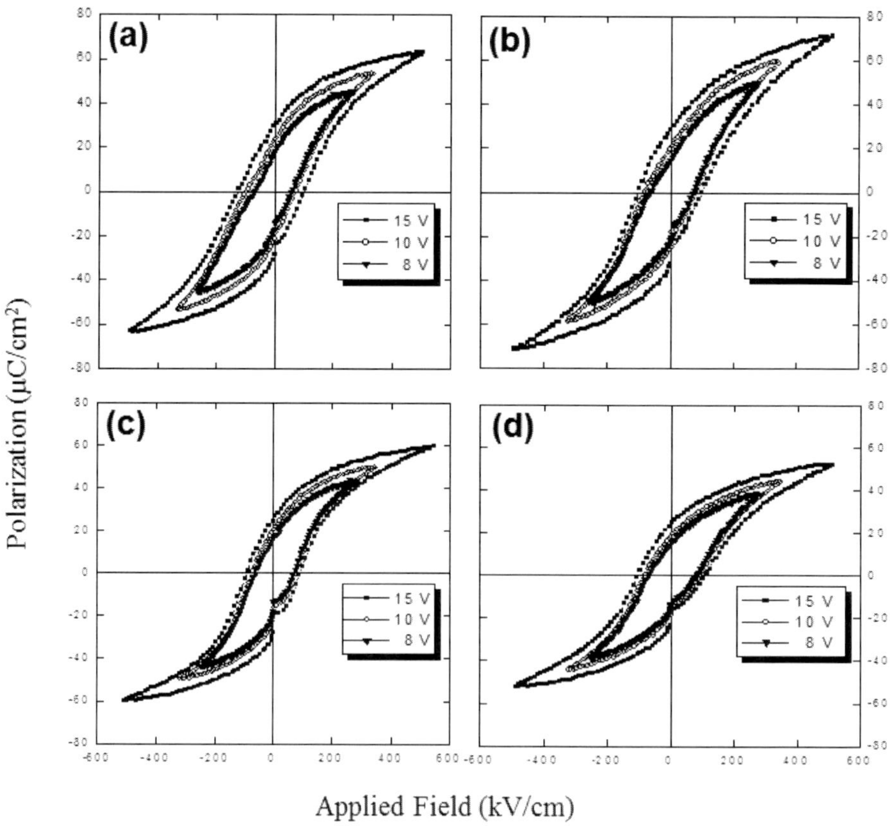

Figure 70. P–E hysteresis loops of the PZT/PLZT multilayer films annealed at 650°C: (a) 4-PZT, (b) PZT/2-PLZT/PZT, (c) PLZT/2-PZT/PLZT and (d) 4-PLZT.

In case of 4-PZT and PZT/2-PLZT/PZT films, the P_r were 30 and 28 $\mu C/cm^2$ an applied voltage of 15 V while the $E_c$were -151, 108 and -105, 98 kV/cm, respectively, with a shift to the left. The shift in E_c is related to the imprint and this can be attributed to the asymmetric distribution of defect dipoles such as oxygen and Pb vacancies at the film/electrode interface, inducing an internal field [110]. However, in case of the 4-PLZT and PLZT/2-PZT/PLZT films, the P–E hysteresis was symmetric.

The measured P_r values of the 4-PLZT and the PLZT/2-PZT/PLZT films are 23 and 25 $\mu C/cm^2$ and the E_c values are 107 and 94 kV/cm, respectively. The introduction of the La^{3+} dopant reduces the P_r and E_c values of the films. This is as expected based on the relative ionic radii, which are 1.32 Å for Pb^{2+} and 1.22 Å for La^{3+} [70] resulting in a decrease in the tetragonal distortion. The crystallographic structure of the PLZT film is transformed from the ferroelectric state to the paraelectric state by doping with La, consequently the P_r and E_c values are decreased [112].

The grain size effect also has to be considered because the small grained PLZT/2-PZT/PLZT film showed saturation behavior due to a small value of the net dipole, P_r, but the large grained PZT/2-PLZT/PZT film showed a larger P_r value. The P_r value of 4-PZT film is larger than that of 4-PLZT film. From the shift in the P–E loop, we can infer an asymmetric distribution of the defect dipoles at the PZT layer/Pt electrode. In the case of the film with the PLZT/Pt interfaces, the P–E hysteresis loop was symmetric. Therefore, at the interface between the ferroelectric and Pt, defect dipoles such as caused by O vacancies are reduced by La doping.

The variation in the current density with applied field for 300 nm thick PZT/PLZT multilayer films is given in Figure 71. In case of the 4-PZT and PZT/2-PLZT/PZT films, the leakage current density with negative bias was a little higher than positive bias. The I–V characteristics of those films show a voltage polarity dependence [147]. The major source of leakage current is crystalline PZT and related compounds is known to be Pb or O vacancies created during the high temperature anneal [29].

In our case, the voltage polarity dependence in the I–V characteristics might be related with asymmetric distribution of space charge density at the PZT/Pt electrode interface [30]. In contrast, in case of the 4-PLZT and PLZT/2-PZT/PLZT films, the leakage current density was lowered and appeared symmetric as shown in Figure 70. This is presumably a consequence of reducing the defect dipole concentration at the film/electrode interface. Recall that these films had characteristically smaller grain sizes (Figure 68). The lower leakage current density of small-grained films might be due to the inter-grain depletion of grain boundary limited conduction (GBLC) [125]. In this model, local space charges near grain boundaries inhibit current flow and the overlapping depletion regions of neighboring grain boundaries are the cause of the low leakage current.

Figure 72 illustrates the change of polarization of the PZT/PLZT multilayer films with the number of switching cycles through a bipolar square wave of -10 V at a frequency of 500 kHz. After 10^9 cycles, the remaining remnant polarizations of 4-PZT, PZT/2-PLZT/PZT, PLZT/2-PZT/PLZT, and 4-PLZT films were about 8, 35, 57, and 50% of the original values respectively. The PLZT and PLZT/2-PZT/PLZT films showed a somewhat improved fatigue resistance. The film/electrode interface was promoted for cyclic electrical fatigue with excess of electrons due to doping with La. Namely, the films with PLZT at the first and the last layers on Pt substrate were improved in fatigue properties. This implies that defects such as O vacancies at film/electrode interface play a significant role in the degradation of fatigue resistance. The possibility of direct-patterning of PZT and PLZT films by removing the area unexposed to UV using hexane was examined and the results are given in Figure 73. As shown in Figure 73(a) and (b), the bright area corresponds to PZT and PLZT films and the dark area to the Si substrate, respectively. These micrographs indicate that direct lithographic patterning of PZT and PLZT films could be accomplished with PMOD. Direct-patterning of PZT and PLZT films by PMOD was accomplished by UV exposure and removal of the unexposed area of the film by rinsing with hexane. In the 4-PZT and PZT/2-PLZT/PZT films an asymmetric behavior of the P–E was observed and defects appear to be the major source of leakage current. In case of the 4-PLZT and PLZT/2-PZT/PLZT films improved, symmetric, P–E characteristics were obtained by reducing the defect-dipole concentration at ferroelectric layer/Pt electrode interface. Lanthanum doping of the PZT results in a lowering of the number of oxygen vacancies in the resultant PLZT.

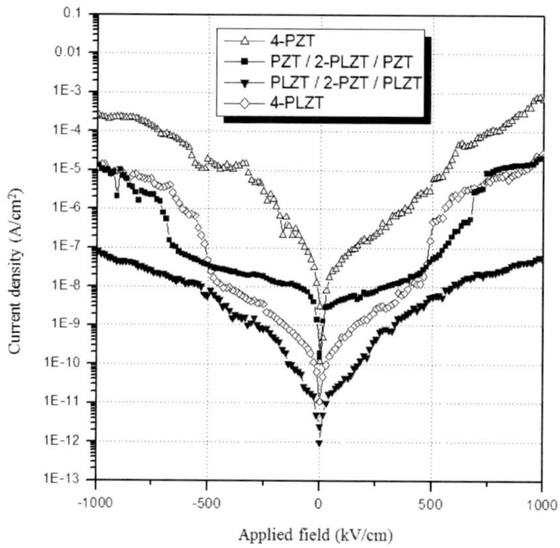

Figure 71. I-V characteristics of the PZT/PLZT multilayer films annealed at 650°C.

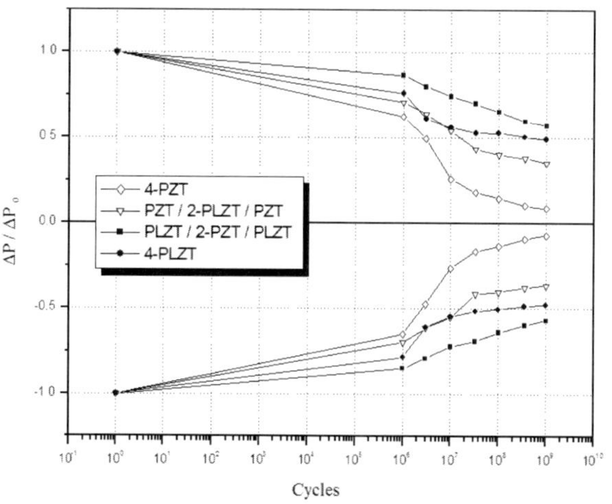

Figure 72. Fatigue characteristics of the PZT/PLZT multilayer films annealed at 650°C.

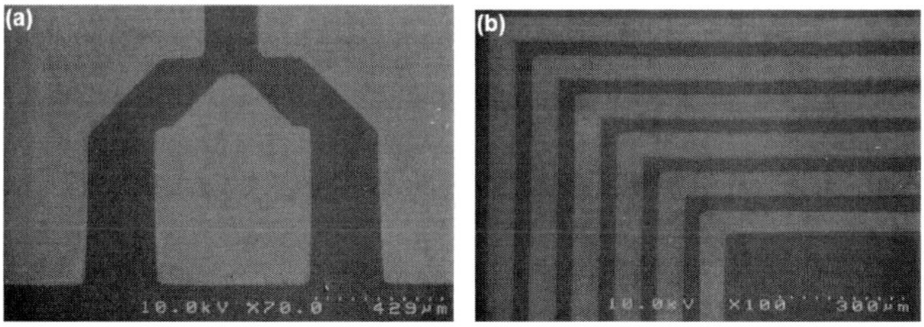

Figure 73. SEM micrographs of UV-direct-patterned (a) PZT and (b) PLZT films.

The lower number of oxygen vacancies in turn has the effect of reducing the number of trapped charges at the interface. Since these trapped charges at the interface contribute to the leakage current, polarization and ultimately loss of ferroelectric behavior the presence of PLZT at the interface mediates these effects. Therefore, the PZT/PLZT multilayer films with PLZT/Pt interfaces show an improvement in the electrical properties, in particular the fatigue resistance.

2.1.6.2. Electric and Ferroelectric Properties of PZT/BLT and PZT/SBT Multilayer Thin Films

Among related materials of interest, PZT has been intensively investigated due to its large P_r and small E_c [148]. However, PZT films with Pt electrodes suffer from fatigue in which the P_r declines after about 10^8 switching cycles [149,150]. On the contrary, a BLT film has become an attractive candidate due to its larger polarization compared to SBT and much better fatigue behavior when compared to conventional lead-based materials [34,49]. So, PZT/BLT multilayer films are of interest in that they may provide improved ferroelectric properties such as large P_r, a small E_c, and excellent fatigue resistance. In this part, a direct-patterning method for the lithographic deposition of PZT/BLT multilayer films formed by PMOD was reported and the effects of stacking order on the electric and ferroelectric properties of the resulting PZT/BLT multilayer films were investigated.

The precursors used for the photochemical production of PZT films were lead 2-ethylhexanoate, zirconyl 2-ethylhexanoate, and titanium isopropoxide. These precursors were dissolved in hexane. The atomic ratio of Zr/Ti was fixed at 52/48. 15 % excess Pb was added for the compensation of Pb-loss during high temperature annealing [25]. Photosensitive starting precursors for the BLT were bismuth 2-ethylhexanoate, lanthanum 2-ethylhexanoate, and titanium isopropoxide. A 10 % excess of bismuth 2-ethylhexanoate was added to compensate for the loss of Bi during high temperature annealing [39]. PZT and BLT precursors were spin-coated at 2000 rpm for 30 s on two different types of substrates. One substrate was chemically cleaned Pt(111)/Ti/SiO_2/Si(100) for measuring the electric and ferroelectric properties and the other substrate was p-Si(100) wafer to demonstrate the direct-patterning process of 200-nm thick PZT and BLT films by PMOD. The precursor films for lithography were exposed for 15 min to the output of a 1000 watt mercury arc lamp passed through a chrome mask. The latent image was developed with a solvent rinse to reveal a negative pattern. The resultant film was annealed for 10 min at 400 and 600°C, respectively. This process was repeated for each layer. The entire structure of the multilayer films was annealed at 700°C for 1 h under ambient O_2. The single layer thickness of PZT and BLT was

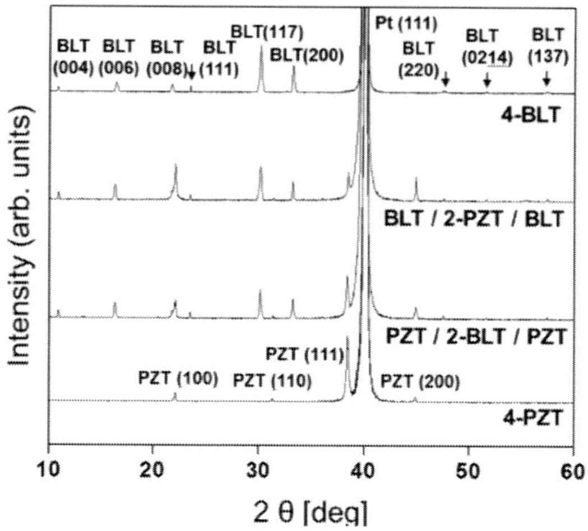

Figure 74. XRD spectra of the PZT/BLT multilayer films annealed at 700°C: 4-PZT, PZT/2-BLT/PZT, BLT/2-PZT/BLT, and 4-BLT films.

around 95 nm measured using an elipsometer and the final thickness of the ferroelectric layers was almost uniform regardless of the order of stacking of the PZT and BLT layers. For the measurement of electric and ferroelectric properties, a 150-nm-thick Pt top electrode with a 200-μm–diameter was sputter-deposited on the 380-nm-thick PZT/BLT multilayer films. The ferroelectric properties of the films were measured using RT66A (Radiant Technology) ferroelectric tester with pulse width of 0.15×10^{-3} s and the I-Vcharacteristics of the films were obtained using a HP 4145B (Hewlett Packard) with holding time of 0.4 s and step of 0.5 V. The direct-patterning image of the films was observed by using a SEMand optical microscopy. The phase formation and crystallinity of the films were studied with an XRD using Cu k_α radiation. XRD spectra of the various PZT/BLT multilayer films coated on Pt(111)/Ti/SiO_2/Si substrates are given in Figure 74.

All the diffraction peaks of the PZT/BLT multilayer films corresponded to PZT, BLT, and Pt. The 4-PZT film was crystallized with a preferred <111> orientation due to (111) plane matching between Pt and PZT. This highly preferred orientation indicates that the nucleation and growth of PZT is strongly affected by the Pt substrate [122].

However, the diffraction intensities of BLT showed a distribution similar to BLT ceramics [41], and Pt (111) was found to not influence the growth orientation of BLT films [42]. An enhancement of the (100) PZT diffraction peak intensity was observed in the PZT/2-BLT/PZT film compared to the 4-PZT film. However, the BLT peak intensity distribution of the PZT/2-BLT/PZT film was not changed from that of the 4-BLT film. These results confirmed that neither the Pt (111) substrate nor the (111)-oriented PZT film showed an influence on the growth orientation of the BLT film. This enhancement of the (100) PZT diffraction peak was from the upper-most top PZT layer because the (100) growth plane of PZT has the lowest activation energy [151].

Also, the BLT/2-PZT/BLT film showed more enhanced (100) PZT diffraction peak intensity compared to the PZT/2-BLT/PZT film because the growth orientation of 2-PZT intermediate layers was not affected by both the bottom BLT layer and the Pt substrate. This result implies that the bottom ferroelectric layer, the firstly deposited layer on the Pt substrate, has a significant influence on the growth orientation of the entire film.

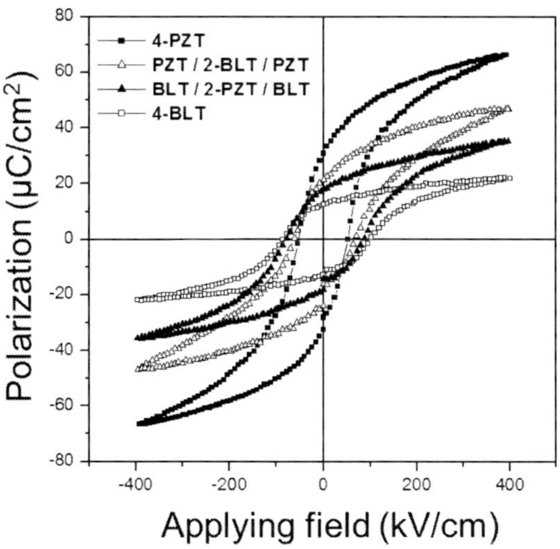

Figure 75. P-E hysteresis loops of the PZT/BLT multilayer films annealed at 700°C: 4-PZT, PZT/2-BLT/PZT, BLT/2-PZT/BLT, and 4-BLT films.

P-E hysteresis loops of the various PZT/BLT multilayer films with an applied voltage of 15 V are shown in Figure 75. In the case of the 4-PZT and 4-BLT films, the P_r values were 31 and 12 μC/cm^2 while the E_c values were 53 and 101 kV/cm, respectively. The 4-PZT film showed higher P_r and lower E_c values than the 4-BLT film because PZT domains are relatively easier to switch than bi-layered perovskite domains [152]. Namely, upon the application of the electric field, the PZT film shows larger displacements of ions and charge defects than those of the BLT film. In the case of the PZT/2-BLT/PZT and BLT/2-PZT/BLT films, the P_r values were 21 and 17 μC/cm^2 while the E_c values were 67 and 84 kV/cm, respectively. In the case of PZT films, it is well known that coexistence of rhombohedral phase and tetragonal phase commonly observed around the morphotropic phase boundary (MPB), the Zr/Ti ratio of 52/48 [118,153]. The MPB composition of PZT films tends to move toward rhombohedral phase, Zr-rich side, under the influence of in-plane film stress, regardless of whether it is tensile or compressive [118,153] and the spontaneous polarization direction of a rhombohedral PZT is [111] [154].

In our case, as shown in Figure 74, the PZT in the PZT/2-BLT/PZT film showed a relatively higher <111> growth orientation than that in the BLT/2-PZT/BLT film. Namely, due to the increase of <111> spontaneous polarization, the P_r value of the PZT/2-BLT/PZT film was higher than that of the BLT/2-PZT/BLT film. In addition, the E_c value of the PZT/2-BLT/PZT film was lower than that of the BLT/2-PZT/BLT film. It might be relatively easy switching of PZT domains compared to BLT domains at the film/electrode interfaces.

The current densities of the 380-nm-thick PZT/BLT multilayer films with an applying field are given in Figure 76.

The leakage current densities of the 4-PZT and PZT/2-BLT/PZT films were 2.1×10^{-5} and 1.5×10^{-5} A/cm^2, respectively, while those of the BLT/2-PZT/BLT and 4-BLT films were 9.1×10^{-7} and 4.6×10^{-7} A/cm^2 at 300 kV/cm, respectively.

Namely, the films with a BLT layer on the Pt substrate had about a two orders of magnitude improved leakage current behavior as compared to the films with a PZT layer on the Pt substrate. The lattice structure of BLT consists of perovskite-like slabs and/or $(Bi_2O_2)^{2+}$

layers [34]. The Bi_2O_2 layers have net electrical charges and their positioning in the lattice is self-regulated to compensate for space charges near the Pt substrate [32].

In case of BLT layers, used for the top and bottom layers in contact with Pt electrodes, the lower number of defects and space charges is expected near the electrodes among all the PZT/BLT multilayer films.

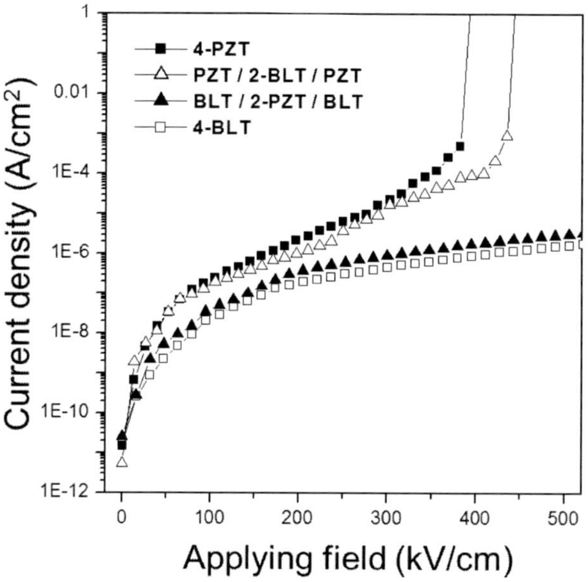

Figure 76. I-V characteristics of various PZT/BLT multilayer films annealed at 700°C: 4-PZT, PZT/2-BLT/PZT, BLT/2-PZT/BLT, and 4-BLT films.

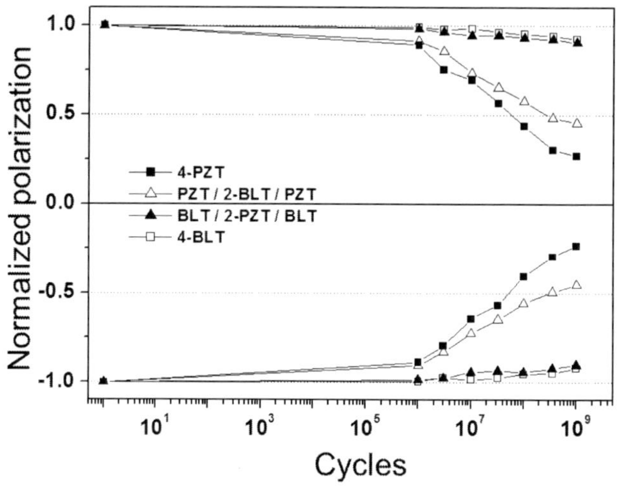

Figure 77. Fatigue characteristics of various PZT/BLT multilayer films annealed at 700°C: 4-PZT, PZT/2-BLT/PZT, BLT/2-PZT/BLT, and 4-BLT films.

Figure 77 illustrates the change of polarization of the PZT/BLT multilayer films as a function of the number of switching cycles through a bipolar square wave of ± 10 V at a frequency of 500 kHz. After 10^9 cycles, the remaining polarizations of the 4-PZT, PZT/2-BLT/PZT, BLT/2-PZT/BLT, and 4-BLT films were about 27, 45, 90, and 92 % of their initial values, respectively. The BLT/2-PZT/BLT and 4-BLT films showed excellent fatigue resistances when compared to the 4-PZT and PZT/2-BLT/PZT films.

It is well known that polarization fatigue of the PZT films on Pt electrodes mainly originate from the accumulation of oxygen vacancies and domain wall pinning [155] while that of the BLT films show fatigue-free behavior due to the $(Bi_2O_2)^{2+}$ layers as a sinker for oxygen vacancies in Bi-layered perovskite structure [34]. In addition, Liu et al. [156] reported that the increase in stress with the repeated switching cycles in the PZT film can be attributed to the lattice distortion during the fatigue process but in the Bi-layered perovskite structure, polarization switching involves little lattice distortion. Because PZT domains are relatively easier to switch than BLT domains, the 4-PZT film could involve more internal stress compared to the 4-BLT film. As a result, the 4-BLT film exhibited the improved fatigue endurance compared to the 4-PZT film. In the case of the PZT/2-BLT/PZT and BLT/2-PZT/BLT films, similar internal stress might be initially involved. However, the Bi_2O_2 layer is known to minimize interlayer stress at Pt electrodes [32]. As a result, the fatigue resistance of the BLT/2-PZT/BLT films was better than that of the PZT/2-BLT/PZT due to the reduced internal stress and decreased charge defects at the film/electrode interfaces. This is why the fatigue endurance was improved when BLT layers were deposited for the top and bottom layers in contact with Pt electrodes.

The possibility of direct-patterning of PZT and BLT films by removing the unexposed area to UV using a solvent was examined and the results are given in Figure 78. In the SEM micrograph (Figure 78(a)), the relatively dark area corresponds to PZT films and the bright area corresponds to the Si substrate. In the optical micrograph (Figure 78(b)), the relatively dark area corresponds to BLT films and the bright area corresponds to the Si substrate. Micron-scaled patterning of PZT and BLT films was successfully obtained by the lithographic patterning using PMOD. From the results of Figure 78, the direct-patterning process can be utilized to fabricate a micro-patterned system without using photoresist or a dry etching process. Various PZT/BLT multilayer films were successfully constructed by PMOD using photosensitive starting precursors. Direct-patterning of the films with widths of several microns was accomplished by UV exposure and removal of the unexposed area of the films using a solvent.

The bottom ferroelectric layer, the firstly deposited layer on the Pt substrate, has a significant influence on the growth orientation of PZT/BLT multilayer films. Though the P_r value of the 4-BLT film was 12 $\mu C/cm^2$, the enhanced P_r value, 17 $\mu C/cm^2$ was obtained with the BLT/2-PZT/BLT film, i.e., by deposition a BLT layer for the top and bottom layers in contact with the Pt electrodes. The films with BLT layers at the top and bottom showed improved leakage current and fatigue endurance compared to the films with PZT layers at the top and bottom due to the reduced internal stress and decreased charge defects at the film/electrode interfaces. From this part, PZT/BLT multilayer films were proven to be effectively adopted for the enhancement of electric and ferroelectric properties, in particular the fatigue resistance.

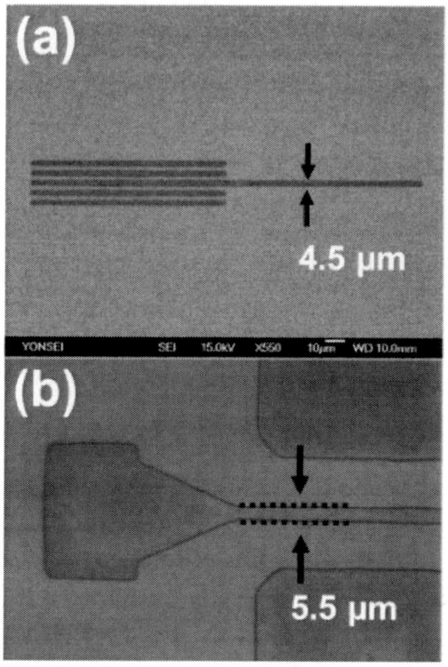

Figure 78. (a) SEM micrograph of direct-patterned PZT film and (b) optical micrograph of direct-patterned BLT film.

We studied PZT/SBT multilayer films also. And the effects of stacking order on the electric and ferroelectric properties were investigated as above PZT/BLT study. The precursors used for the photochemical production of PZT films were lead 2-ethylhexanoate, zirconyl 2-ethylhexanoate, and titanium isopropoxide. These precursors were dissolved in hexane. The atomic ratio of Zr/Ti was fixed at 52/48. 15% excess Pb was added for the compensation of Pb-loss during high temperature annealing. Photosensitive starting precursors for the SBT films were strontium 2-ethylhexanoate, bismuth 2-ethylhexaoate, and tantalum ethoxide which were dissolved in xylene. SBT films with a concentration ratio of Sr/Bi/Ta = 0.8/2.2/2.0 were investigated in this work. PZT and SBT precursors were spin-coated at 2000 rpm for 30 s on two different types of substrates. One substrate was chemically cleaned Pt(111)/Ti/SiO_2/Si(100) for measuring the electric and ferroelectric properties and the other substrate was p-Si(100) to show the generality of the direct-patterning process. Two hundred nanometer thick direct-patterned PZT and SBT films were prepared by PMOD after UV exposure of photosensitive PZT and SBT precursor films. The precursor films for lithography were exposed for 15 min to the output of a 1000W mercury arc lamp passed through a chromium mask. The latent image was developed with a solvent rinse to reveal a negative pattern. The direct-patterning image of the films was observed by using a SEM. The resultant film was annealed for 10 min at 400 and 600 °C, respectively. This process was repeated for each layer. The entire structure of the multilayer films was annealed at 700 °C for 30 min under ambient O_2. The phase formation and crystallinity of the films were studied with an XRD using Cu K_α radiation. A 150-nm thick Pt top electrode with a 100 μm radius was sputter-deposited to obtain electric and ferroelectric properties of the films including I–V and P–E characteristics. X-ray diffraction patterns of the various PZT/SBT multilayer films coated on Pt(111)/Ti/SiO_2/Si substrates are given in Figure 79. All the diffraction peaks of the PZT/SBT multilayer films corresponded to PZT, SBT, and Pt. The

4-PZT film was crystallized with a preferred <111> orientation due to (111) plane matching between Pt and PZT.

This highly preferred orientation indicates that the nucleation and growth of PZT is strongly affected by the Pt substrate [122]. The 4-SBT film was randomly oriented with a relatively high <115> orientation. Lee et al. [157] reported that the influence of Pt substrate orientation on the growth of SBT films was elucidated in terms of atomic matching at the SBT/Pt interface. Neither {115} nor {200} SBT planes can solely match with Pt (111) since the atomic arrangements are appreciably different. Through the reconstruction of atomic positions at the SBT/Pt interface on the Pt (111) substrate, the 4-SBT film was grown towards both <115> and <200> orientations on the Pt (111) substrate.

An enhancement of the (100) PZT diffraction peak intensity was observed in the PZT/2-SBT/PZT film compared to the 4-PZT film. However, the peak intensity distribution of the PZT/2-SBT/PZT film was not changed from that of the 4-SBT film. These results confirmed that neither the Pt (111) substrate nor the (111)-oriented PZT film showed an influence on the growth orientation of the SBT film. This enhancement of the (100) PZT diffraction peak was from the upper-most top PZT layer because the (100) growth plane of PZT has the lowest activation energy [151]. In the case of the SBT/2-PZT/SBT film, the PZT layer showed relatively high <100> growth orientation since the growth orientation of PZT film was not affected by the bottom SBT layer. From these results, it could be said that SBT layer grown on the Pt (111) substrate and the (111)-oriented PZT film and PZT layer grown on the SBT film showed polycrystalline growth behavior. This implies that the growth orientation of the PZT/SBT multilayer film is mainly dependent on the first layer formed on the Pt substrate.

P–Ehysteresis loops of the various PZT/SBT multilayer films with an applied voltage of 15V are shown in Figure 80. In the case of the 4-PZT and 4-SBT films, the P_r values were 30 and 6 μC/cm^2 while the E_c values were 40 and 115 kV/cm, respectively. The 4-PZT film

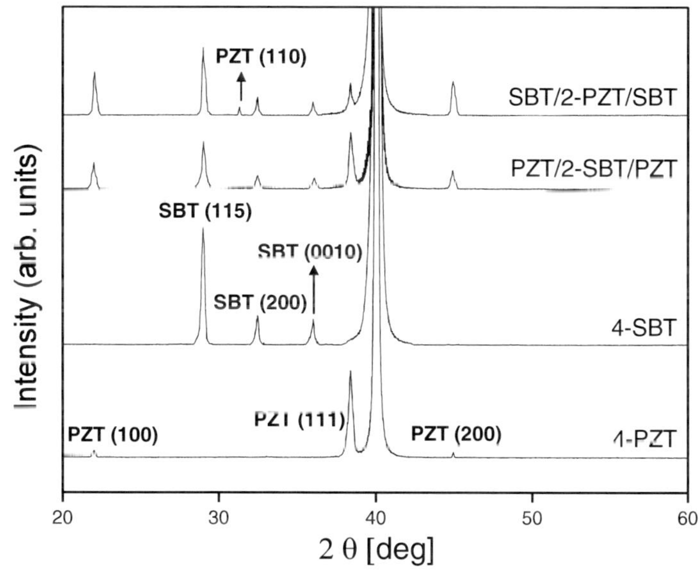

Figure 79. XRD patterns of the PZT/SBT multilayer films annealed at 700°C: 4-PZT, 4-SBT, PZT/2-SBT/PZT, and SBT/2-PZT/SBT films.

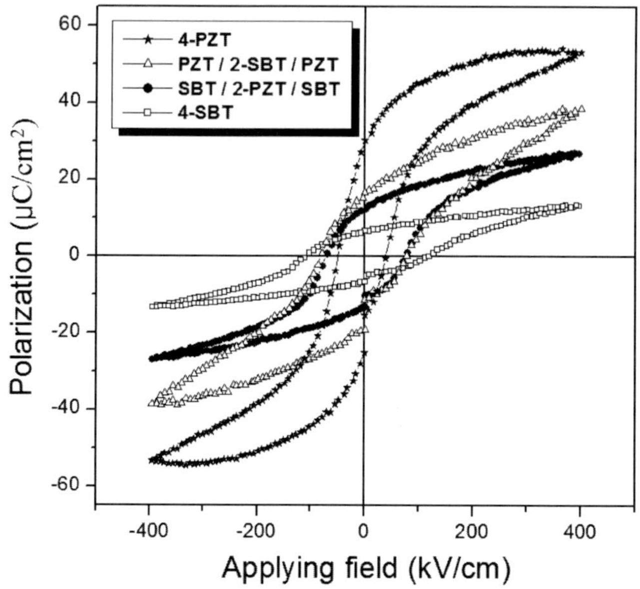

Figure 80. P-E hysteresis loops of the PZT/SBT multilayer films annealed at 700 °C: 4-PZT, PZT/2-SBT/PZT, SBT/2-PZT/SBT, and 4-SBT films.

showed higher P_r and lower E_c values than the 4-SBT film because PZT domains are relatively easier to switch than SBT domains [152].

Namely, upon the application of the electric field, the PZT film shows larger displacements of ions and charge defects than those of the SBT film. In the case of the PZT/2-SBT/PZT and SBT/2-PZT/SBT films, the P_r values were 16 and 12μC/cm^2 while the E_c values were 75 and 78 kV/cm, respectively. As shown in Figure 79, the PZT in the PZT/2-SBT/PZT film showed a relatively higher <111> growth orientation than that in the SBT/2-PZT/SBT film. Since the <100> direction of PZT is perpendicular to the polarization direction, the remnant polarization value of the SBT/2-PZT/SBT film is lower than that of the PZT/2-SBT/PZT film [158]. This is why the PZT/2-SBT/PZT film exhibited a higher P_r value than the SBT/2-PZT/SBT film. The current densities of the 380 nm thick PZT/SBT multilayer films with an applying field are given in Figure 81.

The leakage current densities of the 4-PZT and PZT/2-SBT/PZT films were 2.2×10^{-5} and 1.1×10^{-5} A/cm^2, respectively, while those of the SBT/2-PZT/SBT and 4-SBT films were 9.0×10^{-8} and 5.9×10^{-8} A/cm^2, respectively, at 300 kV/cm. Namely, films with an SBT layer on the Pt substrate had about a three orders of magnitude improved leakage current behavior as compared to the films with a PZT layer on the Pt substrate. The lattice structure of SBT consists of perovskite-like layers with a net composition of $(SrTa_2O_7)^{2-}$ alternating with $(Bi_2O_2)^{2+}$ layers [159]. The Bi_2O_2 layers have net electrical charges and their positioning in the lattice is self-regulated to compensate for space charge near the Pt substrate.

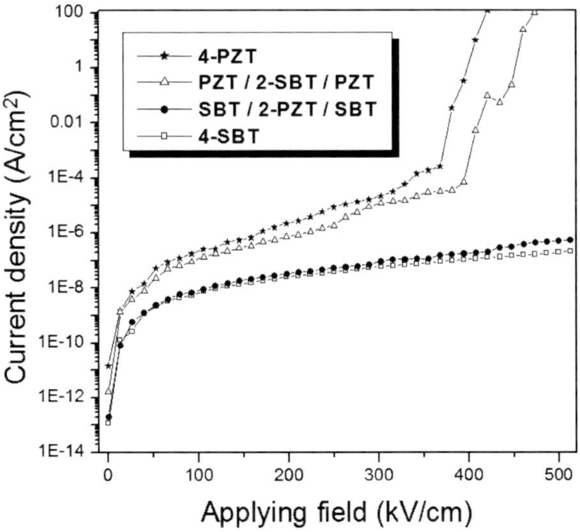

Figure 81. I-V characteristics of various PZT/SBT multilayer films annealed at 700 °C: 4-PZT, PZT/2-SBT/PZT, SBT/2-PZT/SBT, and 4-SBT films.

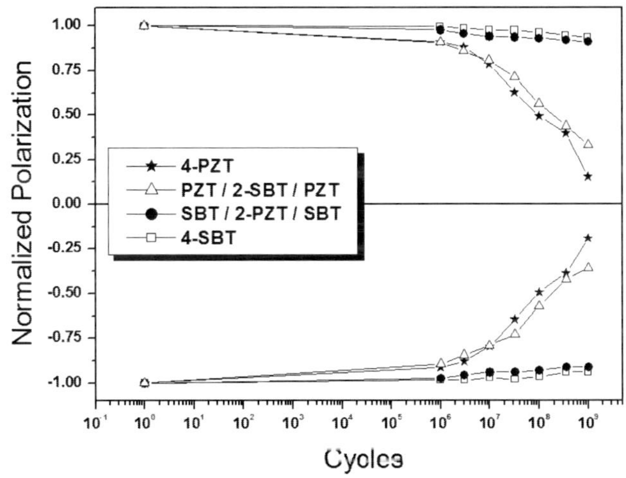

Figure 82. Fatigue characteristics of various PZT/SBT multilayer films annealed at 700 °C: 4-PZT, PZT/2-SBT/PZT, SBT/2-PZT/SBT, and 4-SBT films.

This is why the leakage current density was improved when SBT layers were used for the top and bottom layers in contact with Pt electrodes.

Figure 82 illustrates the change of polarization of the PZT/SBT multilayer films as a function of the number of switching cycles through a bipolar square wave of ±10V at a frequency of 500 kHz. The pulse widths (τ_1), switching period (τ), and duty cycle (τ_1/τ) were 1 μs, 2 μs, and 0.5, respectively. After 10^9 cycles, the remaining polarizations of the 4-PZT, PZT/2- SBT/PZT, SBT/2-PZT/SBT, and 4-SBT films were about 15, 33, 91, and 93% of their initial values, respectively. The SBT/2-PZT/SBT and 4-SBT films showed excellent fatigue resistances when compared to the 4-PZT and PZT/2-SBT/PZT films. This fatigue-free

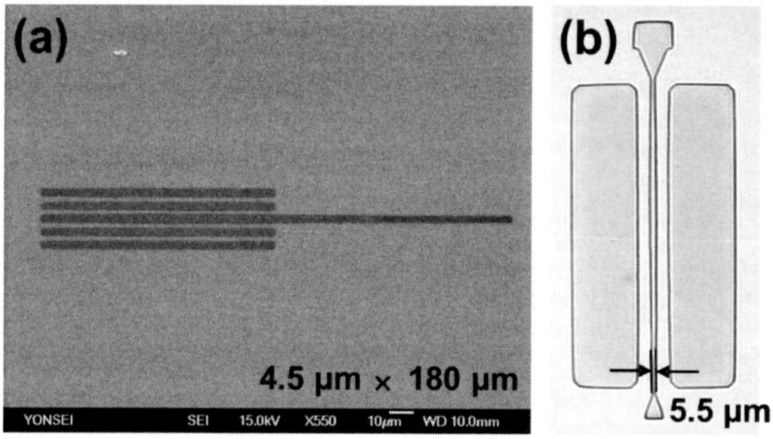

Figure 83. (a) SEM micrograph of PZT and (b) optical micrograph of SBT direct-patterned films.

behavior is predominantly related to the stability in the Bi_2O_2 layers and Sr–Ta–O perovskite blocks [160].

Also, we can observe that the measured P–E hysteresis loops of the various PZT/SBT multilayer films showed significant difference in the value of coercive field. Recall that the E_c value of 4-PZT film was smaller than that of the 4-SBT films as shown in Figure 80. Liu et al. [156] reported that an increase in stress with the repeated switching cycles in PZT film can be attributed to the lattice distortion during fatigue process but in SBT film, polarization switching involves little lattice distortion. Furthermore bismuth oxide is well known to minimize interlayer stress [32].

As a result, the 4-SBT films exhibited an improved fatigue property compared to the 4-PZT films. In the case of the PZT/2-SBT/PZT and SBT/2-PZT/SBT films, similar internal stress might be initially involved. In our case, the Bi_2O_2 layer in the SBT/2-PZT/SBT films might reduce internal stress with Pt substrate.

Also, during the fatigue process, the defect charges in the SBT/2-PZT/SBT films can barely diffuse toward the electrode/film interface due to the barriers formed by the Bi_2O_2 layers and/or stable perovskite slabs.

As a result, the SBT/2-PZT/SBT films showed the similar fatigue-free behavior as the 4-SBT films.

The possibility of direct-patterning of PZT and SBT films by removing the unexposed area to UV using a solvent was examined and the results are given in Figure 83. In the SEM micrograph (Figure 83(a)), the relatively dark area corresponds to PZT films and the bright area corresponds to the Si substrate. In the optical micrograph (Figure 83(b)), the relatively dark area (bluish area) corresponds to SBT films and the bright area (whitish area) corresponds to the Si substrate. Micron-scaled patterning of PZT and SBT films was successfully obtained by the lithographic patterning using PMOD. From the result of Figure 83, the direct-patterning process can be utilized to fabricate a micropatterned system without using photoresist or a dry etching process. In our previous work, direct-patterned PZT multilayer films were applied to micromachined piezoelectric cantilevers without using a dry etching process [65]. A limit of resolution showed to be dependent on the wavelength of the

UV used for direct-patterning, the numerical aperture, and k_1, which is usually referred to as a "k-factor" for the process [50].

Various PZT/SBT multilayer films were successfully constructed by PMOD using photosensitive starting precursors. Direct-patterning of the films with widths of several microns was accomplished by UV exposure for 15 min and removal of the unexposed area of the films using a solvent. Even though the P_r value of the 4-SBT film was relatively small (6 $\mu C/cm^2$), that of the SBT/2-PZT/SBT film was two times larger (12 $\mu C/cm^2$) as a result of using a SBT layer for the top and bottom layers in contact with the Pt electrodes. The films with SBT layers in contact with Pt electrodes showed improved leakage current and fatigue-free behaviors. The low number of defect charges effectively reduced the number of trapped charges at the Pt electrodes due to the presence of self-regulation layers such as $(Bi_2O_2)^{+2}$ in the SBT film. Therefore, the PZT/SBT multilayer films with SBT/Pt interfaces showed improved electric and ferroelectric properties and, in particular, fatigue resistance.

2.1.6.3. Formation of Multilayer Thin Films of $Nd_2Ti_2O_7$ and $Bi_{3.25}La_{0.75}Ti_3O_{12}$ For Applying to Ferroelectric Field Effect Transistor

In this part, improvements in the ferroelectric properties, including a reduction in the leakage current, were studied through the fabrication of a multilayer ferroelectric film for ferroelectric devices. The Aurivillius family, generally represented by $M_2A_{n-1}B_nO_{3n+3}$, contains fluorite-like $[M_2O_2]$ units and perovskite-like $[A_{n-1}B_nO_{3n+1}]$ blocks. In the Aurivillius phase, the perovskite A site with 12-coordination is normally occupied by large cations such as Ca^{2+}, Sr^{2+}, Ba^{2+}, Bi^{3+}, and Ln^{3+}, while the B site contains d^0 cations such as Ti^{4+}, Nb^{5+}, and W^{6+} according to the formula $[M_2O_2][A_{n-1}B_nO_{3n+1}]$, where 'n' is the number of perovskite-like layers and can be 2, 3, 4, or 5. $SrBi_2Ta_2O_9$ and $Bi_4Ti_3O_{12}$ are representative Aurivillius family ferroelectric materials [86,87]. In particular, BLT of the Aurivillius family has been intensively investigated as a material suitable for nonvolatile memory due to its relatively low phase formation temperature, large polarization, and good fatigue properties [88,89]. The ferroelectric material BLT is a strong candidate material for FeFET nonvolatile memory due to its improved ferroelectric properties by elemental substitution [87-89]. However, BLT ferroelectric film does not have a low leakage current suitable for application in microelectronic devices.

Neodymium dititanate (NT) is a member of a small family of rare earth titanates $Ln_2Ti_2O_7$ (Ln = La, Ce, Pr, and Nd) with ferroelectric properties [140]. At room temperature, NT is monoclinic with a $P2_1$ space group and exhibits a layered structure formed by infinite layers of strongly distorted TiO_6 octahedra [134]. The NT has eight sites for Ti atoms, with Ti–O distances ranging from 182 to 223 pm, and Nd atoms are positioned in eight sites, all coordinated by 12 oxygen atoms with Nd–O distances in the range of 240–385 pm. Furthermore, NT ferroelectric material is one of the candidate materials for MFIS-FET memory due to an extremely high Curie temperature $T_C > 1500^oC$, spontaneous polarization $P_S = 5\sim10\ \mu C/cm^2$, and low permittivity $\varepsilon = 31\sim47$ [161,162]. However, a high phase formation temperature and a low remnant polarization are disadvantageous for application to FeFET memory.

Both BLT and NT ferroelectric materials have layered structures, and the materials are capable of reciprocal diffusion of cations such as Bi^{3+}, La^{3+}, and Nd^{3+} at the BLT/NT interface. Therefore, a relatively stable interface can be expected during a phase formation of broad stoichiometry without occurrence of a secondary phase or a disruption in stoichiometry.

Table 2. Precursors of BLT and NT ferroelectric materials

	$Bi_{3.25}La_{0.75}Ti_4O_{12}$	$Nd_2Ti_2O_7$
Precursor	Bismuth 2-ethylhexanoate Lanthanum 2-ethylhexanoate Titanium isopropoxide	Neodymium(III)-acetylacetonate hydrate Titanium isopropoxide
Solvent	Hexane	2-methoxyethanol Acetylacetonate Acetic acid
Solution concentration	0.2 M	0.2 M
Precursor stoichiometry	Bismuth 5 % excess	Nd : Ti = 1 : 1

This part fabricated multilayer ferroelectric thin films using these ferroelectrics in order to improve the properties of the ferroelectric layer in the device structure. Electrical and ferroelectric properties of the multilayer film were compared with those of individual BLT ($Bi_{4-x}La_xTi_3O_{12}$) or NT ($Nd_2Ti_2O_7$) ferroelectric layers. For the characterizations of BLT, NT, and BLT/NT multilayer ferroelectric films, various analyses and measurements were carried out, including XRD, XPS, I-V, and P-E.Ferroelectric films of BLT, NT, and BLT/NT multilayers were formed via chemical solution deposition onto a Pt(111)/Ti/SiO_2/Si(100) substrate. Stock solutions of BLT and NT were prepared using the precursors listed in Table 2. The stock solutions were spin-coated at 3000 rpm for 20 sec onto a chemically cleaned Pt(111)/Ti/SiO_2/Si(100) substrate to measure the electrical properties of the resulting films. During each coating, a film thickness of 40 nm was applied, and the total film thickness was controlled by the number of applied coats.

Spin-coated samples were dried for 2 min on a hotplate at 200°C and then at 400°C to decompose and remove the organics. The thicknesses of the films were 200 nm for the BLT and NT and NT (40 nm)/BLT (120 nm)/ NT(40 nm). The thickness of each film was determined based on the electric field distribution of the permittivity differences. To crystallize the ferroelectrics, BLT, NT, and multilayer films were annealed at 700°C, 800°C, and 800°C, respectively, for 30 min in an O_2 atmosphere in a tube furnace using direct insertion.

The direct insertion annealing was processed to minimize an interface diffusion of BLT/NT film. After annealing, the resulting 150-nm-thick Pt top electrode with a 100 μm radius was sputter-deposited to measure the electrical properties of the films. The phase formations and crystallinities of the films were studied with XRD using Cu k_α radiation. We measured the P-E characteristics with an RT66A. Current-voltage characteristics were measured with an HP4145A at a holding time of 0.5 s and a delay time of 1 s. The composition depth profile of the multilayer structures was analyzed using XPS.

Figure 84 shows the XRD spectra of (a) BLT, (b) NT, and (c) NT/BLT/NT films formed on the Pt(111) substrate. All films were well-crystallized without a secondary phase. The crystal orientation of the BLT film was not affected by Pt substrate, as shown in Figure 84 (a). The NT film showed strong diffraction peaks as (014), (020), (212), and (203). In the case of the NT/BLT/NT multilayer film, diffraction peaks of both BLT and NT films were observed. The BLT film in the NT/BLT/NT multilayer film showed a decrease in the (00*l*) diffraction intensity and increases in the (111) and (117) diffraction intensities, as shown in Figure 84 (c). The difference in BLT growth orientation of the multilayer film was caused by an increase in Nd^{3+} cation substitution in the Bi^{3+} A site due to diffusion at the BLT/NT interface [49,92]. Both the Aurivillius compound $Bi_4Ti_3O_{12}$ and the pseudo pyrochlore structure $Nd_2Ti_2O_7$ ferroelectric materials had similar layered structures.

These materials include Bi^{3+}, La^{3+}, and Nd^{3+} cations between distorted Ti-O octahedral structures, in which the cations are able to achieve reciprocal diffusion at the BLT/NT interface during the annealing process. The diffraction data revealed a small difference in the lattice parameters between the BLT film and the multilayer film.

A decrease in the a - b axes parameters from 5.46 - 5.40 Å to 5.40 - 5.37 Å, and an increase in the c axis parameter from 32.74 Å to 32.88 Å were observed with the BLT in the multilayer film. Multilayer BLT showed a tendency to form tetragonal structures with reduced orthorhombicity due to an increase of Nd^{3+} substitution through the interface diffusion [91,92]. Therefore, formation of a stable interface can be expected due to a possible broad stoichiometry in the multilayer structure without an occurrence of secondary phase, as shown in Figure 84. Therefore, this part analyzed the composition according to the depth of the NT(40 nm)/BLT(120 nm)/NT(40 nm) film in order to determine the interface diffusion state.

Figure 85 presents the stoichiometry of the NT/BLT/NT multilayer film according to etching time(s) as determined by XPS. Bi^{3+} cations were observed at the NT layer region of the multilayer film, while small diffusions of Nd^{3+} and La^{3+} cations were observed in the BLT and NT regions. Due to these diffusions of cations, the amount of $Nd^{3+}$and La^{3+} cations increased to six atomic percent in the BLT layer, corresponding to an increase in x to 1.15 in $Bi_{4-x}(Nd, La)_xTi_3O_{12}$.

This increase in substitution led to a decrease in the (00*l*) diffraction intensity and increases in the (111) and (117) diffraction intensities in the BLT of the multilayer film, as shown in Figure 84 (c). Bi^{3+} cation diffusion allows for not only the substitution of the Nd^{3+} site between the Ti-O octahedral structures in the NT layer, but also the formation of a $[Bi_2O_2]^{2+}$ inter-layer. However, phase transformation from NT to the Aurivillius phase was not observed in the XRD results.

The stable interface with a broad stoichiometry plays an important role in the establishment of stable electrical properties [163]. In this part, the stable interface in a multilayer film was expected to contribute to stable ferroelectric properties. Figure 86 shows P-E hysteresis loops of (a) BLT, (b) NT, and (c) NT/BLT/NT ferroelectric films at an applied voltage of 10 V. The measured remnant polarization and coercive field values of BLT, NT, and NT/BLT/NT films were 7.5, 4, and 3 μC/cm^2 and 130, 300, and 170 kV/cm, respectively, showing that the remnant polarization and coercive field values of the NT/BLT/NT multilayer were 75% and 57%, respectively, of those of the NT layer.

These values correspond to a decrease of 60% and an increase of 30% compared with those of the BLT layer. In spite of a decrease in the c-axis orientation in the low remnant polarization direction and increases in the (111) and (117) diffraction intensities at high remnant polarization, the measured remnant polarization of the multilayer film decreased. This decrease in remnant polarization originates from a reduced ferroelectric distortion represented by a low T_c [81]. The substitution reduces stress in the structure:

by the replacement of Bi^{3+} by Nd^{3+}, La^{3+} would reduce the ion displacement of the A site in the perovskite layer and the remnant polarization decreases [91]. The value of E_c decreased gradually by increasing the substitution contents. Sugita et al. [83] reported that T_c and the crystallization temperature of BLT films decreased with increasing substitution contents. With increased substitution, since the phase transition temperature is reduced with increasing substitution contents, the residual stress decreases during cooling. However, the value of E_c increased for the multilayer film from the high E_c characteristic of NT ferroelectric.

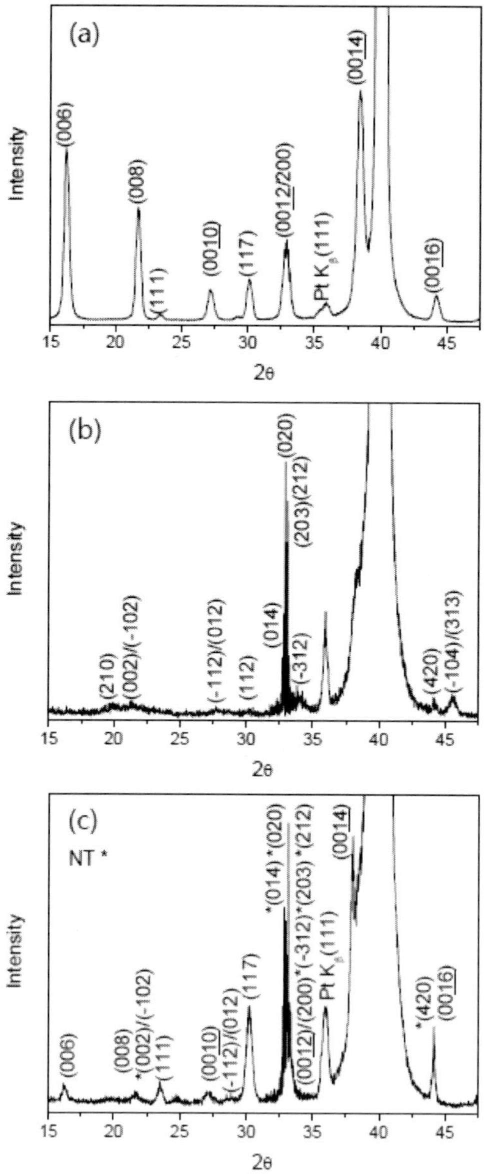

Figure 84. X-ray diffraction patterns of (a) BLT, (b) NT, and (c) NT/BLT/NT films formed on Pt(111) substrate.

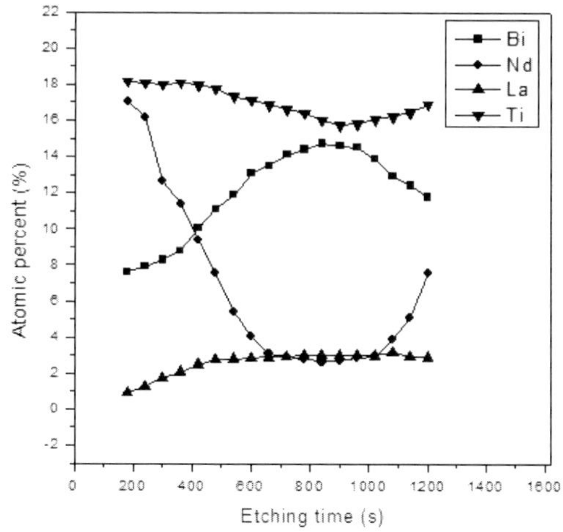

Figure 85. Stoichiometry of the NT/BLT/NT multilayer film as determined using XPS depth profile analysis.

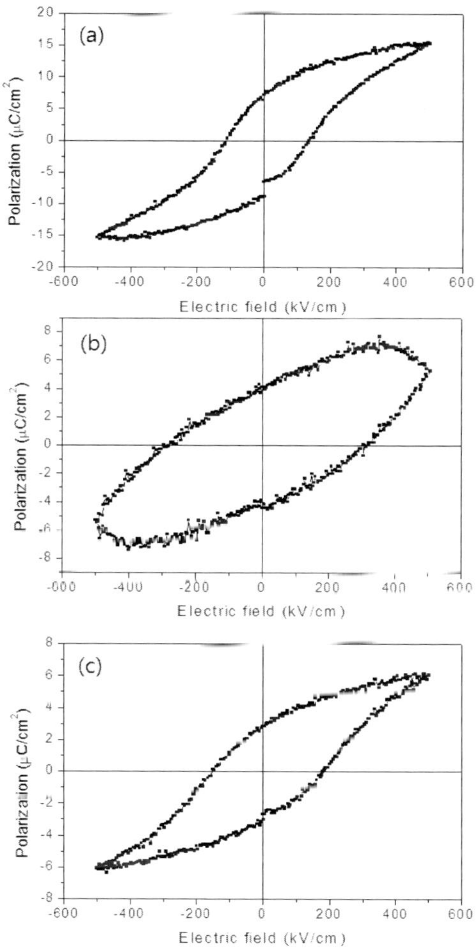

Figure 86. P-E hysteresis loops of (a) BLT, (b) NT, and (c) NT/BLT/NT ferroelectric films at an applied voltage of 10 V.

The current-voltage characteristics of BLT, NT, and NT/BLT/NT films are given in Figure 87. The NT/BLT/NT multilayer film showed a smaller gradient than did the NT single layer film in the Schottky conduction region of the current-voltage curve. However, it shows a transition of the conduction mechanism from Schottky to Poole-Frenkel at 6.2 V. In spite of a decrease in c-axis orientation at low leakage current, the multilayer film showed a low leakage current and a transition in the conduction mechanism from Schottky to Poole-Frenkel at higher voltages. These results are due to an NT layer contact with an electrode and a more stable orbital hybridization between Nd 5d and O 2p levels, which induces a decrease in oxygen vacancy due to an increased substitution in the BLT film. The substitution of BLT with Nd induced a reduction in the permittivity and polarization of BLT, dependent upon the substitutional ratio; sometimes these reductions are smaller than those of NT [81,91]. In our case the gradient of the current-voltage curve of the multilayer film was smaller than that in the NT single layer film in the Schottky conduction region due to an electric field distribution resulting from the permittivity difference. This part showed an improvement in the properties of ferroelectric film through multilayer fabrication between BLT and NT films. An improvement in the leakage current property was obtained to significantly reduce the retention property.

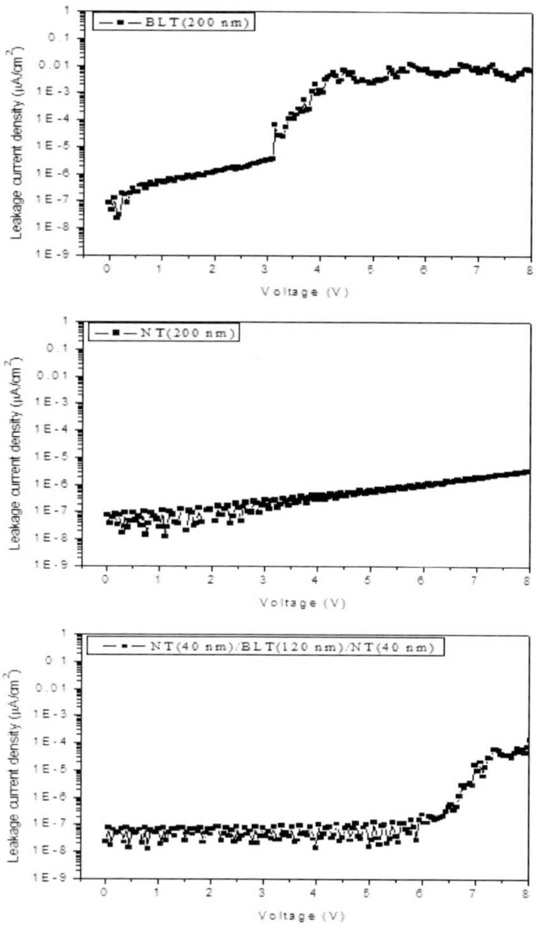

Figure 87. Current-voltage characteristics of BLT, NT, and NT/BLT/NT/Pt films.

A multilayer film of stable interface was fabricated using a possible phase formation at a broad stoichiometry region between BLT and NT ferroelectric materials, without formation of a secondary phase. The multilayer film did not show an enhancement in remnant polarization. On the contrary, the remnant polarization of the multilayer film deceased due to increases in La^{3+} and Nd^{3+} substitutions by reciprocal diffusion in the BLT layer. Even though the multilayer film did not show an improved P-E property, it showed a highly insulating property. As a result, the multilayer ferroelectric films showed an advantage for improving the retention property due to reduced leakage current behavior.

2.2.6.4. Effect of $SrTiO_3$ Buffer Layer on the Phase Formation and Properties of $BiFeO_3$ Thin Films

Recently, multiferroic materials have attracted much attention, due to their abilities to exhibit simultaneous ferroelectric and ferromagnetic properties. Moreover, many studies have described the magnetoelectric effect produced upon the coupling of these two properties [164]. The magnetoelectric effect is expected to form the basis of a new type of memory device, as magnetization can be tuned using an electric field, and, conversely, the ferroelectric polarization can be controlled using a magnetic field [165]. Bulk $BiFeO_3$ (BFO) with a rhombohedrally-distorted perovskite structure in the space group R3c, has a lattice parameter of a = 3.96 Å and α = 89.5 °. In addition, BFO is one of a few multiferroic materials which possess simultaneous ferroelectric and ferromagnetic properties at room temperature. Various studies of BFO have been conducted under high potentiality conditions, a high Curie temperature T_c of about 850°C and a high Néel temperature T_N around 370°C [166,167].

One major problem with BFO is the difficulty in achieving single phase formation. The formation of BFO without impurity phases, such as excess bismuth and iron, is only possible in a narrow pressure-temperature-stoichiometry window [168,169]. A process outside of the optimum window can lead to the formation of a secondary phase with high current leakage and an anomalously increased saturation magnetization, caused by both oxygen vacancies and valence fluctuations of the Fe ion [170].

In this part, direct patternable BFO thin films were fabricated using PMOD and the phase formation, electrical properties of the films were studied. The precursors used for the photochemical production of BFO films and $SrTiO_3$ buffer layer were bismuth 2-ethylhexanoate [$Bi(C_7H_{15}COO)_3$], iron 2-ethylhexanoate [$Fe(C_7H_{15}COO)_3$], strontium 2-ethylhexanoate [$Sr(C_7H_{15}COO)_3$], and titanium isopropoxide [$Ti\{OCH(CH_3)_2\}_4$] [38,90]. The compositions of the BFO precursors were varied, with bismuth excesses including 2, 5, 7.5, and 10% to compensate for the loss of Bi during the annealing process at high temperature. The BFO precursors dissolved in hexane were spin-coated at 3000 rpm for 30 s onto three different substrates. Two substrates, $Pt(111)/Ti/SiO_2/Si(100)$ and $SrTiO_3/Pt(111)/Ti/SiO_2/Si(100)$, were chemically cleaned to check for differences in phase formation and to measure the ferroelectric properties. The other substrate, p-Si(100), was used for characterization of the direct-patterning process. The precursor films were lithographically exposed for 15 min using the output of a 1000 W mercury arc lamp (UV lamp with maximum wavelength of 365 nm) filtered through a chrome mask. The exposed film was rinsed with hexane to develop the latent negative pattern. The resultant films were annealed for 2 min at 200°C and 400°C in air, successively. The films were annealed at 550°C for 20 min under an O_2 and N_2 atmosphere to study an ambient gas effect on the BFO phase formation.

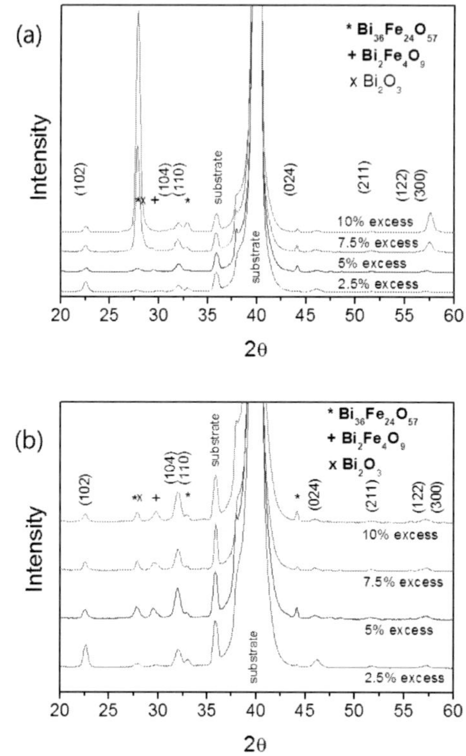

Figure 88. X-ray diffraction patterns of BFO films according to Bi excess after annealing in an (a) N_2 and (b) O_2 atmosphere.

The surface microstructures of the direct-patterned films were observed using optical microscopy. The phase formation and crystallinity of the films were studied with XRD using Cu K_α radiation.

Fourier transform-infrared spectroscopy was used to monitor an evolution of the photosensitive BFO precursors during the process. A 150-nm-thick Pt top electrode with a 100-μm radius was sputter deposited to measure the P-Eand I-Vcharacteristics of the films.

Figure 88 presents the XRD pattern of direct-patternable BFO films on Pt(111)/Ti/SiO_2/Si substrates according to Bi excess. The films were annealed at 550°C for 20 min under an O_2 or N_2 atmosphere. An increase in Bi excess phase was observed distinctly with increased Bi composition in an N_2 atmosphere. The results of the N_2 atmosphere XRD spectra represent a great tendency for compositional deviation in BFO thin films, related to the reactant stoichiometry. During the phase formation at high temperatures, Bi^{3+} is known to evaporate in the form of Bi_2O_3 [91]. Therefore, the O_2 atmosphere was expected to accelerate Bi evaporation; however, a limited amount of Bi excess phase was formed with increasing excess Bi content. To prevent the formation of impurity phases, BFO requires precision control of reactant composition, while considering Bi evaporation. A $BiFeO_3$ thin film with the minimal impurity phase concentration was fabricated with a Bi excess of 2%. However, a Fe excess phase was always observed regardless of Bi content. These results showed that the presence of a secondary phase depended not only on composition, but also the other preparation condition. Therefore in this part, STO buffer layer was used to check the relation between secondary phase formation and the substrate.

$Bi_4Ti_3O_{12}$ material, including volatile Bi, has been used as a representative ferroelectric material of layered perovskite structures. In the Bi excess region of the material, elasticity of stoichiometry is possible by controlling the layer arrangement ratios of the Bi_2O_2 and pseudo-perovskite layers, such as 1:2, 1:1, and 2:1 [171]. An exception is the BFO thin film in this work, which requires an exact stoichiometry. Conversely, the PZT ferroelectric thin film with a perovskite structure, which includes a volatile Pb atom, has been reported to have a deficiency of nucleation sites, related to an induction of the secondary phase [172,173]. In this report, a secondary phase was induced by a deficiency of nucleation sites in the PZT material. From this point of view, an experiment concerning BFO phase formation was performed on STO buffer layer with a perovskite structure of lattice constant 3.905 Å in order to study the substrate effect on phase formation. The STO film was fabricated with a 40 nm thickness on Pt(111) using PMOD and BFO thin films were fabricated on the STO film using the Bi 2% excess solution which showed the least amount of secondary phase.

Figure 89 shows the XRD pattern of direct-patternable BFO films deposited onto STO/Pt(111)/Ti/SiO_2/Si substrates and annealed at 550°C for 20 min under O_2 and N_2 atmospheres. The secondary phase was reduced remarkably and it was found that the STO substrate could prevent the formation of a Fe excess phase. Each step of the BFO film production procedure was monitored using FT-IR spectroscopy to investigate the difference between the Pt(111) and STO substrates. Figure 90 shows the FT-IR spectra results after drying at 200°C and 400°C. With the BFO film formed on the Pt substrate, absorptions by iron oxides such as γ–FeOOH and Fe-O were observed at 640 ~ 670 cm^{-1}, after the 400°C drying process for 2 min [174,175]. However the absorptions were not observed in the BFO film formed on the STO substrate.

These results indicate an induction of a large amount of iron oxides on the Pt(111) substrate. From these results, it can be said that the Pt(111) substrate provides a nucleation site for iron oxides, leading to the nucleation of iron oxides. Normally a formation of FeO and successive island growth of Fe_3O_4 on Pt (111) substrate are occurred from the catalytic effect of Pt [176]. Disproportion of the stoichiometry in BFO thin films was expected from the induction of iron oxides. From this origin, a Fe excess phase was observed regardless of Bi excess content and the secondary phases could be induced from the stoichiometric disproportion, as shown in Figure 88. A 150-nm-thick Pt top electrode with a 100-μm radius was sputter deposited onto a BFO/STO/Pt/Si structure for P-E and I-V characterization of the films. The P-E characteristics of the Pt/BFO/STO/Pt/Si structure with an applied voltage of 11 V are illustrated in Figure 91. The measured P_r and E_c values of the Pt/BFO(200 nm)/STO(40 nm)/Pt/Si structure were 10 μC/cm^2 and 300 kV/cm, respectively. When comparing with the typical values of BFO, these values are relatively small due to a presence of secondary phase as shown in Figure 89. The secondary phase induced oxygen vacancies and valence fluctuations of the Fe ion and BFO became leaky with degraded ferroelectric properties [170].

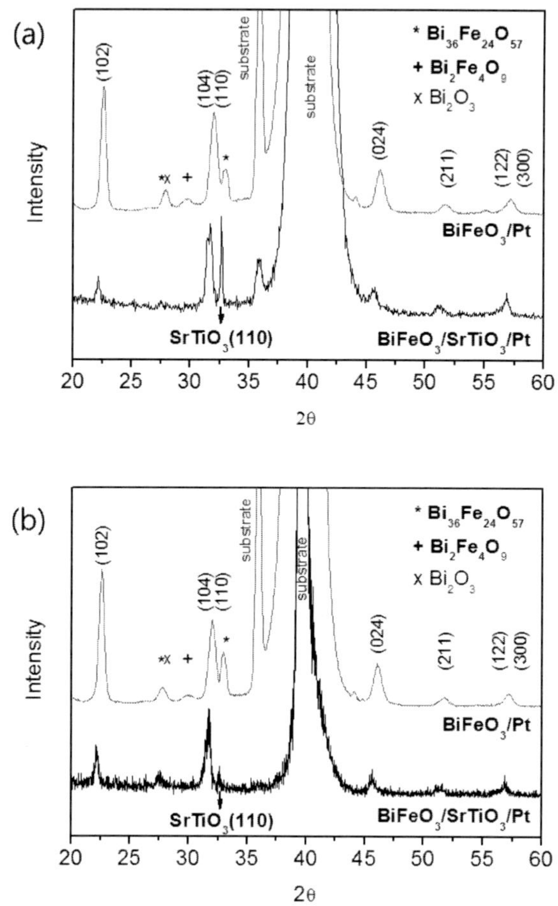

Figure 89. X-ray diffraction patterns of BFO films on Pt(111) and STO/Pt(111) using Bi 2% excess solution after annealing in an (a) N_2 and (b) O_2 atmosphere.

Figure 92 shows an I-V characteristic of the Pt/BFO/STO/Pt/Si structure. The BFO(200 nm)/STO(40 nm) film showed a linear increase of the current density with the current density with the square root of applied electrical field, However at 225 kV/cm of applied electrical field, a sudden increase in the current density was observed due to an occurrence of Poole-Frenkel emission and a breakdown at 360 kV/cm. Leakage current of perovskite BFO film was dominated by Poole-Frenkel emissions as other perovskite ferroelectrics [177,178]. These electrical properties are induced from the structure of Pt/BFO/STO/Pt, because the applied voltage was distributed by the permittivity of each material. Furthermore, the BFO on a Pt(111) substrate did not show ferroelectric behavior due to an insulating property of the secondary phase.

The possibility of direct-patterning of spin-coated BFO precursor films with a 2% Bi excess was examined and the results are given in Figure 93. To obtain an image, a spin-coated BFO film was exposed to UV for 15 min and then rinsed in hexane.

In this case, after the exposure of the precursor film to UV, metal-oxide is remained through the successive metal-charge transfer reaction and the organic ligand was removed through the decomposition. This is the reason why the exposed part of the coated film to UV remains during solvent washing while unexposed part is removed. The basic mechanism of PMOD is addressed in the literature [90].

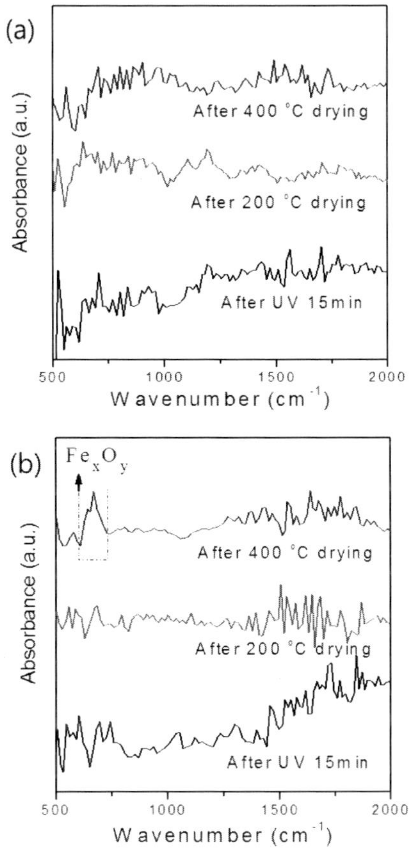

Figure 90. FT-IR spectra of each drying step in BFO film formation using 2% excess Bi solution on (a) STO/Pt(111) and (b) Pt(111).

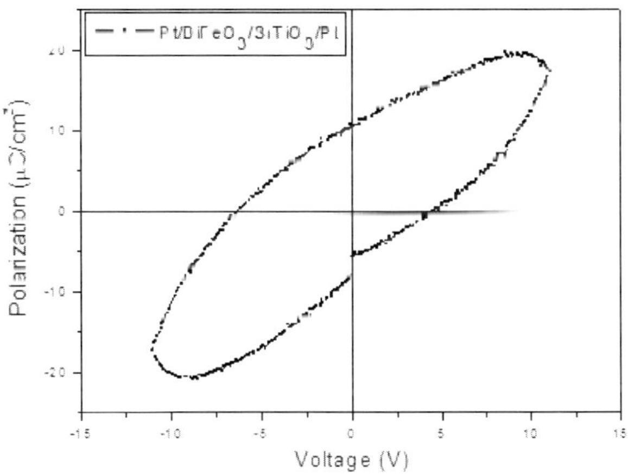

Figure 91. P-E hysteresis loop of BFO (200 nm)/STO (40 nm) film.

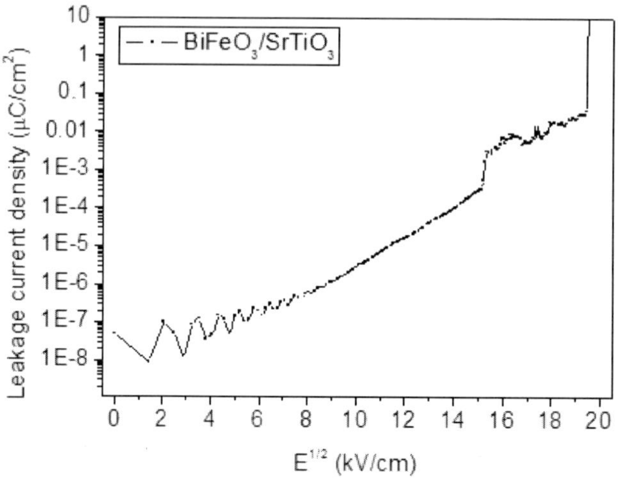

Figure 92. I-V characteristic of the Pt/BFO (200 nm)/STO (40 nm)/Pt(111) structure.

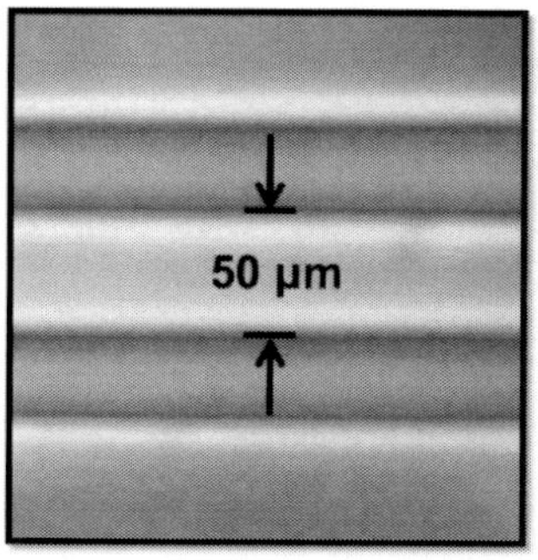

Figure 93. A direct-patterned image of BFO film on p-Si(100) substrate.

As can be seen in Figure 93, the brownish area corresponds to the Si substrate and the bluish area corresponds to the direct-patterned BFO film. Based on the well-defined width in the direct-patterned BFO film, the width of the line pattern was 50 μm and micron-scale patterning was successfully obtained via lithographic patterning using PMOD. From the above results, direct lithographic patterning of a BFO film can be accomplished for ferroelectric device applications using PMOD. Direct patternable BFO thin films were successfully fabricated using PMOD. $BiFeO_3$ was very sensitive to the stoichiometry and formation of a pure BFO phase required precise control of reactant stoichiometry with consideration of Bi evaporation. A preferable formation of iron oxides was observed on a Pt substrate, as demonstrated by the FT-IR results. The Pt(111) substrate led to preferred nucleation of iron oxides due to the presence of nucleation sites and the preferred nucleation of iron oxides was disproportionate to the stoichiometry. The STO buffer layer was found to somewhat reduce the formation of secondary phases. However due to the partial presence of secondary phases, BFO film on STO showed leaky and degraded ferroelectric behaviors. None the less, phase formation of direct-patternable BFO film on an STO buffer layer could be used for micro-patterned systems.

2.2. Study on the Improvement of Retention Property in MFIS-FeFET

A Study of FeFET stays in this research phase. This chapter shows improvement study and the result of leakage current, depolarization field for the improvement of retention property.

2.3.1. Insulator Materials in MFIS for Retention Improvement of FeFET

In case of metal-ferroelectric-metal (MFM) structure, when ferroelectric layer is polarized, a compensating charge is equally developed on each side of electrodes.

Therefore there is no depolarization field in MFM structure. On the other hand, in the stack of MFIS structure, a depolarization field always exists due to a finite dielectric constant of semiconductor. Moreover, a permittivity difference induces an applied voltage drop across buffer layer. As a result, the ferroelectric layer does not have a sufficient electric field for polarization saturation.

The leakage current of the insulator layer compensates for the polarization of the ferroelectric layer by charge trapping at the interface. Improvements in the retention properties of MFIS structures have been achieved through a reduction of the leakage current and permittivity difference (C_{IS}/C_F) between the ferroelectric and insulation barrier materials. As a result, a layered perovskite ferroelectric with low permittivity and high fatigue resistance, and an insulation barrier with low leakage current, high insulation, and high permittivity have been adopted for the construction of FeFET structures [179-181].

In this part, a wide range of three different high-k insulation materials, HfO_2, Y_2O_3, and CeO_2 were adopted to construct an MFIS structure, and their properties, including retention, were compared. These materials have distinctly different characteristics and are suitable for the study of MFIS structures. Table 3 presents the properties of the HfO_2, Y_2O_3, and CeO_2 insulators. $Nd_2Ti_2O_7$ (NT) rare earth titanate is monoclinic with a $P2_1$ space group and exhibits a layered structure formed by infinite layers of strongly distorted TiO_6 octahedra [134,140]. The NT has eight sites for Ti atoms, with Ti–O distances ranging from 182 to 223 pm, and Nd atoms are positioned in eight sites, all coordinated by 12 oxygen atoms with Nd–O distances in the range of 240–385 pm. Furthermore, NT ferroelectric material is one of the candidate materials for MFISFET memory due to an extremely high Curie temperature $T_C >$ 1500°C, spontaneous polarization P_S = 5~10 μC/cm^2, and low permittivity ε = 31~47 [161,162,182]. Two other layered structures of MIM and metal-insulator-silicon (MIS) were fabricated using Pt(111) and silicon substrates to reduce the permittivity in the MFISFET. Various analyses and measurements were carried out to characterize HfO_2, Y_2O_3, and CeO_2 insulators, including XRD, Auger electron spectroscopy (AES), I-V, and C-V including a memory window analysis. CeO_2, Y_2O_3, and HfO_2 insulator films of 20 nm thickness were fabricated on p-type Si(100) substrate by reactive direct current magnetron sputtering. After deposition, the films were annealed at 800°C for 20 min under an O_2 atmosphere.

Table 4 shows the deposition conditions for HfO_2, Y_2O_3, and CeO_2 films. For the construction of MFIS structures, a $Nd_2Ti_2O_7$ ferroelectric film of 240 nm was formed on the insulation films by a chemical solution deposition method. The structure was then annealed at 800°C for 30 min under ambient O_2. Platinum with a 100 μm radius and 150 nm thickness was sputter-deposited as a top electrode. Three types of structures, MIM, MIS, and MFIS were also fabricated with the three different insulation materials to study the effect of silicate formation at the interface on the electrical properties of the MIM, MIS, and MFIS structures.

Table 3. Properties of Y_2O_3, CeO_2, and HfO_2 insulator films

	Y_2O_3	CeO_2	HfO_2
Crystal structure	Bixbyite	Fluorite	Fluorite
Crystal system	cubic	cubic	monoclinic
Dielectric constant	12.03	20.05	22.29

Table 4. Deposition conditions for Y_2O_3, CeO_2, and HfO_2 films

	Y_2O_3	CeO_2	HfO_2
Target	Y(99.99 %)	Ce(99.99 %)	Hf(99.99 %)
Sub. temp.	30°C		
Sputtering gas $[O_2/(Ar+O_2)]$	20 %		
Gas pressure	2.2×10^{-3} Torr		
Deposition rate	4.55 Å/min	4.85 Å/min	5 Å/min

The phase formation and crystallinity of the films were studied with an XRD using Cu kα radiation. A 150 nm thick Pt top electrode with a 100 μm radius was sputter deposited to measure the retention and I-V characteristics of the films. The current-voltage characteristic was measured with an HP4145A at a holding time of 0.5 s and delay time of 1 s. The composition profiles of the layered structures were analyzed by depth profile analysis using AES. The memory window of the samples was measured using an HP4284A and HP4280A. The amplitude of the alternating current signal was 30 mV with a sweep rate of 0.2 V/s at an applied voltage of 5 V. The retention property was investigated by measuring the reduction in permittivity after repeated poling with an applied voltage of -5 V for 10 s for enough poling of domains.

X-ray diffraction spectra of Y_2O_3, CeO_2, and HfO_2 films on /Si substrates are shown in Figure 94. Y_2O_3 showed only (111) diffraction peaks, and this (111) preferred growth plane of yttrium oxide is known to have no orientational relationship with the underlying substrate as has been previously reported [184-186]. CeO_2 showed a strong (111) directional peak with (200) and (220) peaks. The growth direction of CeO_2 is known to be affected not only by substrate orientation but also by deposition method. For example, CeO_2 films prepared by the sputtering technique showed various growth orientations [187-189]. In the case of HfO_2, the polycrystalline nature of growth orientation was observed [190].

C-V characteristics for Y_2O_3, HfO_2, and CeO_2 films are given in Figure 95. Normally, insulation layers undergo an extra annealing step for 30 min at 800°C for the phase formation with $Nd_2Ti_2O_7$ ferroelectrics in MFIS structures. Diffusion of silicon into the insulation layers and the formation of silicon oxides and silicates would be expected from this process, which should affect the properties of the MFIS structure. Thus, the C-V characteristics of the insulation layers annealed at 800°C for 20 min or 50 min were measured and compared as shown in Figure 95. The permittivities of Y_2O_3, HfO_2, and CeO_2 films were reduced by 13.5%, 32.4%, and 31.5%, respectively after an extra annealing for 30 min. These results indicated a progressive interfacial reaction between the insulation layers and the silicon substrate with increased annealing time. For a comparison of interfacial stability, equations representing the thermodynamic reactions between the insulator layer materials and silicon at 298 °K are given below [192].

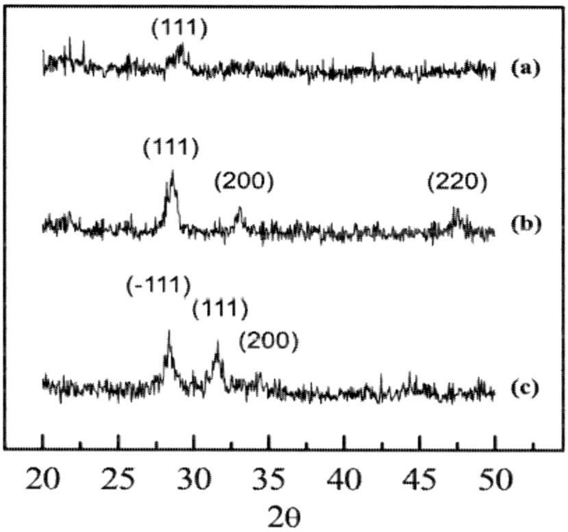

Figure 94. XRD patterns of 20 nm-thick (a) Y_2O_3, (b) CeO_2, and (c) HfO_2 thin films annealed at 800° C for 20 min in ambient O_2.

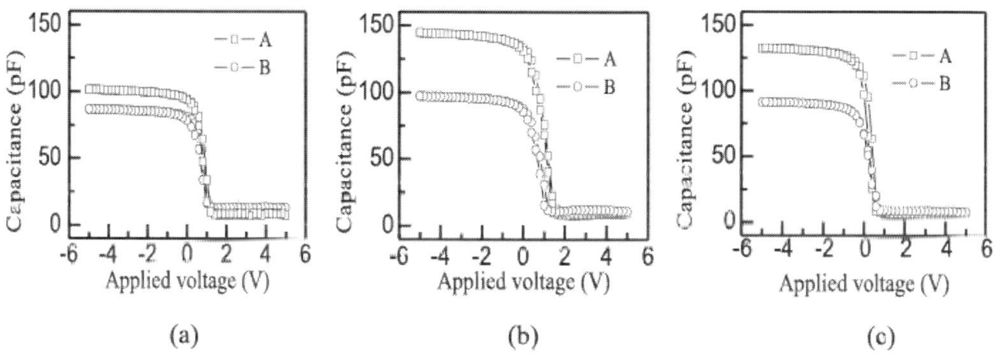

Figure 95. C-V characteristics of 20 nm thick (a) Y_2O_3, (b) CeO_2, and (c) HfO_2 thin films after annealing at 800° C for A; 20 min and B; 50 min in ambient O_2.

$$Ce + SiO_2 \rightarrow CeO_2 + Si \quad \Delta G = -178.3 \text{ kJ/mol}$$

$$4Y + 3SiO_2 \rightarrow 2Y_2O_3 + 3Si \quad \Delta G = -1075 \text{ kJ/mol}$$

$$Hf + SiO_2 \rightarrow HfO_2 + Si \quad \Delta G = -47 \text{ kJ/mol}$$

Yttrium oxide, the most stable of the metal oxides in the presence of silicon, was found to effectively suppress the formation of silicates. The thermodynamic reaction values match well with the reduction in permittivity of the insulator materials during annealing, the process responsible for the reduction of insulator materials and the formation of metal-silicates. Depth profiling analyses were carried out using AES to examine the diffusion barrier capability of the insulation layers against the $Nd_2Ti_2O_7$ ferroelectric layer and the interfacial stability with Si by checking for Si diffusion into the insulator layers. The results, given in Figure 96, show that all of the insulator layers effectively prevented diffusion into the $Nd_2Ti_2O_7$ ferroelectric layer. However, in the case of HfO_2, a heavy inter-diffusion of Si into the insulation layer was observed, and this observation is in good agreement with a report that HfO_2 forms silicates with Si in a wide range of compositions [193, 194]. The formation of silicates at the interface may induce an increase of the depolarization field due to the relatively lower permittivity of silicates compared to other insulator material, which is disadvantageous for the retention property.

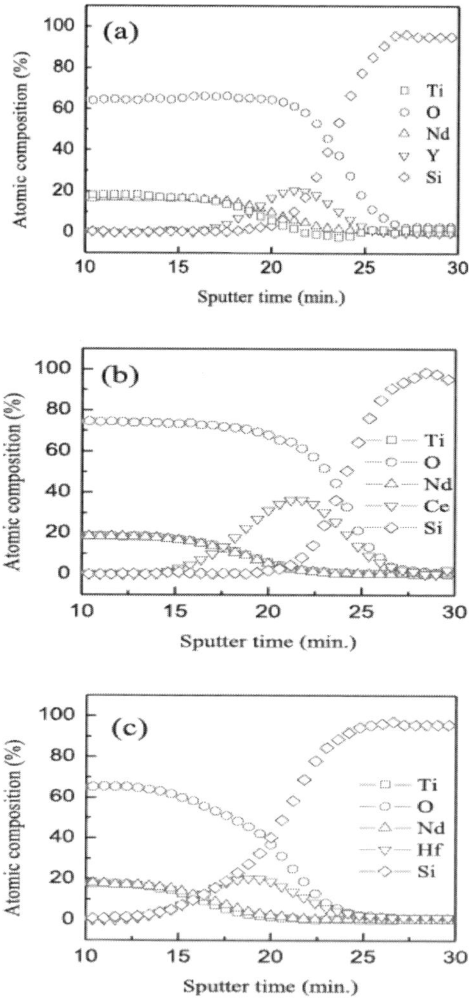

Figure 96. AES depth profiles of $Nd_2Ti_2O_7$ film deposited on (a) Y_2O_3, (b) CeO_2, and (c) HfO_2/Si substrate after annealing at 800° C for 30 min.

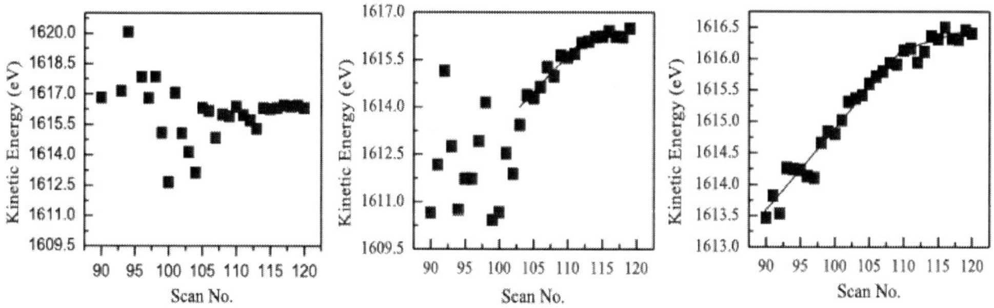

(a) Y (b) Ce (c) Hf

Figure 97. Variation of Si position in the AES spectrum. (a) NT/Y_2O_3, (b) NT/CeO_2, and (c) NT/HfO_2.

However, on the other hand, electrically stable silicates may induce an improvement in the retention property by the reduction of leakage current. The kinetic energy of the Si $_{KLL}$ Auger line was monitored within the MFIS structure, as shown in Figure 97, to investigate the chemical state of the silicon: metal-silicate; silicon-oxides; metallic silicon. Normally, the kinetic energy of Si $_{KLL}$ changes from 1616 eV to 1609 eV when Si is oxidized. In the formation of metal-silicate, an intermediate kinetic energy value is observed [54, 195]. In the case of Y_2O_3, a convergence of the Si $_{KLL}$ kinetic energy to 1616 eV was observed after scan number 107, but only a scattering of kinetic energy values was found before the silicon substrate, which is an indication of an inhomogeneous and trivial formation of yttrium-silicates and of the irregular presence of unavoidable silicon-oxides [196]. The MFIS structure of CeO_2 showed a relatively gradual formation of cerium-silicates at the interface from scans 105 to 115; however, HfO_2 formed gradual hafnium-silicates in a wide range of interfaces in good agreement with the above observations.

The highest reduction in the permittivity of HfO_2 after an extra annealing at 800°C was due to the gradual formation of hafnium-silicates, which have a low permittivity at the interface with the silicon substrate.

The current densities of MFIS structures with applied voltages are given in Figure 98. In the cases of HfO_2 and CeO_2, a transition of the conduction mechanism from Schottky to Poole-Frenkel was observed. The Pt/NT/Y_2O_3/Si structure showed a soft breakdown, which might be due to an unstable interface in the Y_2O_3-MFIS structure. The highest breakdown voltage for the HfO_2-MFIS structure was the result of uniform and gradual formation of hafnium-silicates which have a relatively high breakdown voltage of 10 MV/cm [163].

The memory windows of the three MFIS structures containing HfO_2, Y_2O_3, or CeO_2 were measured and are given in Figure 99. The values did not differ greatly, 1.44~1.62 V at 5 V of applied voltage, and this similarity was the result of a decrease in the permittivity of the insulator layer region due to the formation of metal-silicates with low permittivity between the insulators and silicon substrate.

Figure 100 reports the retention properties of the three MFIS structures with the HfO_2-MFIS structure having the best retention property.

The capacitance values of MFIS structures dropped sharply about 15–25% after writing. After initial decrease of the capacitance, decreasing rates became very small and the retention property was stabilized.

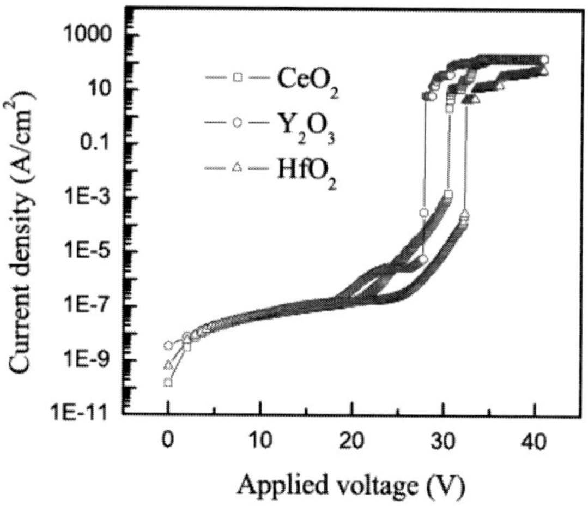

Figure 98. I-V characteristics of $Nd_2Ti_2O_7$ thin film deposited on Y_2O_3, CeO_2, and HfO_2 insulator in the MFIS structure.

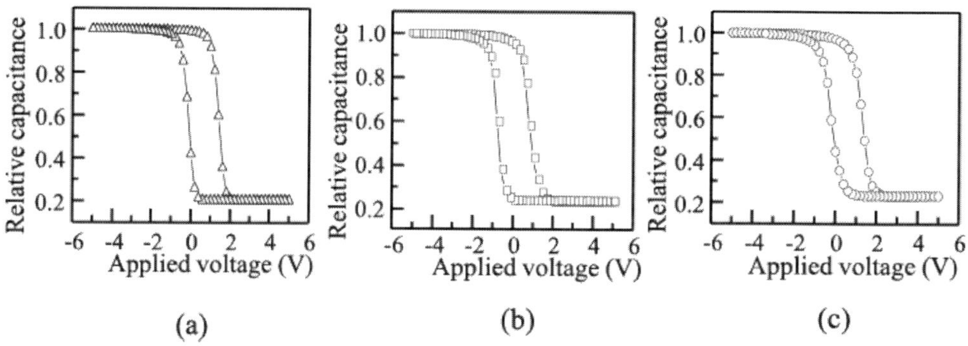

Figure 99. Memory window of $Nd_2Ti_2O_7$ thin films deposited on (a) Y_2O_3, (b) CeO_2, and (c) HfO_2 insulator in the MFIS structure.

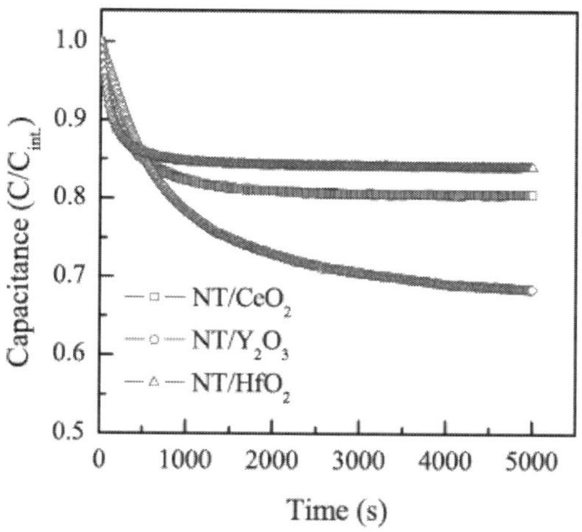

Figure 100. Retention property of $Nd_2Ti_2O_7$ thin film deposited on Y_2O_3, CeO_2, and HfO_2 insulator in the MFIS structure.

The HfO_2-MFIS structure showed the smallest initial decrement and the shortest saturation time. The Y_2O_3-MFIS structure shows a continuous decrease of capacitance after an initial decrement in the retention property. The ferroelectric retention is the loss of polarization as a function of time from a given-poled state. A retention tendency of the three MFIS structures seemed not to be related with the C_I/C_F of the MFIS structures. The retention property might depend mostly on the leakage current behavior because C_I/C_F of the MFIS structures was not very different each other. The unstable interface of the Y_2O_3-MFIS structure induced the worst retention property. As shown in Figure 98, leakage current density of three insulator layers showed almost same values up to 15 V and this means that Schottky current characteristics were not different. However the electrically unstable interface of Y_2O_3-MFIS corresponded to a high Frenkel defect density state. A discharge of trapped charge in the defect induced the retention behavior by long poling time [197].

From these results, it can be concluded that a minimization of leakage current through the formation of stable interfacial metal-silicates is more important to the retention property than the simple adoption of insulator material with high permittivity and shows the advantages of minimization of the depolarization field and effective distribution of the applied field.

Among the Y_2O_3, CeO_2, and HfO_2 insulator materials, the most thermodynamically stable Y_2O_3 formed the least stable interface with the silicon substrate after annealing at 800°C for 50 min with the inclusion of inhomogeneous, trivial yttrium-silicates and irregular, unavoidable silicon-oxides. In contrast, the thermodynamically least stable HfO_2 formed gradual hafnium-silicates with low permittivity with the silicon substrate. This gradual and uniform metal-silicate between the insulator and silicon substrate induced a stable and homogeneous interface and proper leakage current behavior even with a reduction in permittivity. Because high temperature annealing is necessary for the preparation of MFIS, the stable formation of interfacial metal-silicates is more important from the point of view of retention property than is the simple adoption of insulator material with high permittivity and high thermodynamic stability. The results of this part indicate the advantages of minimization of the depolarization field and stable interface in MFIS structure.

2.2.2. Ferroelectric Materials in MFIS for Retention Improvement of FeFET

As mentioned above, MFIS structures have two major problems: retention and disturbance. The retention problem must first be resolved to successfully fabricate non-volatile MFISFET memory. Reduction in retention time is mainly caused by the depolarization field and leakage current in the insulator layer [10]. If the permittivity of the buffer layer differs greatly from that of the ferroelectric film, the ferroelectric layer will have a great depolarization field. Most of the applied voltage drop occurs across the buffer layer. As a result, the ferroelectric layer does not have a sufficient electric field for polarization saturation. The leakage current of the insulator layer compensates for the polarization of the ferroelectric layer by charge trapping at the interface. Actually, an improvement of retention property of MFIS structure has been achieved by the works to reduce the leakage current and the permittivity difference (C_{IS}/C_F) between ferroelectrics and insulators. For example, a layered perovskite ferroelectric with low permittivity, high fatigue resistance, and low leakage current and an amorphous HfO_2 with high permittivity were adopted for FeFET structure [179-181]. In this work, wide range of ferroelectric characteristic materials with three different typical structures, PZT of Perovskite, SBT of layered-Perovskite and NT of pseudo-pyrochlore were adopted to construct an MFIS structure and their ferroelectric

Table 5. Properties of PZT, SBT, NT ferroelectric films

	$Pb(Zr_{0.52}Ti_{0.48})O_3$	$SrBi_2Ta_2O_9$	$Nd_2Ti_2O_7$
Crystal structure	Perovskite	Layer-perovskite	Pseudo-pyrochlore
Dielectric constant	1100	180	45
Spontaneous polarization	~52 $\mu C/cm^2$	~10 $\mu C/cm^2$	~9 $\mu C/cm^2$
Coercive field	~42 kV/cm	~100 kV/cm	~200 kV/cm

properties including retention were compared. These materials have distinctly different ferroelectric characteristics and are suitable for the study of MFIS structure.

Table 5 represents the properties of PZT, SBT, and NT ferroelectrics. Y_2O_3 with a high dielectric constant was used as an insulator layer to minimize the formation of low permittivity silicate [162,198]. Two other layer structures of MFM and MFIM were fabricated using Pt(111) substrate for comparison studies. For the characterization of PZT, SBT, and NT ferroelectrics, various analyses and measurements were carried out including XRD, AES, SEM, I-V, and P-Eincluding memory window analysis.

Y_2O_3 insulator was deposited on p-type silicon substrate with a reactive direct current magnetron sputtering method and annealed at 800°C for 30 min under O_2 ambient [162,198]. The insulator layer thickness was 20 nm with a dielectric constant of 12. Ferroelectric films of PZT, SBT, and NT were formed by chemical solution deposition on the Y_2O_3/Si substrate. Three types of structures, MFIM, MFM, and MFIS with three different ferroelectric materials were fabricated to investigate the role of the insulator layer and to compare the ferroelectric properties of MFIS, MFM, and MFIS structures.

Stock solutions of PZT, SBT, and NT were prepared using the precursors in Table 6. 2-methoxyethanol was used as a common solvent for each stock solution and extra catalyst was not used because acetic acid was used as a solvent. Extra water was not added due to esterification. By making the solution preparation consistent, we minimized the property changes due to sol composition and preparation procedure. Spin-coated samples were dried for 2 min on a hotplate at 200°C and then at 500°C to decompose and remove the organics. The thicknesses of PZT, SBT, and NT films were 200 nm, 210 nm, and 200 nm, respectively. To crystallize ferroelectrics, PZT, SBT, and NT films were annealed at 600°C, 800°C, and 800°C, respectively for 30 min under O_2 ambient in a tube furnace [183,199].

After annealing, the resulting 150 nm thick Pt top electrode with 100 μm radius was sputter-deposited to measure the electrical properties of the films. The images of the surface microstructure were observed using SEM. The phase formation and crystallinity of the films were studied with XRD using Cu k_α radiation. We measured P-E characteristics with RT66A. Current-voltage characteristics were measured with the HP4145A at a holding time of 0.5 s and delay time of 1 s.

Table 6. The precursors of PZT, SBT, NT stock solutions

	$Pb(Zr_{0.52}Ti_{0.48})O_3$	$SrBi_2Ta_2O_9$	$Nd_2Ti_2O_7$
Precursor	Lead acetate tri-hydrate Zirconium n-butoxide Titanium isopropoxide	Strontium acetate Bismuth 2-ethylhexanoate Tantalum ethoxide	Neodymium(III) acetylacetonate hydrate Titanium isopropoxide
Solvent	2-methoxyethanol, acetylacetonate, acetic acid		
Solution concentration	0.2 M	0.2 M	0.2 M
Precursor stoichiometry	Pb:Zr:Ti = 1:0.4:0.6	Sr:Bi:Ta = 0.8:2.2:2	Nd:Ti = 1:1

The composition profile of the layer structures was analyzed using an AES depth profiler. The memory window of the samples was measured using an HP4284A and HP4280A. The amplitude of the alternating current signal was 30 mV with a sweep rate of 0.2 V/s at an applied voltage of 5 V. The retention time was measured from the reduction in permittivity after poling at an applied voltage of 5 V for 10 s. XRD spectra of PZT, SBT, and NT films formed on Y_2O_3/Si substrate are shown in Figure 101. From these results, it is apparent that all three films were well crystallized without the formation of a secondary phase. The preferred growth orientation of PZT film, (110)/(101) was observed and from this result, it can be said that there was no influence from the Y_2O_3/Si substrate [123]. SBT film was crystallized with a preferred (115) growth orientation the same as in a previous report [199]. For NT film, diffraction peaks of (h00)/(0k0)/(hk0) were relatively stronger than the other diffraction peaks of NT due to possible lattice matching between the c-axis of NT(7.679 Å) and interplanar spacing of (111) Y_2O_3 (7.443 Å), i.e., a small lattice mismatch of 3.17%. For SBT and NT, relative diffraction intensities along with preferred growth orientation were not very strong because they have only one directional matching with Y_2O_3 substrate. Figure 102 shows the hysteresis loops of PZT, SBT, and NT in MFM and MFIM structures. The MFIM structure was formed to investigate the effect of Y_2O_3 insulator layer on the P-E characteristics of the ferroelectric films because with the MFIS structure, formation and presence of yttrium-silicate and silicon-dioxide also affect the measured properties. With the MFM structure, an excellent hysteresis loop was found for all the ferroelectrics but the polarization property was not maintained with the MFIM structure because the performance of the ferroelectrics decreased by an increase of permittivity difference with insulator layer due to reduced effective applied voltage to the ferroelectrics [10]. With the MFIM structure, SBT and NT ferroelectrics were revealed to have better retention properties because retention property is also affected by a coercive field value. The coercive field value is a characteristic of maintaining a polarization in ferroelectrics against an applied field on the reverse direction from the polarization, for example, depolarization field. In FeFET structure, a depolarization field is developed from a difference in the charge compensation which is resulted from the

C_{IS}/C_F difference [10]. Then a low coercive field value can affect to the retention property disadvantageously.

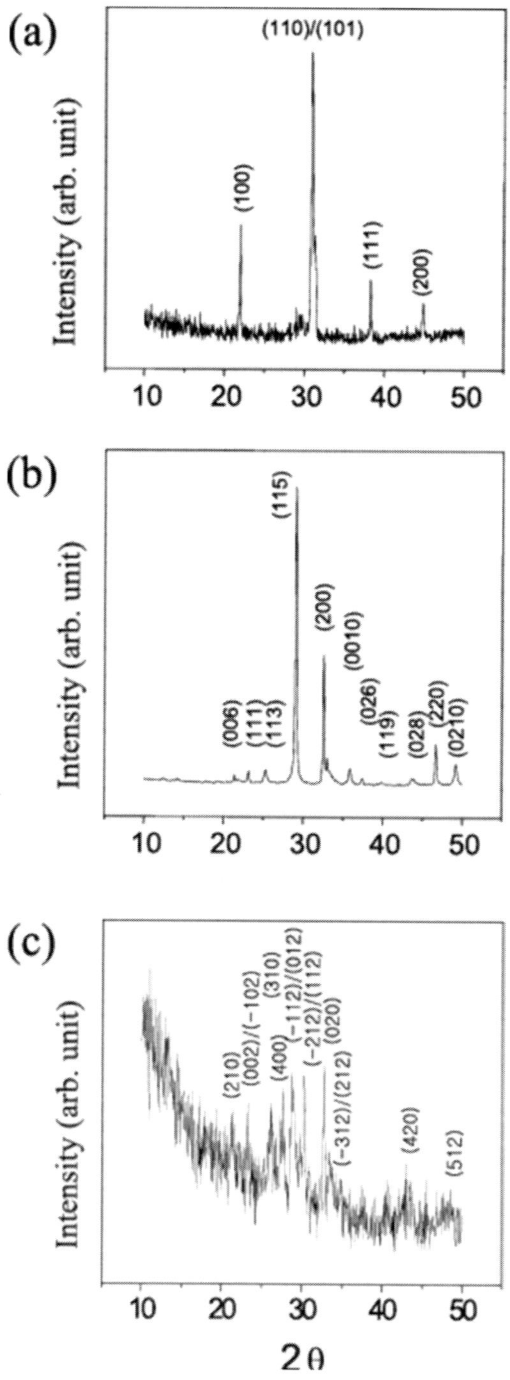

Figure 101. XRD patterns of (a) PZT (b) SBT (c) NT films deposited on Y_2O_3/Si substrate after anneal at each temperature for 30 min.

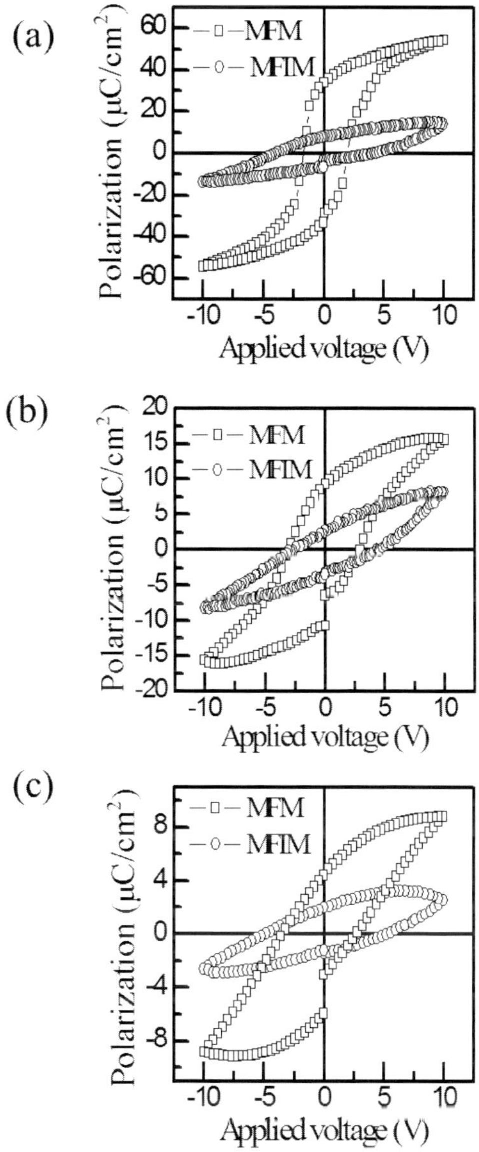

Figure 102. P-E hysteresis loop of MFM and MFIM structure of (a) PZT, (b) SBT, and (c) NT.

Figure 103 represents I-V characteristics of three ferroelectric films in the MFIS structure. A small difference in the I-V characteristics might result from the generation of SiO_2 and Y-silicate in the MFIS structures. A relatively high Schottky conduction was observed in the PZT-MFIS because the applied voltage distribution on the SiO_2 and Y-silicate was effectively increased due to the relatively high dielectric constant of PZT. Furthermore a transition from Schottky conduction to Poole-Frenkel conduction was observed for PZT-MFIS in the lowest voltage region due to a relatively high concentration of defects. From these results, it can be said that proper amounts of SiO_2 and Y-silicate are necessary to reduce leakage current density in the MFIS structure and this observation agreed well with a previous report: leakage current could be effectively reduced when a SiO_2 layer greater than 7 nm thick was formed [179].

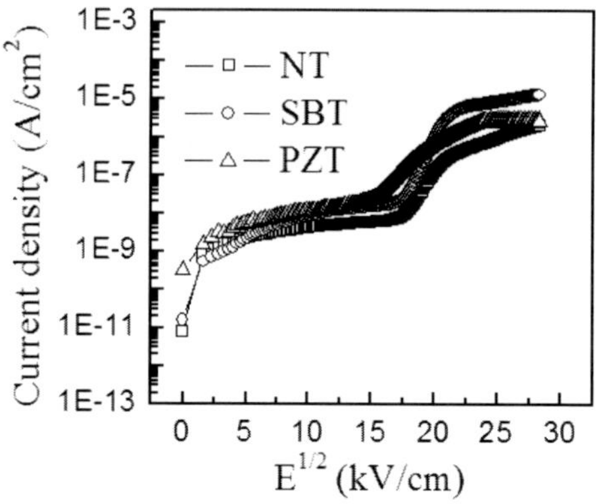

Figure 103. I-V characteristics of MFIS structure.

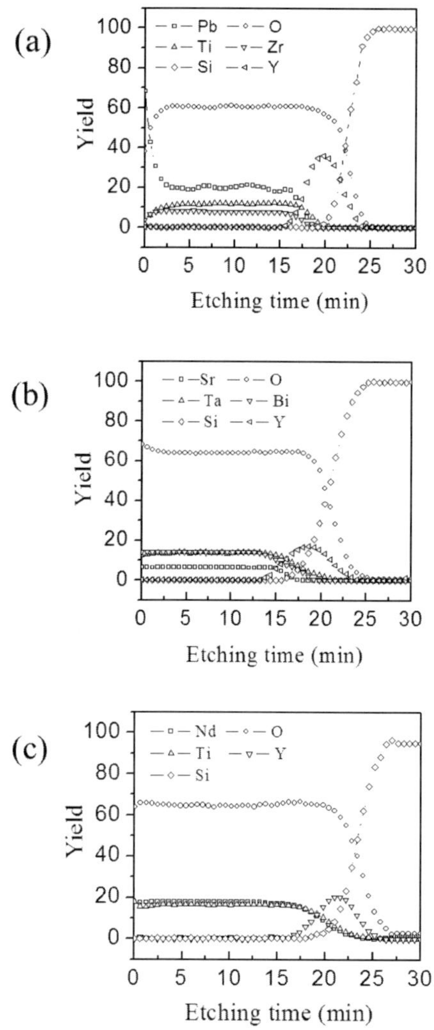

Figure 104. AES depth profiles of (a) PZT, (b) SBT, (c) NT films deposited on Y_2O_3/Si substrate after anneal at 600° C, 800° C, 800° C for 30 min.

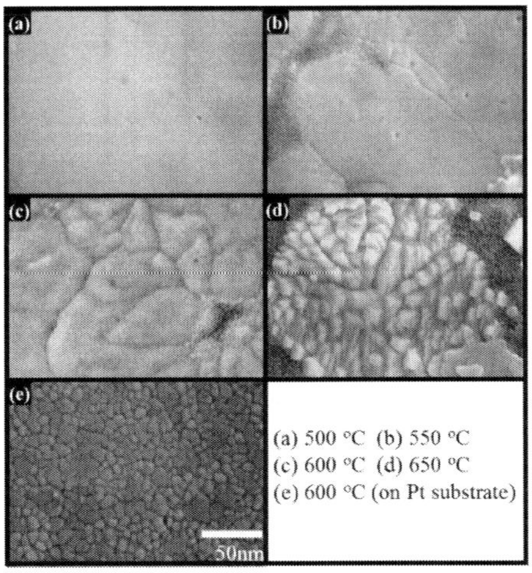

Figure 105. SEM micrographs of PZT films deposited on Y_2O_3/Si with annealing.

Depth profiling analyses were done for three FIS structures after anneal to monitor the inter-diffusion between layers and the formation of SiO_2 and Y-silicate by using AES. The results are given in Figure 104. As shown in the figure, there was almost no difference in the profiling results between SBT and NT ferroelectrics. With PZT-FIS, a large deviation from the designed composition was found at the surface region due to a high volatility of Pb [29]. At the region of overlap with Y_2O_3, for example after argon ion etching for 16 min, an insufficiency of Pb from the designed stoichiometry was observed around 15.8 % in the cation ratio of A-site to B-site, i.e., Pb at % to Zr at % plus Ti at %. The possibility of substitution of Pb with Y is less than 1% because the difference in the ionic radii is above 26 % [200]. The SEM images in the surface microstructure of PZT on Y_2O_3 after anneal at various temperatures and on Pt at 600°C were given in Figure 105(a-d) and (e), respectively. With a high temperature anneal (above 550°C) of PZT on Y_2O_3, the generation of secondary phase with rosette structure was observed due to an excess concentration of surface Pb and insufficient nucleation sites for PZT [201,202]. The generation of the rosette structure became important as annealing temperature increased. Rosette structure was not observed when PZT was formed on the Pt substrate even after anneal at 600°C as shown in Figure 105(e).

Normally PZT is easily crystallized into a columnar structure when it is formed on Pt(111) substrate [158]. From these results, it can be said that nucleation of PZT on Y_2O_3 is more difficult than on Pt substrate. These kinds of surface and interface problems could be confirmed with a large shift in threshold voltage in the memory window of PZT-MFIS as given in Figure 106(a).

The threshold voltage shift of the PZT-MFIS structure could be expected from the generation of fixed charge, trapped charge and interfacial charge trapping. In this case, from the stability of the inversion region of the PZT-MFIS structure as shown in Figure 106(a), the threshold shift is not due to fixed charge or interface charge trapping at the interface [203]. Because the direction of the threshold voltage shift was negative, the shift could be expected from the break in charge neutrality from cation diffusion and deficiency of surface oxygen as shown in Figure 104.

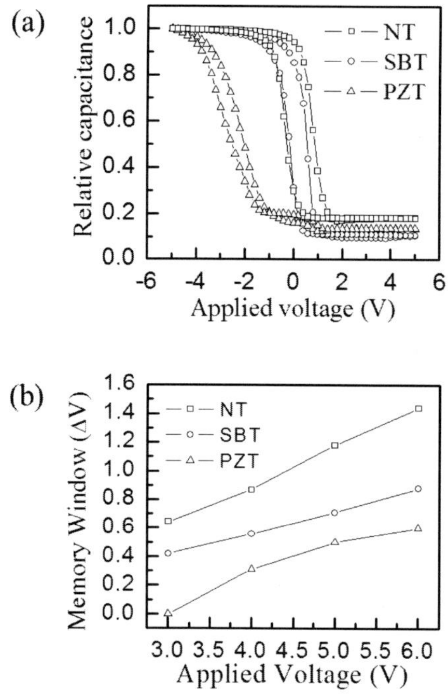

Figure 106. (a) C-V characteristics and (b) the values of the memory window with applied voltage of PZT, SBT, and NT films deposited on Y_2O_3/Si substrate.

In Figure 106(a), NT shows a relatively wide memory window of 1.18 V at 5 V of applied voltage. The memory window of SBT was 0.71 V. These two ferroelectrics showed the same lattice structure profile. The difference in memory window is induced by the difference in permittivity which results in difference effective applied voltage. Normally the effective applied voltage of the ferroelectric layer on the MFIS structure depends on the difference in permittivity between ferroelectric and insulator layers [182]. The SBT film layer received a smaller applied electrical field than the NT layer due to its relatively high permittivity. It also showed a smaller memory window than NT. The width of the memory window increased with increasing applied voltage as shown in Figure 106(b). The differences among memory windows also increase with increasing applied voltage.

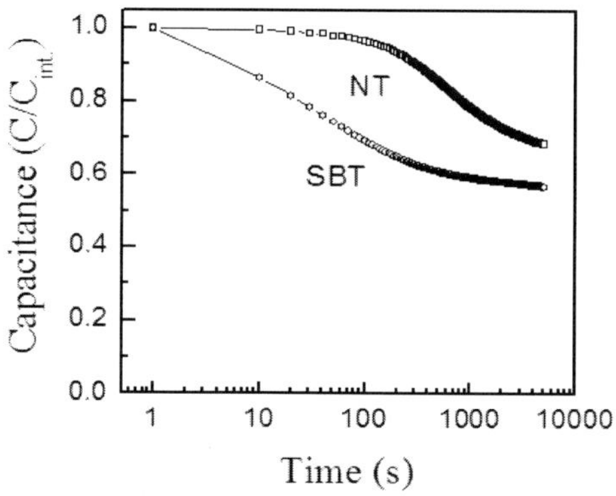

Figure 107. Retention property of SBT and NT films on Y_2O_3/Si substrate.

We measured retention properties of MFIS of SBT and NT, and the results are given in Figure 107. The PZT-MFIS was excluded from the measurement due to its high threshold voltage shift. The NT in the MFIS structure receives higher effective applied voltage than SBT due to its low permittivity and lower depolarization field due to a small difference in permittivity from the insulator, Y_2O_3. Furthermore NT has a larger coercive field. From these observations, it can be said that NT ferroelectric film showed the best retention property with MFIS structure among the above three typical ferroelectrics in this experimental condition.

PZT-MFIS structure showed a shift in threshold voltage resulting from generated defects in PZT due to non-stoichiometric diffusion from the interface with Y_2O_3 to the surface of PZT. With SBT-MFIS, layer and interface structure were found to be stable but showed relatively short retention time due to a small memory window, large depolarization field, and unsaturated polarization from a relatively large difference in the permittivity compared with Y_2O_3. NT-MFIS showed relatively beneficial properties in its memory window as well as its retention time from a large coercive field and low permittivity compared with other ferroelectric films. From the results of this study, it can be concluded that ferroelectric material for MFIS structure is required low permittivity, high coercive field, stable interface, and phase formation without compositional deviation to have large memory window and good retention properties. $Nd_2Ti_2O_7$ is the best ferroelectric material for Fe FETs of the three studied.

CONCLUSION

In NVMs research field, it is true that material holds the key. We have been studied an improvement in the properties of ferroelectric material for FeFET memory, as shown above. First, direct-patternable ferroelectric materials are studied for an application to device process. And then, for improvement of their properties as fatigue, leakage current, permittivity, phase formation temperature and remnant polarization, we studied substitution, multilayer and fabrication process by using represent ferroelectric materials as perovskite, layered perovskite, pseudo pyrochlore. In conclusion, insulator material of high permittivity, ferroelectric material of low permittivity and their low leakage current properties are very important factors for MFIS FeFET device. To date, many studies were focused on retention property as shown above, and there were many achievements. However, there are still disturbance for FeFET device as above mentioned. In fact, this disturbance is related to retention in a sense. Ferroelectric material of unselected FeFET cells has to endure the disturbed electric field in write process. For the endurance, ferroelectric material has to show hysteresis loop of a rectangular shape. However, commonly observed hysteresis loop in ferroelectric materials cannot endure field stress of the disturbance reliably. In other words, there should be no loss of remnant polarization of ferroelectric material in the repeated field stress below the coercive field and depolarization field of MFIS structure. Polarized domains of ferroelectric material should endure up to coercive field. And inversion of polarization has to be generated in the E_c value. However, ideal hysteresis loop of a rectangular shape is difficult from piezo property and the inversion mechanism of ferroelectric material in reality. In spite of so many studies, we could not pass these difficult barriers, clearly. In addition, during a study for the retention and disturbance of FeFET, competitiveness of FeRAM is reduced from the rapid progress of flash memory and DRAM. Nevertheless, steady study of FeFET has been progressed. As a result, new application of ferroelectric NAND (Not AND) flash memory is designed from applied FeFET to NAND structure, recently [204]. It is same as the operating process of NAND flash memory.

As it is well known, under the 20 nm scaling in NAND flash memory, a disturbance between cells can be generated from trap charge in floating gate. Therefore, a study to overcome the limit of scaling is currently underway by the FeFET. New applications as the ferroelectric NAND flash memory are awaking the potential of ferroelectric material.

REFERENCES

[1] J. F. Scott, Ferroelectric Memories. (Springer-Verlag. New York, 2000).

[2] H. Maurice, Francombe, D. J. Taylor (Ed.), Handbook of Thin Film Devices. Ferroelectric Film Devices. (London, Academic Press, 2000).

[3] K. Sugibuchi, Y. Kurogi, and N. Endo, *J. Appl. Phys.,* 1975, 46, 2877.

[4] S. L. Miller and P. J. McWhorter, *Appl. Phys. Lett.,* 1992, 72, 5999.

[5] E. Tokumitsu, R. Nakamura, and H. Ishiwara, *Electron Device Lett.* 1997, 18, 160.

[6] S. K. Lee, Y. T. Kim, S.-I. Kim, and C. E. Lee, *J. Appl. Phys.*, 2002, 91, 9303.

[7] S. Sakai, R. Ilangovan and M. Takahashi, Jpn. *J. Appl. Phys.*, 2004, 43, 7876.

[8] C. K. Lee, W. S. Kim, H. H. Park, H. T. Jeon and Y. H. Pae, *Thin Solid Films,* 2005, 473, 335.

[9] W. S. Kim, S. M. Ha, J. K. Yang, and H. H. Park, *Ferroelectrics*, 2002, 271, 333.

[10] T. P. Ma, and J. P. Han, *IEEE Electron Device Lett.*, 2002, 23, 386.

[11] E. Tokumitsu, G. Fujii and H. Ishiwara, *Jpn. J. Appl. Phys.*, (2000), 39, 2125.

[12] D. S. Shin, S. T. Parka, H. S. Choia, I. H. Choia and J. Y. Lee, *Thin Solid Films,* 1999, 354, 251.

[13] S.Wang, M. Takahashi, Q.-H. Li, K. Takeuchi, and S. Sakai, Semicon. *Sci. Technol.*, 2009, 24, 105029.

[14] T. Hatanaka, M. Takahashi, S. Sakai, and K. Takeuchi, IEICE Trans. *Electron.*, 2011, E94-C, 4, 53.

[15] B. J. Palmer, A. Becalska, and R. H. Hill, *J. Photoch. Photobio. A: Chem.*, 1991, 57, 457.

[16] A. Becalska, J. D. Debad, H. K. Sanati, and R. H. Hill, *J. Photoch. Photobio. A: Chem.,* 1991, 60, 279.

[17] W. Xia and R.H. Hill, *Polyhedron,* 1992, 11, 1319.

[18] T. W. H. Ho, S. L. Blair, R. H. Hill, and D. G. Bickley, *J. Photoch. Photobio. A: Chem.,* 1992, 69, 229.

[19] M. Gao and R. H. Hill, *J. Photoch. Photobio. A: Chem.,* 1996, 97, 73.

[20] Y. Shi, G. Li, and R. H. Hill, *Mat. Sci. Semicon. Proc.*, 1999, 2, 297.

[21] W. L. Law, Photochemical metal organic deposition of metal oxides, Simon Fraser University, Ph.D. thesis, (1995) p. 6.

[22] J. P. Bravo-Vasquez, Solid-state photochemistry of thin films: Metal organic deposition from mesomorphous and amorphous films of inorganic complexes, Simon Fraser University, Ph.D. thesis (2001) p. 7.

[23] H. Funakubo, K. Tokita, T. Oikawa, M. Aratani, and K. Saito, *Appl. Phys. Lett.*, 2002, 92, 5448.
[24] J. Hong, H. W. Song, S. Hong, H. Shin, and K. No, J. *Appl. Phys.*, 2002, 92, 7434.
[25] S. B. Krupanidhi, H. Hu, and V. Kumar, J. *Appl. Phys.*, 1992, 71(1), 376.
[26] S. M. Ha, W. S. Kim, H. H. Park, and T. S. Kim, Ferroelectrics 2002, 273, 35.
[27] S. Y. Chen and I. W. Chen, J. *Am. Ceram. Soc.*, 1998, 81, 97.
[28] C. R. Cho, L. F. Francis, and M. S. Jang, *Jpn. J. Appl. Phys.*, Part 2, 1999, No. 7A, L751.
[29] J. K. Yang, W. S. Kim, and H. H. Park, *Thin Solid Films*, 2000, 377-378, 739.
[30] I. Stolichnov, A. Tagantsev, N. Setter, S. Okhonin, P. Fazan, J.S. Cross, and M. Tsukada, *J.Appl. Phys.*, 2000, 87, 1925.
[31] S. M. Ha, D. H. Kim, H. H. Park, and T. S. Kim, *Thin Solid Films* 1999, 355-356, 525.
[32] C. A Araujo, J. D. Cuchiaro, L. D. McMillan LD, M. C. Scott, and J. F. Scott JF, *Nature*, 1995, 374, 627.
[33] A. Z. Simões, B. D. Stojanovic, M. A. Ramirez, A. A. Cavalheiro, E. Longo, and J. A. Varela, *Ceram Int.,* 2008, 34, 257.
[34] B. H. Park, B. S. Kang, S. D. Bu, T. W. Noh, J. Lee, and W. Jo, *Nature*. 1999, 401, 682.
[35] J. K. Lee, T. Kim, I. Chung, and S. B. Desu, *Appl. Phys. Lett.*, 1999, 75, 334.
[36] C. Soyer, E. Cattan, D. Remiens, and M. Guilloux-Viry, *J. Appl. Phys.*, 2002, 92, 1048.
[37] K. T. Kim, and C. I. Kim, *Microelectron Eng.*, 2004, 71, 266.
[38] H. H. Park, H. H. Park, and *R. H. Hill*, 2006, 132, 429.
[39] H. Y. Chen, J. Lin, K. L. Tan, and Z. C. Feng, *Thin Solid Films*, 1996, 289, 59.
[40] H. Ke, Y. Zhou, D. C. Jia, W. Wang, and X. Q. Xu, J. Sol-Gel. *Sci. Techn.,* 2005, 34, 131.
[41] J. S. Kim, C. W. Ahn, H. J. Lee, S. Y. Lee, S. H. Kang, I. W. Kim, and K. Lee, *J. Korean. Phys. Soc.*, 2005, 46, 147.
[42] B. T. Liu, W. T. Zhang, Y. Wang, H. L. Sun, F. Li, Z. Yan, J. Du, and Q. X. Zhao, *Mater Lett.*, 2007, 61, 1933.
[43] S. Ryu, N. Lee, W. Lee, and S. Yoon, Jpn J Appl Phys., 2004, 43, 6594.
[44] K. T. Kim, C. I. Kim, D. H. Kang, I. W. Shim, J. Korean. *Phys. Soc.*, 2002, 41, 1003.
[45] W. Yunyi, Z. Duanming, Y. Jun, W. Yunbo, *Mat. Sci. Eng. B.*, 2008, 149, 34.
[46] H. H. Park, W. S. Kim, J. K. Yang, H. H. Park, and R. H. Hill, *Microelectron Eng.*, 2004, 71, 215.
[47] N. Zhong, and T. Shiosaki. *Mater. Lett.*, 2007, 61, 2935.
[48] H. S. Yoo, Y. H. Son, T. W. Hong, S. C. Ur, S. L. Ryu, and S. Y. Kweon. *Mater. Sci. Forum.*, 2007, 534-536, 565.
[49] H. N. Lee, D. Hesse, N. Zakharov, and U. Gösele. *Science*, 2002, 296, 2006.
[50] B. J. Lin, P. Rai-Choudhury, H. J. Levinson, W. H. Arnold, M. A. McCord, M. J. Rooks, F. Cerrina, R. D. Allen, W. E. Conley, R. R. Kunz, J. G. Skinner, T. R. Groves, A. Novembre, H. Pfeiffer, R. Singh, L. J. Lauchlan, D. Nyyssonen, N. Sullivan, A. R. Neureuther, C. A. Mack, M. C. Peckerar, F. K. Perkins, E. A. Dobisz, O. J. Glembocki, in: P. Rai-Choudhury (Ed.), Handbook of microlithography, micromachining, and microfabrication, Vol. 1, SPIE optical engineering press, *Bellingham,* (1997) p. 24-38.
[51] Z. Huang, Q. Zhang, and R. W. Whatmore, *J. Appl. Phys.*, 1999, 85, 7355.
[52] M. Alexe, C. Harnagea, D. Hesse, and U. Gösele, *Appl. Phys. Lett.* 1999, 75, 1793.
[53] T. J. Zhu and L. Lu, J. *Appl. Phys.*, 2004, 95, 241.

[54] D. Briggs and M. P. Seah, Practical surface analysis, 2nd edn. (John Wiley and Sons press, New York, 1990), pp. 623.
[55] H.-H. Park, H.-H. Park, and R. H. Hill, *Appl. Surf. Sci.*, 2004, 237, 427.
[56] A. Jeyakumar, C. L. Henderson, P. Roman, and S. Suh, *J. Vac. Sci. Technol. B,* 2003, 21, 3157.
[57] M. S. M. Saifullah, K. R. V. Subramanian, E. Tapley, D.-J. Kang, M. E. Welland, and M. Butler, *Nano Lett.*, 2003, 3, 1587.
[58] Y. L. Tu and S. J. Milne, *J. Mater. Res.*, 1995, 10, 3222.
[59] A. D. Cristoforo, P. Mengucci, G. Majni, F. Leccabue, B. E. Watts, and G. Chiorboli, *Mat. Sci. and Eng.*, 1997, B47, 263.
[60] Z. Huang, Q. Zhang, and R. W. Whatmore, *J. Mat. Sci. Lett.,* 1998, 17, 1157.
[61] F. Yang, W. D. Fei, and Q. Sun, *J. Mater. Process. Tech.*, 2009, 209, 220.
[62] A. Q. Jiang, Y. Y. Lin, and T. A. Tang, *J. Appl. Phys.*, 2007, 101, 056103.
[63] A. Q. Jiang, Y. Y. Lin, and T. A. Tang, *J. Appl. Phys.*, 2007, 101, 104105.
[64] A. Q. Jiang, Y. Y. Lin, and T. A. Tang, *J. Appl. Phys.*, 2007, 102, 074118.
[65] G. Y. Kang, S.-W. Bae, H.-H. Park, and T. S. Kim, *Appl. Phys. Lett.*, 2006, 88, 042904.
[66] E. B. Araújo and J. A. Eiras, *J. Mat. Sci. Lett.*, 1999, 18, 1679.
[67] Q. Zou, H. E. Ruda, B. G. Yacobi, K. Saegusa, and M. Farrell, *Appl. Phys. Lett.,* 2000, 77, 1038.
[68] [68] V. Kumar, Y. Ohya, and Y. Takahashi, *Jpn. J. Appl. Phys.*, 1998, 37, 4477.
[69] J. Y. Kim, R. Khamankar, B. Jiang, P. Maniar, R. Maniar, R. Moazzami, R. E. Jones, and J. C. Lee, Mater. Res. SOC. *Symp. Proc.*, 1995, 361, 409.
[70] S. H. Kim, J. G. Hong, S. K. Streiffer, and A. I. Kingon, *J. Mater. Res.*, 1999, 14, 1371.
[71] G. Arlt, *Ferroelectrics*, 1990, 104, 217.
[72] T. S. Kim, D. J. Kim, J. K. Lee, and H. J. Jung, *J. Vac. Sic. Technol.*, 1997, A 15, 2831.
[73] T. Hase, T. Sakumma, Y. Miyasaka, K. Hiraja, and N. Hosokawa, *Jpn. J. Appl. Phys.,* 1993, 32, 4061.
[74] I. K. Yoo, S. B. Desu, and J. Xing, *Mater. Res. SOCP. roc.*, 1993, 310, 165.
[75] E. L. Colla, A. K. Tagantsev, D. V. Taylor, and A. L. Kholkin, *Intern. Ferroelectric.,* 1997, 18, 19.
[76] J. H. Joo, J. M. Seon, Y. C. Jeon, K. Y. Oh, J. S. Roh, and J. J. Kim, *Appl. Phys. Lett.*, 1997, 70, 3053.
[77] S. T. Kim, H. H. Kim, Y. I. Kim, M. Y. Lee, and W. J. Lee, Jpn. *J. Appl. Phvs.,* 1997, 36, 5663.
[78] T. Takenaka and K. Sakata, *Ferroelectrics*, 1981, 38, 769.
[79] J. C. Bae, S. S. Kim, E. K. Choi, T. K. Song, W.-J. Kim and Y.-I. Lee, *Thin Solid Films*, 2005, 472, 90.
[80] T. Watanabe, H. Funakubo, M. Mizuhira and M. Osada, *J. Appl. Phys.*, 2001, 90, 6533.
[81] D. Wu, A. Li, T. Zhu, Z. Li, Z. Liu and N. Ming, *J. Mater. Res.*, 2001, 16, 1325-1332.
[82] H. Wang and M.F. Ren, *Mat. Sci. Eng. B-Solid.,* 2005, 122, 201.
[83] N. Sugita, M. Osada and E. Tokumitsu, *Jpn. J. Appl. Phys.*, 2002, 41, 6810.
[84] J. Shi-Zhao, J.E. Lim, M.J. Cho, C. S. Hwang and S.-H. Kim, *Appl. Phys. Lett.,* 2002, 81, 1477.
[85] Y.-F. Tsun and C. C. Chou, *Jpn. J. Appl. Phys.*, 1999, 38, 3585.
[86] J. F. Dorrian, R. E. Newnham, D. K. Smith, and M. I. Kay, *Ferroelectrics*, 1971, 3, 17.
[87] Y. Noguchi, *Phys Rev B.* 2001, 63, 214102.

[88] D. Wu, A. Li, and N. Ming, *J. Appl. Phys.,* 2004, 95, 4275.
[89] T. Kojima, T. Sakai, T. Watanabe, and H. Funakubo, *Appl. Phys. Lett*., 2002, 80, 2746.
[90] A. A. Avey, and R. H. Hill, *J. Am. Chem. Soc.,* 1996, 118, 237.
[91] K. Liang, X. He, Y. Qi, and C. Lu, 2008, 310, 2471.
[92] M. K. Jeon, Y. I. Kim, S. H. Nahm, J. M. Sohn, C. H. Jung, and S. I. Woo, *J. Phys. D, Appl. Phys.,* 2007, 40, 4647.
[93] M. Takahashi, Y. Noguchi, and M. Miyayama, *Jpn. J. Appl. Phys*., 2002, 41, 7053.
[94] M. Iwata, C. H. Zhao, R. Aoyagi, M. Maeda, and Y. Ishibashi, *Jpn. J. Appl. Phys.,* 2007, 46, 5894.
[95] M. Iwata, C. H. Zhao, Y. Suzuki, R. Aoyagi, M. Maeda, and Y. Ishibashi, *Jpn. J. Appl. Phys*., 2007, 46, 7155.
[96] U. Chon, H. M. Jang, M. G. Kim, and C. H. Chang, *Phys. Rev. Lett*., 2002, 89, 087601.
[97] W. Sakamoto, M. Yamada, N. Iizawa, Y. K. Mizutani, D. Togawa, K. Kikuta, T. Yogo, T. Hayashi, and S. I. Hirano, *Journal of Electroceramics.,* 2004, 13, 339.
[98] T. Watanabe, T. Kojima, T. Sakai, and H. Funakubo H, *J. Appl. Phys*., 2002, 92, 1518.
[99] U. Chon, J. S. Shim, H. M. Jang, *Solid State Commun*. 2004, 129, 465.
[100] Y. Wang, G. Xu, X. Zhang, Y. Feng, W. Tang, G. Cheng, and Y. Zhu, Mater. *Lett.,* 2004, 58, 813.
[101] R. E. Melgarejo, M. S. Tomar, S. Bhaskar, P. S. Dobal, and R. S. Katiyar, *Appl. Phys. Lett*., 2002, 81, 2611.
[102] M. W. Chu, M. Ganne, M. T. Caldes, and L. Brohan, *J. Appl. Phys*., 2002, 91, 3178.
[103] F. M. F. de Groot, M. Grioni, J. C. Fuggle, J. Ghijsen, G. A. Sawatzky, and H. Petersen, *Phys. Rev. B.,* 1989, 40, 5715.
[104] S. J. Kim, C. Moriyoshi, S. Kimura, and Y. Kuroiwa, *Appl. Phys. Lett.,* 2007, 91, 062913.
[105] F. M. F. de Groot, J. Faber, J. J. M. Michiels, M. T. Czyżyk, M. Abbate, and J. C. Fuggle, *Phys. Rev. B*. 1993, 48, 2074.
[106] T. Haccart, W. R. Cook, and H. Jaffe, *Thin Solid Films*. 2003, 423,235.
[107] B. Jaffe, W. R. Cook, and H. Jaffe: Piezoelectric Ceramics. London and New York: Academic Press.
[108] Q. Zou, H. Ruda, B. G. Yacobi, and M. Farrell, *Thin Solid Films*. 2002, 402, 65.
[109] J. F. Chang, and S. B. Desu, *J. Mater. Res*., 1994, 9, 955.
[110] S. B. Majumder, B. Roy, R. S. Katiyar, and S. B. Krupanidhi, *Appl. Phys. Lett.,* 2001, 79, 239.
[111] N. Husam,A. Shareef, and D. Dimos, *J. Am. Ceram. Soc*., 1997, 80, 3127.
[112] S. M. Mah, N. W. Jung, J. H. Park, D. S. Paik, and C. Y. Park, *Mater. Res. Bull.,* 2000, 35, 1113.
[113] B. Yang, T. K. Song, S. Aggarwal, and R. Ramesh, *Appl. Phys. Lett*., 1997, 71, 3578.
[114] [114] C. Sudhama, J. Kim, J. Lee, V. Chikarmane, W. Shepherd, and E.R. Myers, *J. Vac. Sci. Technol*., 1993, B11(4), 1302.
[115] H. D. Chen, K. R. Udayakumar, C. J. Gaskey, and L .E. Cross, *Appl. Phys. Lett.,* 1995, 67, 3411.
[116] R. C. Evans, An Introduction to Crystal Chemistry (Cambridge Univ. press, London, 1966), p. 38.
[117] K. Lee and S. Baik, *Appl. Phys. Lett*., 2005, 86, 202901.
[118] S. H. Oh and H. M. Jang, *Phys. Rev*., 2000, B62, 14757.

[119] G. H. Haertling and C. E. Land, *J. Am. Ceram. Soc.,* 1971, 54, 1.
[120] Y. Honda, S. Yokoyama, H. Funakubo, and K. Saito, Mater. *Sci. Eng. B,* 2005, 120, 161.
[121] T. Ikeda, *J. Phys. Soc. Japan*, 1959, 14, 1290.
[122] H.-H. Park, S. Yoon, H.-H. Park, and R.H. Hill, *Thin Solid Films*, 2004, 447–448, 669.
[123] C. K. Kwok and S. B. Desu, *J. Mater. Res*., 1993, 8, 339.
[124] B. D. Cullity, Elements of X-ray Diffraction (Addison-Wesley Pub. Co., *Notre Dame*, 1978), p. 127.
[125] H. Hu and S.B. Krupanidhi, *J. Mater. Res*., 1994, 9, 1484.
[126] S. H. Hu, G. J. Hu, X. J. Meng, G. S. Wang, J. L. Sun, S. L. Guo, J. H. Chu, and N. Dai, *J. Cryst. Growth*, 2004, 260, 109.
[127] J.-K. Yang, W. S. Kim, and H.-H. Park, *Appl. Surf. Sci*., 2001, 169–170, 544.
[128] H. B. Sharma, H. N. K. Sarma, and A. Mansingh, *J. Appl. Phys*., 1999, 85, 341.
[129] M. Noda, H. Sugiyama, and M. Okuyama, *Jpn. J. Appl. Phys*., 1999, 38, 5432.
[130] Y. Fujimori, N. Izumi, T. Nakamura, and A. Kamisawa, *Jpn. J. Appl. Phys*. 1998, 37, 5207.
[131] W. S. Kim, S.-M. Ha, S. Yun, and H.-H. Park, *Thin Solid Films*, 2002, 420-421, 575.
[132] L. H. Brixner, *Inorg. Chem*., 1964, 3, 1065.
[133] Powder Diffraction File, Joint Committee on Powder Diffraction Standards, ASTM, Philadelphia, PA, 1967, Card 280517.
[134] K. Scheuneman, and H. Muller-Buschbaum, *J. Inorg. Nucl. Chem*., 1975, 37, 2261.
[135] J. A. Rodriguez, X. Wang, J. C. Hanson, and G. Liu, *J. Chem. Phys*., 2003, 119, 5659.
[136] K. Shibata, Y. Rikiishi, T. Hosokawa, Y. Haruyama, Y. Kubozono, S. Kashino, T. Uruga, A. Fujiwara, H. Kitagawa, T. Takano, and Y. Iwasa, *Phys. Rev. B,* 2003, 68, 094104.
[137] A.V. Soldatov, T.S. Ivanchenko, S.D. Longa, A. Kotani, Y. Iwamoto, and A. Bianconi, *Phys. Rev. B,* 1994, 50, 5074.
[138] F. Klose, O. Schulte, F. Rose, and W. Felsch, *Phys. Rev. B*, 1994, 50, 6174.
[139] S. Nanamatsu, M. Kimura, K. Doi, S. Matsushita, and N. Yamada, *Ferroelectrics*, 1974, 8, 511.
[140] [140] M. Kimura, S. Nanamatsu, T. Kawamura, and S. Matsushita, Jpn. *J. Appl. Phys.,* 1974, 13, 1473.
[141] Y. Fujimori, N. Izumi, T. Nakamura, A. Kamisawa, and Y. Shigematsu, Jpn. *J. Appl. Phys.,* 1997, 36, 5935.
[142] A. Kerber, E. Cartier, G. Groeseneken, H.E. Maes, and U. Schwalke, *J. Appl. Phys.,* 2003, 94, 6627.
[143] S.S. Ang, Y.J. Shi, and W.D. Brown, *J. Appl. Phys.,* 1996, 79, 1968.
[144] J. Nuffer, D.C. Lupascu, and J. Rodel, *Appl. Phys. Lett*., 2002, 80, 1049.
[145] I. Boerasu, L. Pintilie, and M. Kosec, *Appl. Phys. Lett*., 2000, 77, 2231.
[146] M. Es-Souni, M. Abed, A. Piorra, S. Malinowski, and V. Zaporojtchenko, *Thin Solid Films*, 2001, 389, 99.
[147] F. Yan, P. Bao, X.B. Chen, J.S. Zhu, and Y. N. Wang, *Integr. Ferroelectr*. 2001, 33, 379.
[148] Y. J. Song, N. W. Jang, D. J. Jung, H. H. Kim, H. J. Joo, S. Y. Lee, K. M. Lee, S. H. Joo, S. O. Park, and K. Kim, Jpn. *J. Appl. Phys*., 2002, 41, 2635.
[149] A. K. Tagantsev, I. Stolichnov, E. L. Colla, N. Setter, *J. Appl. Phys*., 2001, 90, 1387.

[150] C. Z. Pawlaczyk, A. K. Tagantsev, K. Brooks, I. M. Reaney, R. Klissurska, and N. Setter, *Integr. Ferroelectr.*, 1995, 8, 293.
[151] K. Kim, and Y. J. Song, Microelectron. *Reliab.*, 2003, 43, 385.
[152] Q.-M. Wang, Y. Ding, Q. Chen, and M. Zhao, *J. Cheng, Appl. Phys. Lett.,* 2005, 86, 162903.
[153] R. R. Das, S. B. Majumder, R. S. Katiyar, Integr. Ferroelectr. 2002, 42, 323.
[154] H.-H. Park, H.-H. Park, R. H. Hill, *J. Electroceram.* 2006, 17, 135.
[155] X.-H. Du, J. Zheng, U. Belegundu, K. Uchino, *Appl. Phys. Lett.*, 1998, 72, 2421.
[156] W. L. Warren, D. Dimos, B. A. Tuttle, R. D. Nasby, and G. E. Pike, *Appl. Phys. Lett.*, 1994, 65, 1018.
[157] J. S. Liu, S. R. Zhang, L. S. Dai, and Y. Yuan, *J. Appl. Phys.*, 2005, 97, 104102.
[158] W. S. Kim, J.-K Yang, and H.-H. Park, *Appl. Surf. Sci.* 2001, 169-170, 549.
[159] J. F. Scott, C. A. Paz de Araujo, *Science*, 1989, 246, 1400.
[160] H. Z. Jin, and J. Zhu, *J. Appl. Phys.*, 2002, 92, 4594.
[161] G. Winfield, F. Azough and R. Freer, *Ferroelectrics*, 1992, 133, 181.
[162] S. K. Kang, D. H. Ko, E. H. Kim, M. H. Cho and C. N. Whang, *Thin Solid Films,* 1999, 353, 8.
[163] G. D. Wilk, R. M. Wallace and J. M. Anthony, *J. Appl. Phys.*, 2001, 89, 5243.
[164] G. A. Smolenskii and I. Chupis, *Sov. Phys. Usp.*, 1982, 25, 475.
[165] T. Zhao, A. Scholl, F. Zavaliche, K. Lee, M. Barry, A. Doran, M. P. Cruz, Y. H. Chu, C. Ederer, N. A. Spaldin, R. R. Das, D. M. Kim, S. H. Baek, C. B. Eom and R. Ramesh, *Nature Materials*, 2006, 5, 823.
[166] F. Kubel and H. Schimid, Acta Crystallogr. B, 1990, 46, 698.
[167] P. Fischer, M. Polomska, I. Sosnowska and M. Szymanski, *J. Phys. Solid State Phys.,* 1980, 13, 1931.
[168] I. Sosnowska, T.P. Neumaier and E. Steichele, *J. Phys.*, 1982, C15, 4835.
[169] M.M. Kumar, V.R. Palkar, K. Srinivas and S.V. Suryanarayana, *Appl. Phys. Lett.,* 2000, 76, 2764.
[170] Y. P. Wang, L. Zhou, M. F. Zhang, X. Y. Chen, J. M. Liu and Z. G. Liu, *Appl. Phys. Lett.*, 2004, 84, 1731.
[171] M. W. Chu, M. Ganne, M. T. Caldes and L. Brohan, *J. Appl. Phys.*, 2002, 91, 1048.
[172] G. A. C. M. Spierings, M. J. E. Ulenaers, G. L. M. Kampschaer, H. A. M. van Hal and P. K. Larsen, *J. Appl. Phys.*, 1991, 70, 2290.
[173] Z. Song and C. Lin, *Appl. Surf. Sci.*, 2000, 158, 21.
[174] E. A. Deliyanni, D. N. Bakoyannakis, A. I. Zouboulis, K. A. Matis and L. Nalbandian, Micropor. *Mesopor. Mat.*, 2001, 42, 49.
[175] Y. Wang, A. Muramatsu and T. Sugimoto, Colloids and Surfaces A: Physicochemical and Engineering Aspects, 1998, 134, 281.
[176] G. Ketteler and W. Ranke, *J. Phys. Chem. B*, 2003, 107, 4320.
[177] G. W. Pabst, L. W. Martin, Y. H. Chu, and R. Ramesh, *Appl. Phys. Lett.,* 2007, 90, 072902.
[178] P. Zubko, D. J. Jung, and J. F. Scott, *Appl. Phys. Lett.*, 2006, 100, 114112.
[179] S. Sakai, and R. Ilangovan, IEEE Electr. Device L., 2004, 25, 369.
[180] K. Takahashi, K. Aizawa, B. E. Park, and H. Ishiwara, *Jpn. J. Appl. Phys.*, 2005, 44, 6218.
[181] Q. H. Li, and S. Sakai, *Appl. Phys. Lett.*, 2006, 89, 222910.

[182] C. K. Lee, W. S. Kim, and H. H. Park, *Thin Solid Films*, 2004, 464-465, 155.

[183] W. S. Kim, S. M. Ha, J. K. Yang, and H. H. Park, *Thin Solid Films*, 2001, 398-399, 663.

[184] H. Fukumoto, and M. Morita, *J. Appl. Phys*. 1989, 65, 5210.

[185] K. Onisawa, M. Fuyama, K. Tamura, K. Taguchi, T. Naguchi, and Y. A. Ono, *J. Appl. Phys*., 1990, 68(2), 719.

[186] E. K. Evangelou, C. Wiemer, and M. Fanciulli, *J. Appl. Phys*., 2003, 94(1), 318.

[187] T. Inoue, Y. Yamamoto, S. Koyama, S. Suzuki, and Y. Ueda, *Appl. Phys. Lett.*, 1990, 56, 1332.

[188] T. Inoue, Y. Yamamoto, and M. Satoh, *Jpn. J. Appl. Phys*., 1996, 35, 1685.

[189] T. Inoue, M. Osonoe, H. Tohda, M. Hiramatsu, Y. Yamamoto, A. Yamananka, and T. Nakayama, *J. Appl. Phys*., 1991, 69, 8313.

[190] C. Tian, Y. Du, and S. W. Chan, J. Vac. Sci. Technol. A, 1997, 15, 85.

[191] G. He, Q. Fang, M. Liu, L.Q. Zhu, and L.D. Zhang, *J. Crystal Growth*, 2004, 268, 155.

[192] K. J. Hubbard, and D. G. Schlom, J. Mater. Res., 1996, 11(11), 2757.

[193] A. Callegari, E. Cartier, M. Gribelyuk, H. F. Okorn-Schmidt, and T. Zabel, *J. Appl. Phys*., 2002, 91(9), 6173.

[194] G. D. Wilk, R. M. Wallace, and J. M. Anthony, *J. Appl. Phys*., 2000, 87(1), 484.

[195] S. Bender, R. Franke, E. Hartmann, V. Lansmann, M. Jansen, and J. Hormes, *J. Non-Crystalline Solids*, 2002, 298, 99.

[196] D. Niu, R. W. Ashcrft, M. J. Kelly, J. J. Chambers, T. M. Klein, and G. N. Parsons, *J. Appl. Phys*., 2002, 91(9), 6173.

[197] I. Daisuke, F Norifumi, Y. Takeshi, and I. Taichiro, *J. Appl. Phys*., 2003, 94(4), 4036.

[198] W. S. Kim, M. G. Kang, J. K. Yang, and H. H. Park, *Ferroelectrics*. 2001, 259, 299.

[199] R. R. Das, P. Bhattacharya, R. S. Katiyar, *J. Appl. Phys*., 2002, 92, 6160.

[200] W. D. Kingery, H. K. Bowen, D. R. Uhlmann, in introduction to ceramics, Wiley, 1976.

[201] K. H. Yoon, J. H. Shin, J. H. Park, D. H. Kang, *J. Appl. Phys*., 1998, 83, 3626.

[202] R. D. Klissurska, T. Maeder, K. G. Brooks, N. Setter, *Microelectron. Eng*., 1995, 29, 297.

[203] E. H. Nicollian, J. R. Brew, in MOS Physics and Technology. New York, John Wiley and Sons, 1982.

[204] S. Sakaia, M. Takahashia, K. Takeuchib, L. Qiu-Honga, T. Horiuchia, K. Y. Shouyu Wanga, M. Takamiyac, T. Sakuraic, Non-Volatile Semiconductor Memory Workshop, 2008 and 2008 International Conference on Memory Technology and Design. NVSMW/ICMTD 2008. *Joint*, 2008, 18-22, 103.

In: Ferroelectrics: New Research
Editors: Xui Li Huang and Su Lyn Ma

ISBN: 978-1-61942-282-7

Chapter 2

RESONANT AND NON-RESONANT MICROWAVE ABSORPTION STUDIES IN MULTIFERROIC MATERIALS

G. Alvarez[1,*], H. Montiel[2] and R. Zamorano[1]

[1] Sección de Estudios de Posgrado e Investigación, ESFM-IPN, U.P. Adolfo López Mateos Edificio 9, Av. Instituto Politécnico Nacional S/N, San Pedro Zacatenco, México

[2] Departamento de Tecnociencias, Centro de Ciencias Aplicadas y Desarrollo Tecnológico de la Universidad Nacional Autónoma de México, Apartado Postal, México

ABSTRACT

In this work, the effects of temperature and dc magnetic field on resonant and non-resonant microwave absorption measurements at X-band (8.8-9.8 GHz), in the polycrystalline multiferroics $Pb(Fe_{1/2}Nb_{1/2})O_3$ (PFN), $Pb(Fe_{1/2}Ta_{1/2})O_3$ (PFT), $YMnO_3$ (YM) and $YCrO_3$ (YC), are presented. Three techniques are used: electron paramagnetic resonance (EPR), low-field microwave absorption (LFMA) and magnetically modulated microwave absorption spectroscopy (MAMMAS). These studies are performed in the 77-520 K temperature range. Resonant and non-resonant microwave absorption can distinguish between different dissipative dynamics of microwave absorbing centers; these responses showed distinctive features associated with microwave absorption by magnetic moments and/or electric dipoles, to different temperature ranges. Where these measurements are a manifestation of a similar response to electromagnetic absorption, in which similar physical processes take place.

* Corresponding author, E-mail: memodin@yahoo.com.

1. Introduction

The term multiferroic is associated with phenomena in which at least two of the three properties ferroelectricity, ferromagnetism, and ferroelasticity occur in the same material [1-4]. In particular, the multiferroic materials in which ferroelectricity coexists with a magnetic order are called magnetoelectrics [5,6]; representing a kind of material in which magnetization can be induced by an electric field and electrical polarization by a magnetic field [7-10]. These materials have received much attention in recent years because of the possibility of new device applications envisaged from a strong coupling between these parameters [11,12]. Some magnetoelectric materials exhibit a ferroelectric transition at a relatively high temperature and a magnetic transition at lower temperature [13-16].

Evidence of a coupling between the ferroelectric and magnetic ordering in these materials has been previously reported [14,17,18]. The study of these materials may offer an insight into the conditions of occurrence and interdependence of ferroelectricity and magnetism by examining the coupling between the two ordering phenomena. This coupling can result in the so-called magnetoelectric effect, where the magnetic (dielectric) properties of the multiferroic material may be altered by the onset of the electric (magnetic) transition or by the application of an electric (magnetic) field.

The phenomenology of the coupling can be viewed as follows: a change in electrical ordering caused by the ferroelectric phase transition or an external electric field leads to a redistribution of the electron spins; this produces a change in the magnetic moments, and therefore a variation in magnetic properties [19]. Additionally, the ferroelectric phase transition is strongly dependent on structural changes; and these are able to modify the magnetic exchange interactions and therefore the long-range magnetic order. The change in the magnetic ordering caused by the magnetic phase transition, or the application of a magnetic field, will also lead to changes in ferroelectric and dielectric behaviors of the material, via a magnetostrictive effect or electron-phonon interaction mechanisms; where the symmetry and spatial location of ions in the structure are affected.

To investigate this coupling, the material is excited with a magnetic (electric) field and the observation of a typical electric (magnetic) response such as dielectric permittivity (magnetic susceptibility) is carried out [17,20]; for example, anomalies in the dielectric constant and loss tangent have been observed near the Néel temperature [13,14,18], when the material is driven through magnetic transition.

Electric dipoles and magnetic moments are strong absorbers of microwaves [21]. For a pure dielectric material, only the dielectric losses contribute to microwave absorption energy, while for a pure magnetic material, the effect of the magnetic losses dominates over the dielectric loss. However, for magnetoelectric materials, the microwave absorption is due to a competition between the dielectric and magnetic losses, and which are originated from the combination of both ferroelectric and magnetic phases. Information on absorption processes of each kind of center absorbing and their coupling is expected to be obtained from microwave experiments, especially when the material undergoes a phase transition.

On the other hand, by more of sixty years, since its discovery in 1944 by E. K. Zavoisky, the electron paramagnetic resonance (EPR) spectroscopy has been exploited as a very sensitive technique to study the resonant microwave absorption for different kinds of paramagnetic species in solids; as is the particular case of the transition metal ions. EPR is the most powerful spectroscopic method available to unambiguously determine the valence state of paramagnetic ions [22-24], local structural information, and symmetry of paramagnetic ions incorporated in the structure [24-26]. This technique allows the investigation of the nature of magnetic phases in materials at different temperatures [23,27-29]; and recently, the technique has also allowed the investigation of the ferroelectric transition [30-32].

Additionally, we have implemented two experiments to measure the non-resonant microwave absorption (NRMA) as a function of temperature or dc applied magnetic field [33-35]. They are known as magnetically modulated microwave absorption spectroscopy (MAMMAS) and low-field microwave absorption (LFMA), respectively. Recently, these techniques have been used to detect the magnetic and electric transitions in multiferroic materials [31,32,34,36,37]. Additionally, these techniques provide information on the temperature or dc magnetic field dependence of NRMA in each material. More importantly, these techniques can distinguish between different dissipative dynamics of microwave absorbing centers [23,27,29,31-37]. Also, the magnetoelectric materials have shown to have a coupling between the electric and magnetic phenomena [14,17,18,38]; the NRMA is therefore expected to be due to magnetic moments and/or electric dipoles contained in these materials.

In this work, the changes in the EPR lineshape are studied for polycrystalline samples of $Pb(Fe_{1/2}Nb_{1/2})O_3$ (PFN), $Pb(Fe_{1/2}Ta_{1/2})O_3$ (PFT), $YMnO_3$ (YM) and $YCrO_3$ (YC) in the 77-520 K temperature range; these changes are quantified by means of the following parameters: the peak-to-peak linewidth (ΔH_{pp}), the integrated intensity (I_{EPR}) and the g-factor (or the resonance field, H_{res}) as a function of temperature. Also, we present MAMMAS spectra of phase transitions on some materials, showing distinctive features associated with the microwave absorption by magnetic moments and/or electric dipoles. LFMA technique confirms our assignment for the transition temperatures and the involved absorption processes.

2. Experimental Methods

In this work, the resonant and non-resonant microwave absorptions are studied. The measure of the non-resonant microwave absorption as a temperature function or as a function of an applied dc magnetic field are experimentally denominated as MAMMAS and LFMA, respectively. Additionally, the resonant microwave absorption measurement as a function of a magnetic field employed the EPR technique.

2.1. Resonant Microwave Absorption Measurement (EPR Technique)

The basic components of a standard EPR spectrometer are shown in Figure 1 [25,26]. In this technique, the magnetic field on the sample, placed at the center of a microwave cavity (see Figure 2), is increased until that the energy difference between the spin-up and spin-down orientations, which are also affected by the local environment, match the microwave frequency of the power supply; and an intense absorption is then detected.

The EPR spectrometer plots the derivative of the microwave absorption as a function of the magnetic field.

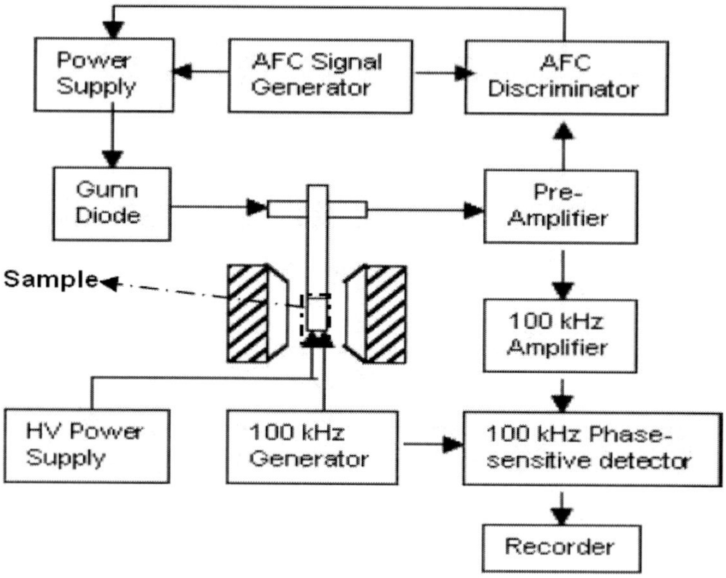

Figure 1. Block diagram of a standard EPR spectrometer [25,26].

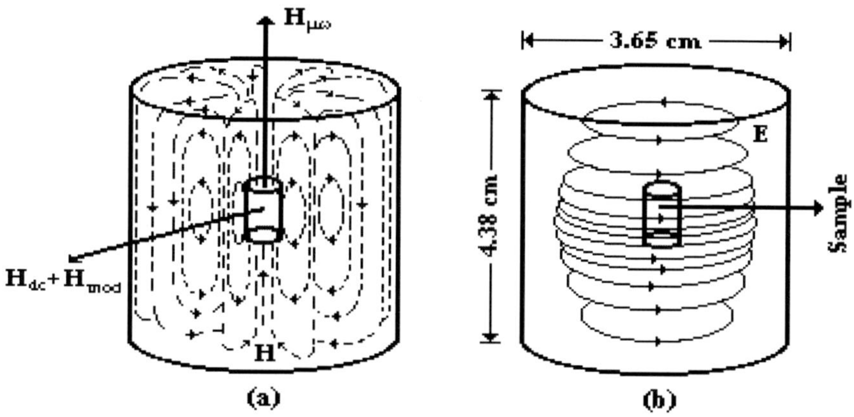

Figure 2. Distribution of (a) magnetic and (b) electric fields, and the sample location inside the TE_{011} cylindrical cavity of the JEOL JES-RES 3X spectrometer.

Next we will give a description more detailed of the EPR spectrometer (Jeol JES-RES3X). In a magic-T bridge, the microwaves (H_{mw}), with power P_0 and frequency ν_{mw}= 8.8-9.8 GHz (X-band), are generated by a JEOL/ES-HX3 microwave unit and are fed to the cylindrical cavity through a rectangular wave-guide; the coupling adjuster of the cavity must be adjusted so that no wave is reflected from the cavity. The absorption of the microwave energy by a sample generates a change in the quality factor, Q, of the cavity; due to this change, the microwave bridge becomes non-balanced, causing a wave that is reflected from the cavity. The cavity Q-factor change as result of changes in the energy absorption resonant by spins. The reflected waves from the cavity (P_{ref}) with the information of the microwave absorption by the sample are directed towards a detector crystal, which was previously biased to a 10% of the incident power in order to work in the linear regime; with this method of detection-homodyne with lock-in amplification, a very high sensitivity in the measurements is guaranteed.

The sample is subjected to a dc magnetic field (H_{dc}), that it is produced by a conventional electromagnet with truncated pole pieces, and a weak ac magnetic field (H_{mod}) is superimposed to H_{dc}. The H_{mod} is achieved by placing small Helmholtz coils on each side of the cavity along the axis of the static field, which are fed and controlled by a sign generator. The amplitude of this field goes from 0.002 Gauss to 20 Gauss with a modulation frequency of 100 kHz, thus allowing, the microwave absorption registration at the modulation frequency. The EPR measurements were carried out in the 77-500 K temperature range, and in EPR spectra, H_{dc} could be varied from 0 to 8000 Gauss.

2.2. Non-Resonant Microwave Absorption Measurements (MAMMAS and LFMA)

Additionally, the Jeol JES-RES3X spectrometer (see Figure 3), was modified connecting the output of a digital voltmeter (signal Y) to a PC enabling digital data acquisition. The electrical signal Y is proportional to NRMA from the sample. The signal Y is fed to a 7½ digits - Keithley DMM-196 voltmeter. Hence, the reading of this voltmeter (V_y) carries the information of the microwaves absorption by the sample.

MAMMAS technique allows measurement as a function of temperature, giving information on the temperature profile of NRMA response of each material and can also provide valuable information about the nature of magnetic and/or electric ordering in the

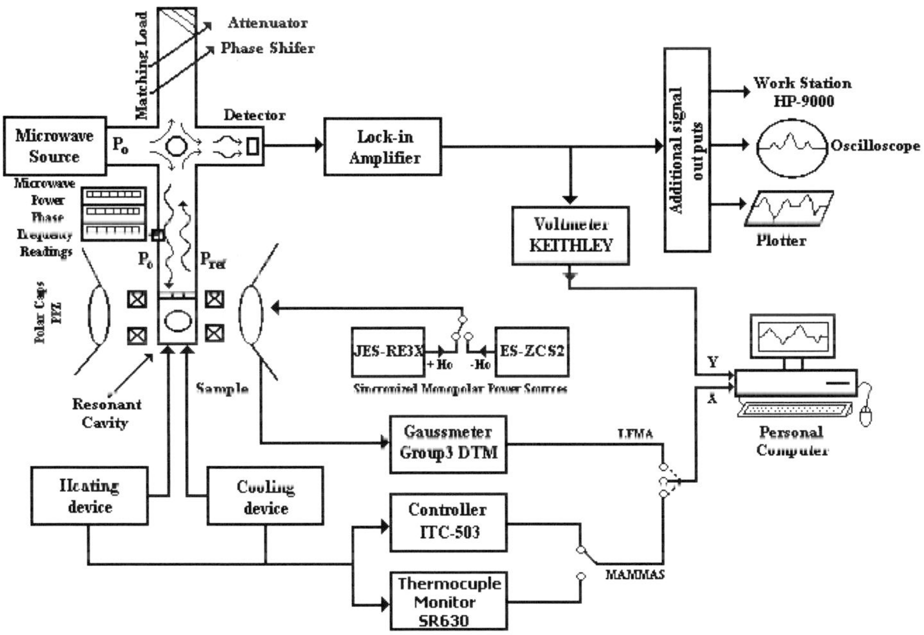

Figure 3. Block diagram of MAMMAS and LFMA techniques.

materials. The temperature of the sample is slowly varied (~1 K/min) and controlled by flowing N_2 gas through a double walled quartz tube, which is inserted in the center of the microwaves cavity; with an operation temperature range of 77-520 K.

The temperature is measured by a copper constantan thermocouple placed inside the sample tube just outside the cavity. Its output signal is further digitized by means of the Stanford SR630 thermocouple monitor. In this technique, the temperature is plotted along the X-axis, and the microwaves absorbed by the sample are collected as V_y and is plotted along the Y-axis.

LFMA technique measures the NRMA as a function of H_{dc}, this uniform field is produced by the same electromagnet, but which receives current regulated from two power supplies (JEOL JES-RE3X and ES-ZCS2); these are synchronized in such a way as obtain a true zero value of magnetic field between the pole caps. The Jeol ESZCS2 zero-cross sweep unit compensates digitally for any remanence in the electromagnet, with a standard deviation of the measured field of less than 0.2 Gauss, allowing measurements to be carried out by cycling the H_{dc} about its zero value continuously from -1000 Gauss to +8000 Gauss. Hence, symmetric field-sweeps from ±0.1 Gauss to ±1000 Gauss are available and asymmetric field-sweeps up to -1000 Gauss $\leq H_{dc} \leq$ 8000 Gauss are also available in order to detect possible hysteresis for NRMA signal, which would point out to irreversible processes of microwave energy absorption. With this technique, the sample is zero field cooled or heated to the fixed temperature. For the present studies, the temperature is maintained fixed with a maximum deviation of 1 K during the whole LFMA measurement (sweeping <8 min). The magnetic field is swept following a cycle; the field sweep schemes have their analog in the magnetic hysteresis measurements. GPIB port of a PC receives the magnetic field coming from the Group3 DTM-141 teslameter, and is displayed as the X-axis on the plot of the data being acquired; meanwhile, the voltmeter V_y receives the signal of NRMA. Then, what is measured is not the microwave absorption itself, but rather its derivative with respect to magnetic field. This allows us to distinguish the field-sensitive part of the microwave absorption from the part that does not depend on magnetic field, and record only the first; and also, to use narrow-band amplifier to enhance the signal, which greatly increases the signal-to-noise ratio.

3. Multiferroic Materials Studies

In this section, the results of studies on magnetoelectric materials using EPR, MAMMAS and LFMA techniques are described. These examples include the multiferroics $Pb(Fe_{1/2}Nb_{1/2})O_3$ (PFN), $Pb(Fe_{1/2}Ta_{1/2})O_3$ (PFT), $YMnO_3$ (YM) and $YCrO_3$ (YC), highlighting their main characteristics and illustrating how magnetic and/or electric phase transitions are manifested in these measurements.

3.1. Magnetoelectric PFN

3.1.1. EPR Technique

Figure 4(a) shows EPR spectra (dP/dH vs. magnetic field) recorded in the temperature range 90-300 K for PFN powders. One single broad symmetric Lorentzian line is observed, in

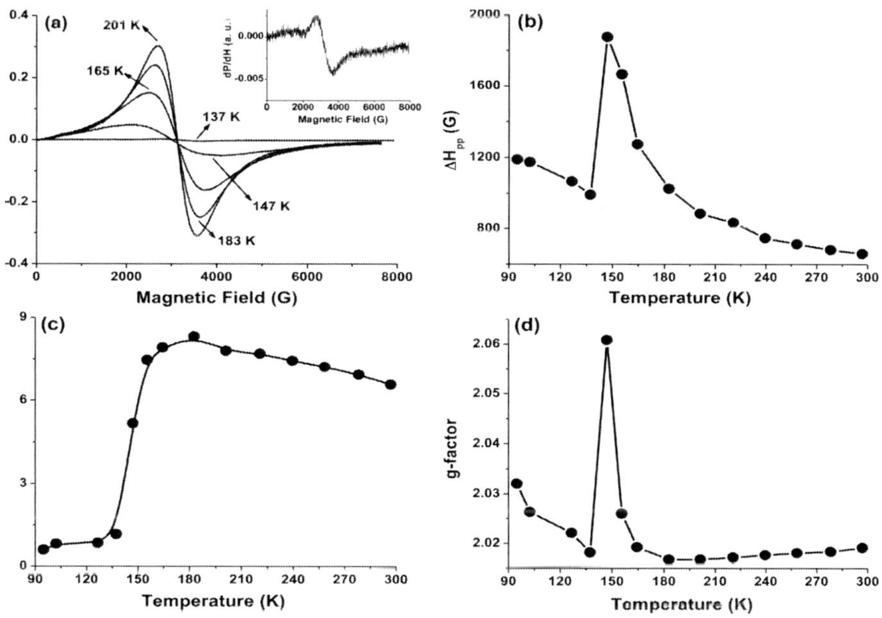

Figure 4. (a) EPR spectra of $Pb(Fe_{0.5}Nb_{0.5})O_3$ for selected temperatures; the inset shows an extended scale of EPR spectrum at 137 K. Temperature dependence of (b) the peak-to-peak linewidth -ΔH_{pp}, (c) the integral intensity -I_{EPR} and (d) the g-factor. Curves connecting points are only guides for the eye.

the entire temperature range, due to the spin of Fe^{+3} ions. When the temperature is decreased below 147 K a weak absorption line has been clearly observed, see inset of Figure 4(a).

Figure 4(b-d) show the temperature dependence of three main parameters deduced from the EPR spectra: ΔH_{pp}, I_{EPR} and g-factor, respectively. ΔH_{pp} increases when the temperature is lowered from 300 K to 147 K, see Figure 4(b), this gradual increase is found to be common to various antiferromagnetic materials [39].

The further increase in linewidth can be due to build-up of magnetic correlations preceding the transition to the long-range antiferromagnetic ordering at Néel temperature (T_N). A peak is observed at 147 K, and then decreases approximately 50% in only 10 K; this is quite unusual, and it is most unusually observed in antiferromagnetic materials. A common feature of the EPR spectra in antiferromagnetic materials is that approaching the T_N, ΔH_{pp} gradually increases and close to T_N it sharply diverges [39]. As temperature decreases down to 95 K, ΔH_{pp} increases again, but now with a much smaller rate of growth.

I_{EPR} increases continuously with the decrease in temperature, having a maximum at ~183 K, followed by a rapid decrease until 137 K; then, for T<137 K a slow decrease is observed.

The g factor shows a weak decrease in the 297-183 K range and then the g-factor increases when the temperature continues to decrease, reaching a maximum at 147 K. It is remarkable that the g-factor increases significantly from 183 K to 147 K. This sharp increase can be due to magnetic fluctuations, i.e., fluctuations in the establishment of the long-range order that precedes the transition to the antiferromagnetic order at T_N. For 90K<T≤137 K an increase in the g-factor is also observed.

The changes in the parameters ΔH_{pp}, I_{EPR} and g-factor in the EPR spectra for PFN at T_N~145 K, are interpreted as a paramagnetic-antiferromagnetic transition. The residual resonant signal observed below T_N, can be interpreted in terms of a resulting magnetic moment, i.e., a canting of Fe^{+3} ion sublattices in the antiferromagnetic matrix.

On the other hand, one single broad symmetric Lorentzian line is also observed for 300-500 K temperature range, due also to the spin of the Fe^{+3} ions. The temperature dependences of the EPR parameters are plotted in Figure 5(a-c).

Figure 5(a) shows the temperature dependence of ΔH_{pp} for this temperature range. Starting from room temperature (300 K), as temperature increases, ΔH_{pp} decreases continuously until 403 K. In this region of temperature, the material is paramagnetic [14,17,40-42], and the paramagnetic absorption decreases as temperature increases [25,26]. As temperature increases further, T>403 K, ΔH_{pp} shows a weak increase. This behavior is associated with the structural change, which accompanies the ferro-paraelectric phase transition [14,40,41,43]; this structural change creates variations in the vicinity of the Fe^{+3} ions, which result in a reorganization of the spin-spin and spin-magnetic field interactions.

The temperature dependence of I_{EPR}, which in the paramagnetic phase is directly proportional to the static spin susceptibility [25,26], is shown in Figure 5(b). In this figure are established two behavior regions, which are limited to the 390-403 K temperature range. This feature is associated with the ferro-paraelectric phase transition. The change in electrical ordering caused by the ferro-paraelectric phase transition leads to a redistribution of the electron spins, which in turn, produces a change in the magnetic moments. Also, this phase transition is in concomitance with the structural change, affording the variation of the Fe-O-Fe angle that defines these magnetic interactions.

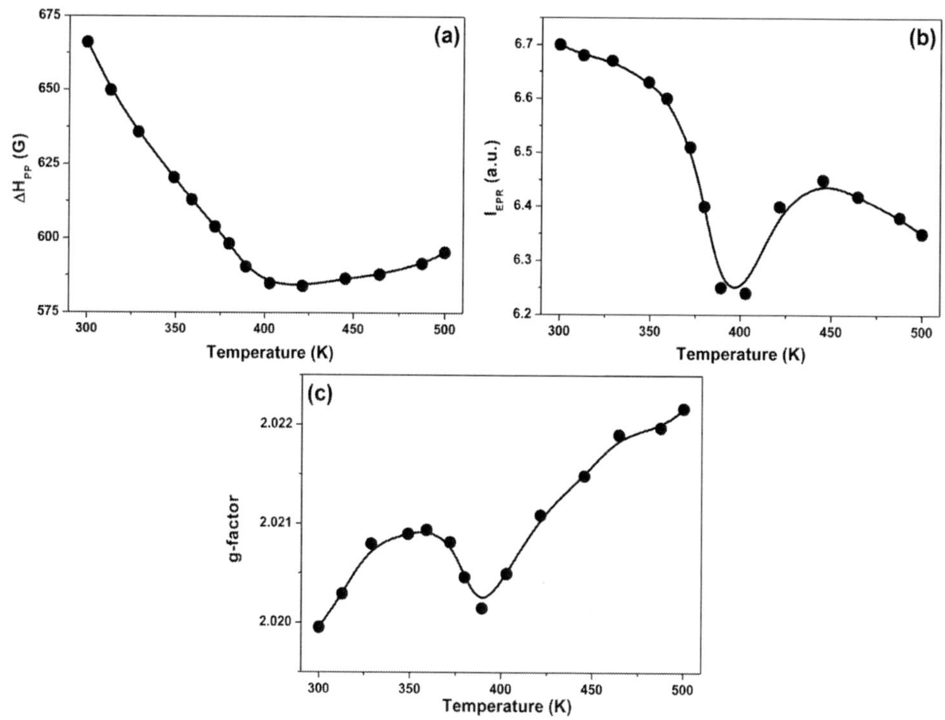

Figure 5. Temperature dependence of (a) the peak-to-peak linewidth -ΔH_{pp}, (b) the integral intensity - I_{EPR} and (c) the g-factor of $Pb(Fe_{0.5}Nb_{0.5})O_3$. Solid lines are guides for the eye only.

Figure 5(c) shows the behavior of the g-factor vs. temperature. The g-factor shows an increase in the 300-359 K range, and as the temperature increases further, the g-factor decreases, with a minimum at 390 K. This behavior is associated with the diffuse character of the ferro-paraelectric transition [43], which is detected starting from 359 K. This characteristic can be ascribed to the random distribution of Fe^{+3} and Nb^{+5} ions, which leads to the fluctuations of the composition, thus forming many microscopic regions with slightly different Curie temperatures. The existence of these microscopic regions broadens the transition over a wide temperature range [43,44]. The additional g-factor increase (from 390 K to 500 K) is associated with the paramagnetic behavior of the Fe ions.

It is necessary to mention that the Curie temperature obtained by the EPR technique (between 390 K and 403 K) is in a good agreement with the reported values [42,43].

3.1.2. MAMMAs Technique

Figure 6 shows the MAMMAS response for the PFN powders, where the non-resonant microwave absorption was taken for both the heating and cooling directions. As the signal does not show thermal hysteresis in all the temperature range, and for simplicity it will only be described the cooling spectrum.

MAMMAS response recorded in the cooling run for the magnetoelectric PFN, shows an increase monotonically as temperature decreases from 300 K to ~184.6 K, following an approximate exponential law. In this temperature range the material is paramagnetic [14,17,40,41], and MAMMAS measurement is in agreement with the interpretation that this first absorbing process is due to paramagnetic moments.

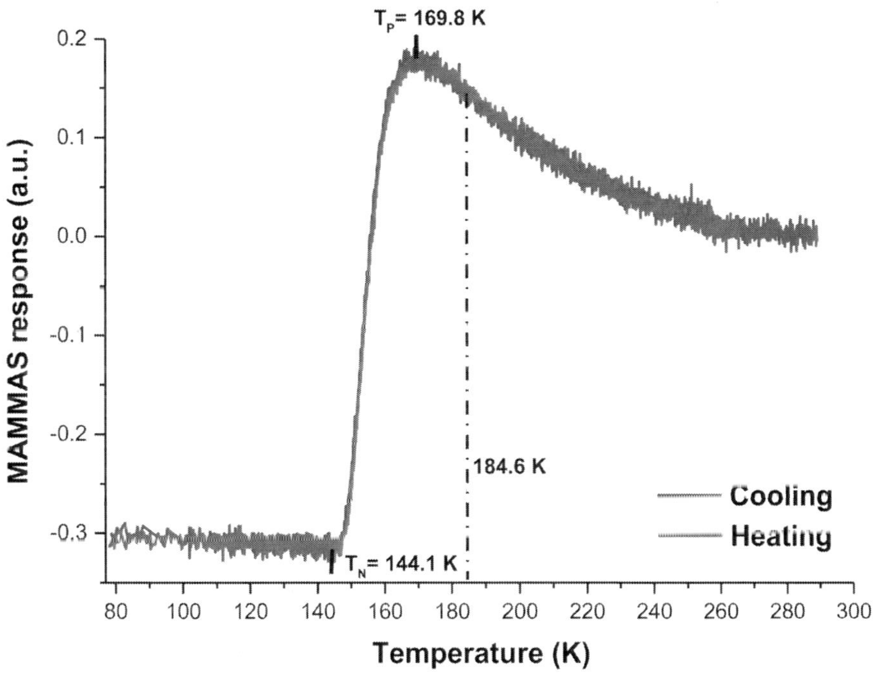

Figure 6. MAMMAS response of magnetoelectric Pb(Fe0.5Nb0.5)O3 in the 77-300 K temperature range, during a heating and cooling cycle.

As the temperature is decreased further, T<184.6 K, another magnetic process sets-in which strongly slows down the previous absorbing behavior. This process can be due to magnetic fluctuations, i.e., fluctuations in the establishment of the long-range order that proceeds to a magnetic transition. MAMMAS response reaches a maximum at T_P=169.8 K, after that, the absorption decreases very fast with decreasing temperature until T_N=144.1 K; in this region the quantity of absorbing centers diminishes considerably due to a process of antiparallel spin alignment, and which is associated with a G-type antiferromagnetic order in this material [14,17,41]. Also, the absence of thermal hysteresis is indicative of a second-order transition, as is expected for para-antiferromagnetic transition [45]. Below T_N, the MAMMAS response remains approximately lineal with a small slope; this behavior is a signature of weak ferromagnetism in this region of temperature [28,34], which can be associated with the magnetoelectric effect [46,47].

MAMMAS response for the magnetoelectric PFN mainly is due to the para-antiferromagnetic behavior of the Fe^{+3} spins. The whole MAMMAS profile depends on the thermal dependence of the spin dynamics prior to their antiparallel alignment at T_N; and the intensity of the signal follows the variations on the number of absorption centers, which in turn is controlled by the establishment of the antiferromagnetic order and by the presence of a weak ferromagnetism at low temperature.

The microwave absorption in magnetic materials is originated from magnetic losses such as reorientation of magnetic domains, domain wall resonance and magnetic resonance. The time-averaged power (P), dissipated by a material magnetic in an alternate magnetic field, $H = H_{mw}\, e^{-i\omega t}$, can be expressed as [35]:

$$P = \tfrac{1}{2}\, \omega\, \mu_0\, \mathrm{Im}(\mu)\, H_{mw}^2 \qquad (1)$$

where μ_0 is magnetic permeability of free space, $\mathrm{Im}(\mu)$ is the imaginary part of the magnetic permeability, and $\omega=2\pi f$, where f is the frequency. All the variables in the above equation are quite sensitive to temperature; therefore, for a constant frequency and as a temperature function the microwave absorption should follow a behavior similar to $\mathrm{Im}(\mu)$.

Additionally, the magnetoelectric PFN has a diffuse phase transition from a ferroelectric phase with rhombohedral structure to the paraelectric phase with cubic structure, which is detected at a temperature about 380 K [14] or 387 K [17,40,41]. Next it is shown as the ferro-paraelectric transition is detected by means of MAMMAS technique. Figure 7 shows the MAMMAS response recorded in the heating run for the magnetoelectric PFN.

Starting from room temperature, as the temperature is increased, the microwave absorption decreases continually. MAMMAS response also follows an exponential law, being a continuation of the paramagnetic process to high temperature; this profile starts to differ from the exponential law about T^*=363.3 K. Such deviation indicates the presence of an additional absorption process, which it is proposed as originated also by the paramagnetic moments, with a strong dipole-dipole interaction caused by the high concentration of Fe spins. When the temperature continues to decrease, MAMMAS signal reaches a minimum value at T_{min}=398.1 K, then the microwave absorption increases and asymptotically reaches a constant value, as shown in Figure 7; this is indicative of the gradual vanishing of this last process.

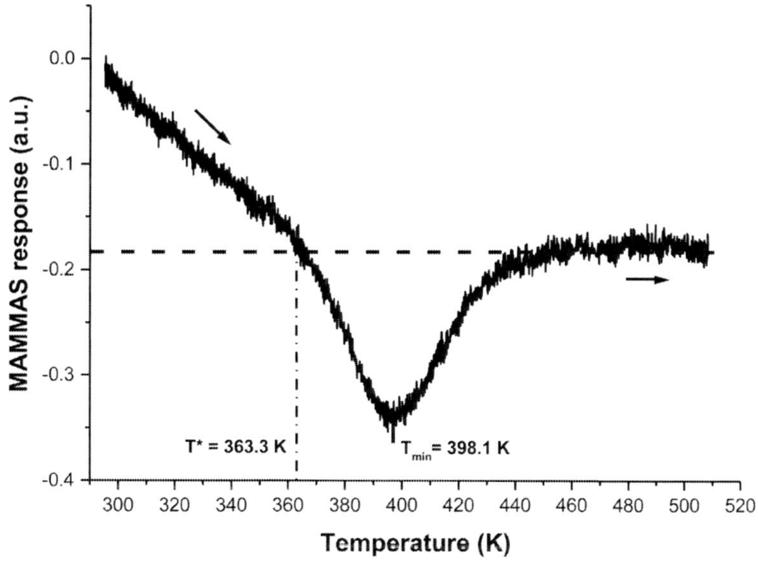

Figure 7. MAMMAS response of $Pb(Fe_{0.5}Nb_{0.5})O_3$ recorded for increasing temperature; with H_{dc} = 600 G and H_{mod} = 4 G.

The ferroelectric Curie temperature (T_{min}= 398.1 K) obtained by the technique MAMMAS is in a good agreement with the reported values [42,43]. The appearance of an inverted peak in the microwave absorption is opposite to that observed in other materials [29,48-52]; this is a signature of the absorption dynamics by paramagnetic moments around a ferro-paraelectric phase transition. Additionally, the absorption process also shows characteristics of a diffusive phase transition over a wide temperature range. The diffusive characteristics can be ascribed to the random distribution of Fe^{+3} and Nb^{+5}, which leads to the fluctuations of the composition, thus forming many microscopic regions with different Curie temperatures. It is just these microscopic regions that broaden the transition temperature over a wide temperature region [42-44].

3.1.3. LFMA Technique

Finally, the LFMA spectra (dP/dH vs. magnetic field, around zero field) recorded at selected temperatures for PFN powders are shown in Figure 8(a). These curves have a linear behavior with a positive slope and non-hysteretic traces. The positive slope implies that this microwave absorption is a minimum around zero magnetic field and it increases with applied magnetic field; in other words, this is a magnetic field-dependent absorption.

When the temperature is reduced below 145 K (G-type antiferromagnetic order), a horizontal and non-hysteretic LFMA line is observed. Additionally, LFMA line shows the absence of any irreversible microwave energy-absorption process in the ordered and disordered phases. This behavior for the magnetoelectric PFN contrasts strongly with that shown by high-Tc superconductors [33], ferrimagnetic [27] and ferromagnetic [53] materials.

The LFMA line can be described by the experimental correlation:

$$dP/dH = S(T)\, H \qquad (2)$$

where S(T) depends only on temperature.

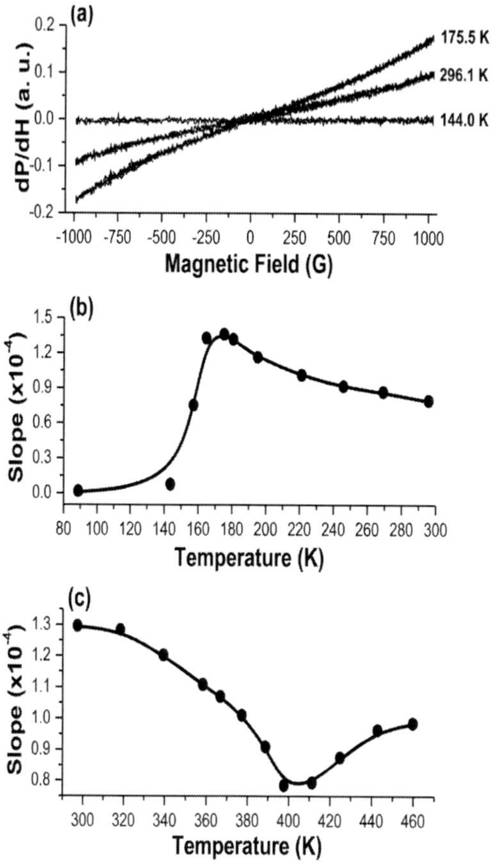

Figure 8. (a) LFMA spectra of PFN for selected temperatures; with H_{mod}= 0.63 G, microwave power 1 mW (cooling) and 2mW (heating). The temperature dependence of the absorption line slope for temperature intervals of (b) 90-300 K and (c) 300-460 K; the solid lines are only guides for the eyes.

Figs. 8(b,c) show the behavior of the slopes of the LFMA line for the 90-300 K and 300-460 K temperature ranges, respectively. We observe that the profile obtained is similar to the shape of the corresponding MAMMAS response. This clearly points to a common origin for both measurements, and allows us to confirm the assignment of the temperature values for the magnetic and ferroelectric transitions.

Additionally, it gives evidence that LFMA spectra are generated by the same absorption processes described in the previous section, and which are due to the absorption dynamics of magnetic moments at all the temperatures. Also, as the MAMMAS responses are similar to corresponding plots of the slope vs. temperature, settling down an experimental correlation between the MAMMAS response and the second derivative microwave absorption (d^2P/dH^2); i.e. for low magnetic field (≤1000 G) the MAMMAS response is given as:

$$\text{MAMMAS response} \propto \frac{d^2P}{dH^2} = \frac{1}{2}\omega\mu_0 H_{mw}^2 \frac{d^2\,\mathrm{Im}(\mu)}{dH^2} \qquad (3)$$

where the amplitude of the microwave magnetic field, H_{mw}, is constant.

3.2. Magnetoelectric PFT

3.2.1. EPR Technique

In Figure 9, we show EPR spectra recorded in the 120-300 K temperature range. It can be observed that resonance spectra exhibit a single broad symmetric Lorentzian line along the entire temperature range, due to the spin of the Fe^{+3} ions.

The derivative of microwave power absorption with respect to the static field (dP/dH) is fitted into the two-component Lorentzian equation accounting for the contributions from the clockwise and anticlockwise rotating components of the microwave magnetic field [54,55], as follows:

$$\frac{dP}{dH} \propto \frac{d}{dH}\left(\frac{\Delta H_{pp}}{(H_{dc} - H_{res})^2 + \Delta H_{pp}^2} + \frac{\Delta H_{pp}}{(H_{dc} + H_{res})^2 + \Delta H_{pp}^2} \right) \tag{4}$$

where H_{dc} is the applied magnetic field and H_{res} is resonant field. The temperature dependences of the EPR parameters obtained from these fits are plotted in Figure 10.

Figure 10(a) shows the temperature dependence of ΔH_{pp} for PFT powders. ΔH_{pp} increases slightly as temperature decreases from 300 K to 170 K. As temperature continues to decrease, T< 170 K, ΔH_{pp} increases continuously, but this time with a higher change rate.

In Figure 10(b), we show the behavior of the H_{res} vs. temperature. For simplicity, we use the H_{res} instead of g-factor. H_{res} increases continuously with the temperature decrease, exhibiting a maximum at 170 K, and followed by a fast decrease until 120 K. For 170-300 K temperature region the material is paramagnetic [56,57], and microwave absorption increases with the decrease in temperature [25,26]; and it would explain the behavior of ΔH_{pp} and H_{res} in this temperature interval. On the other hand, the sudden shift in H_{res} and the broadening in ΔH_{pp}, as the temperature is lowered (T<170 K), can be attributed to a long-range antiferromagnetic order at T_N=170 K, as result of the linear superexchange interaction between Fe ions.

The temperature dependence of I_{EPR} is shown in Figure 10(c). Starting from 300 K, I_{EPR} increases continuously with the temperature decrease, exhibiting a maximum at T_{max}= 240K; this behavior indicates the paramagnetic nature of this signal [25,26]. After T_{max}, when the temperature continues to decrease, I_{EPR} diminishes, where an inflection point at T_N= 170 K is clearly observed. We associate the maximum, T_{max}, with the onset of the ferro-paraelectric phase transition and which is accompanied with a structural change [58]. This structural change leads to a redistribution of the Fe^{+3} ions, and it generates a reorganization of all the magnetic interactions, affording the variation of the Fe O Fe angle that defines these interactions.

The Curie temperature (T_{max}) obtained by the EPR technique is in a good agreement with the reported values [58]. For low temperature, the decrease in I_{EPR} can be associated with the decrement of the quantity of absorbing paramagnetic centers, due to the process of antiparallel spin alignment and which is associated with the antiferromagnetic order. The change in slope at T_N, as is shown in Figure 10(c), can be associated with the onset of antiferro-paramagnetic transition, in a good correspondence with plots of H_{res} and ΔH_{pp}.

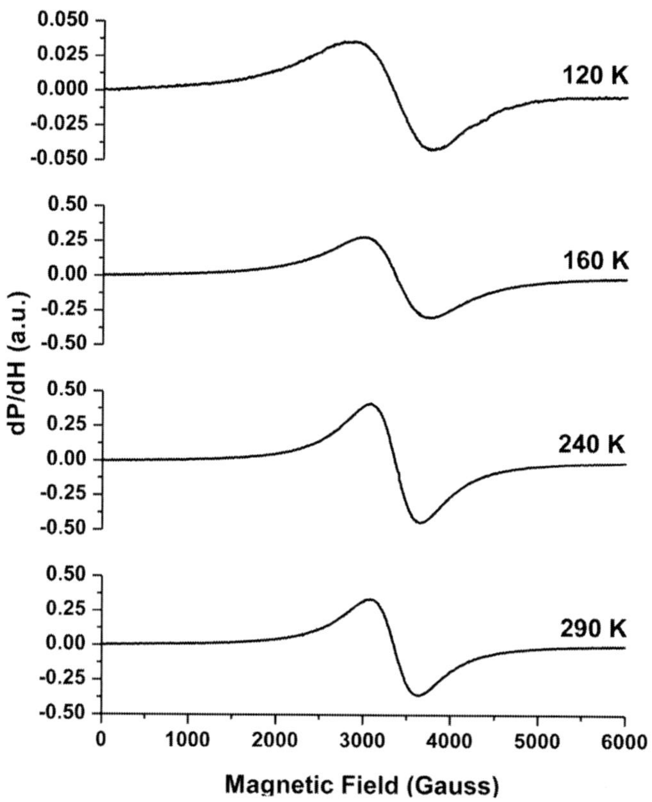

Figure 9. EPR spectra of the magnetoelectric PFT for selected temperatures.

3.2.2. LFMA Technique

Figure 11(a) shows LFMA spectra in the magnetoelectric PFT for selected temperatures. The curves exhibit a linear behavior with a positive slope and non-hysteretic traces. The positive slope implies that this non-resonant microwave absorption is a minimum around zero magnetic field and that it increases with applied magnetic field. LFMA lines show the absence of any irreversible microwave energy-absorption process in the ordered and disordered phases.

Figure 11(b) shows the behavior of the slope of the LFMA line in the 90-290 K temperature range. The slope increases monotonically when the temperature decreases from 290 K, reaching a maximum value at 160 K. This increasing behavior is characteristic of a paramagnetic phase [23,34], in good agreement with the EPR results. Additionally, a change in slope at 240 K is observed, this inflection point is more apparent in the plot d(Slope)/dT vs. T and it is shown as a maximum at 240 K; see the inset of Figure 11(b).

This change indicates the presence of an additional absorption process, which we propose as originated by the ferro-paraelectric phase transition; where the ferroelectric transition originate a change in the relative orientation of magnetic moments due to structural change, affecting the superexchange interactions between Fe ions, and it generate a change in the dynamics of microwave absorption.

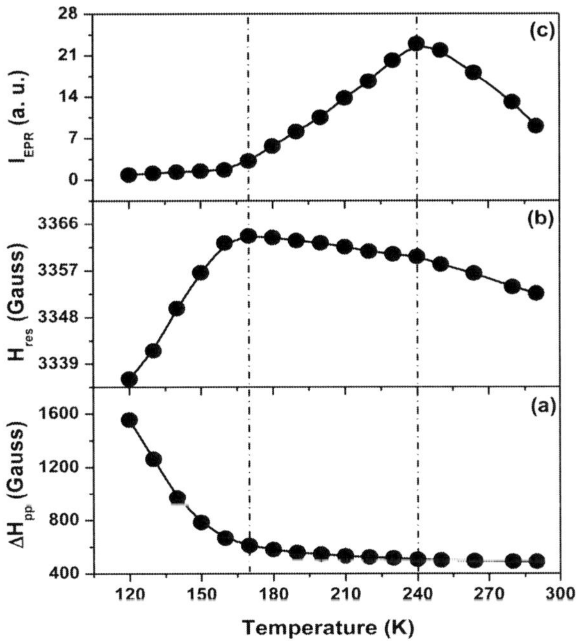

Figure 10. Temperature dependence of (a) the peak-to-peak linewidth -ΔH_{pp}, (b) the resonant field -H_{res} and (c) the integral intensity -I_{EPR}. Solid lines are guides for the eye only.

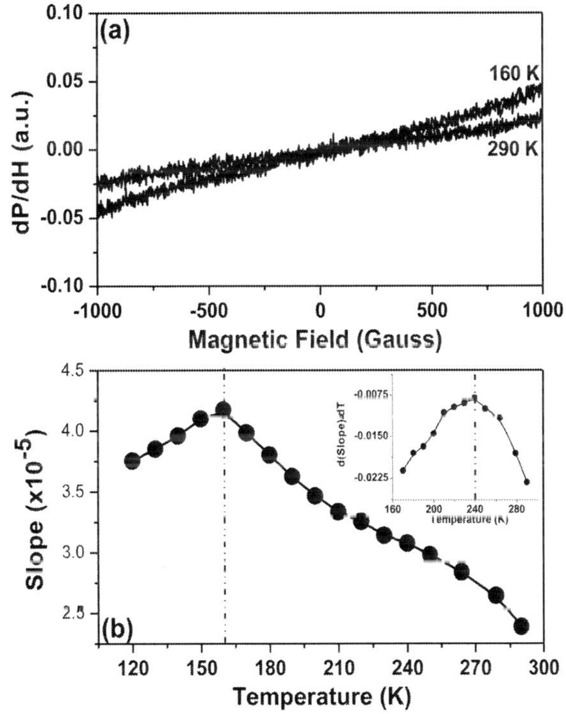

Figure 11. (a) LFMA spectra of PFT for 290 K and 160 K, with H_{mod}= 0.63 G and microwave power of 1 mW. (b) The temperature dependence of the absorption line slope for temperature interval of 120-290K. The inset of (b) shows the temperature dependence of d(Slope)/dT in the 160-290 K temperature range. The curves connecting points are only guides for the eye.

As the temperature is decreased further, T<160 K, the slope slowly decreases and another magnetic process sets-in which we propose as originated by long-range magnetic correlations; in this region the quantity of absorbing centers diminishes, due to the process of antiparallel spin alignment. The whole profile of the slope of the LFMA line depends on the thermal dependence of the spin dynamics prior to their antiparallel alignment at T<160K; and its intensity follows the variation in the number of absorption centers, which in turn is controlled by the establishment of the antiferromagnetic order at low temperature.

3.3. Magnetoelectric YM

3.3.1 EPR Technique

Figure 12 shows EPR spectra for the magnetoelectric YM, at a few selected temperatures in the 100-300 K range. EPR spectra show two symmetric Lorentzian lines along the entire temperature range, and they are labeled as 1 and 2. The broad signal 1 suggests a strong dipole-dipole interaction, due to the high concentration of Mn^{+3} spins in the YM.

The inset of Figure 12 displays the g-factor as function of temperature for signal 1; it decreases continuously with the decrease in temperature. The linewidth of signal 1 broadens as the temperature is lowered, suggesting the existence of short-range magnetic correlations in the paramagnetic phase. This temperature dependence of our EPR data can be understood in terms of a universal behavior for the spin dynamics observed in many classes of manganites [59], and which reveal a strong magnetic coupling well above magnetic order. This behavior is similar to that found in the paramagnetic phase of systems exhibiting spin-freezing phenomena at low temperature (spin glasses), and can be understood as a manifestation of the presence of short-range correlations at $T \geq T_N$.

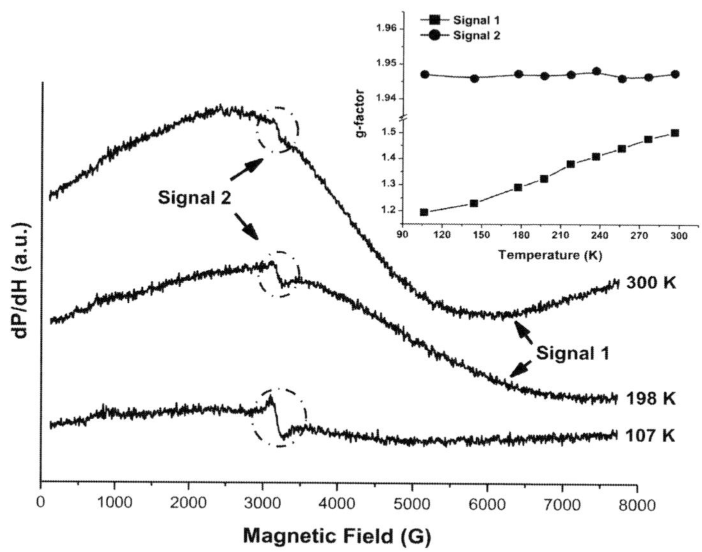

Figure 12. EPR spectra of the magnetoelectric YM for selected temperatures in the 100-300 K range. The inset shows the temperature dependence of the g-factor in signals 1 and 2 for the same temperature range; curves connecting points are only guides for the eye.

For all the temperatures, the thin signal 2 is located approximately in the same resonant field, ~3254 Gauss, where this absorption mode is due to the presence of Mn^{+2} ions [24-26], and it can be assigned at impurity traces from the starting reagents. The inset of Figure 12 also shows the g-factor vs. temperature for the signal 2, where the g-factor is independent of the temperature, and this behavior confirms the paramagnetic nature of this signal [24-26,29].

3.3.2. MAMMAS Technique

Figure 13 shows the MAMMAS response for the magnetoelectric YM that decreases monotonically as temperature decreases from 300 K, following an approximate decay exponential (which we call absorbing process A); reaching a minimum value at T_{min} = 112.4K. To better understand this behavior, we compare the MAMMAS response of a well known ferroelectric material, $BaTiO_3$ powders for the same temperature range, see inset in Figure 13. This profile shows a continuous decrease of exponential type with decreasing the temperature. Since $BaTiO_3$ has only electric dipoles, all its microwave absorption is due to dielectric losses. For the behavior similarity between the MAMMAS profiles, we propose that the process A is due to the electric dipoles with long-range co-operative interactions in the magnetoelectric YM; these dipoles can be driven by a microwave field by increasing their oscillating energy and therefore leading to absorption. When temperature decreases, the thermal excitation also decreases, and the electrostatic interactions increase. Electric dipoles have a diminished freedom to move, and therefore, absorption decreases.

As the temperature is decreased further into the 77 K<T<T_{min} range, the MAMMAS response increases and another microwave energy-absorbing process sets-in, different from the previous behavior.

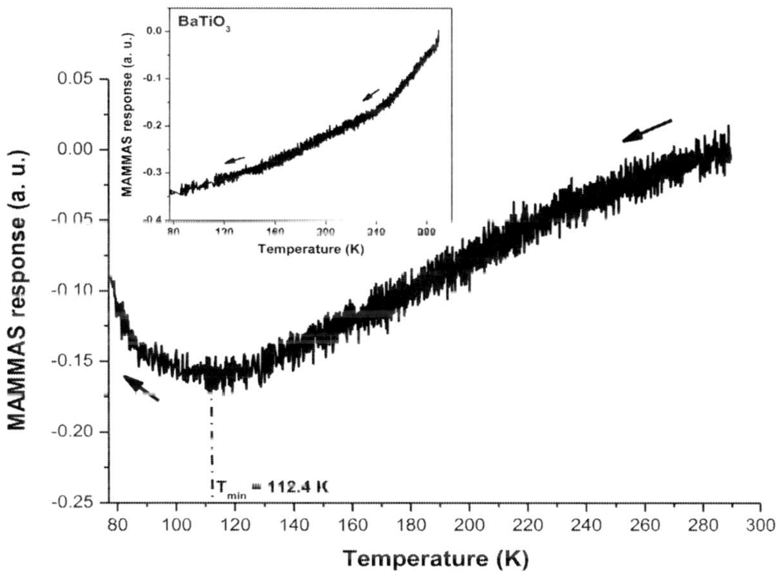

Figure 13. MAMMAS response of the magnetoelectric YM for decreasing temperature; the inset shows the MAMMAS response of $BaTiO_3$ powders in the 77-300 K temperature range.

This process follows an approximate growing exponential function (absorbing process B). We propose that the process B is originated mainly by the magnetic moments in paramagnetic state, since they have effectively exhibited this behavior [29,34]; then for low temperature, the effect of magnetic loss is dominant over the one from dielectric loss.

On the other hand, the microwave absorption in magnetoelectric materials can be originated from magnetic losses or dielectric losses, such as reorientation of magnetic domains and electric charge displacement, respectively. The time-averaged power, P, dissipated in a system carrying a current density j in an alternating electric field $E=E_0\, e^{(-i\omega t)}$ or by its magnetic component in an alternating magnetic field, $H=H_{mw}\, e^{(-i\omega t)}$, can be expressed as:

$$P = \tfrac{1}{2}\,\omega\,(\varepsilon_0\, \mathrm{Im}(\varepsilon) E_0^{\,2} + \mu_0\, \mathrm{Im}(\mu)\, H_{mw}^{\,2}) \tag{5}$$

where ε_0 and μ_0 are the dielectric constant and the permeability of free space, respectively; $\mathrm{Im}(\varepsilon)$ and $\mathrm{Im}(\mu)$ are the imaginary part of the dielectric constant and the magnetic permeability, respectively, and $\omega = 2\pi f$, where f is the frequency.

It can be observed that the microwave power absorption varies linearly with frequency, while the imaginary part of both the magnetic permeability and the dielectric constant depend on the square of the intensity of the magnetic and electric fields. For a constant frequency (and amplitude) of applied fields, the microwave power absorption should follow a behavior similar to $\mathrm{Im}(\mu)$ or $\mathrm{Im}(\varepsilon)$ when temperature is varied.

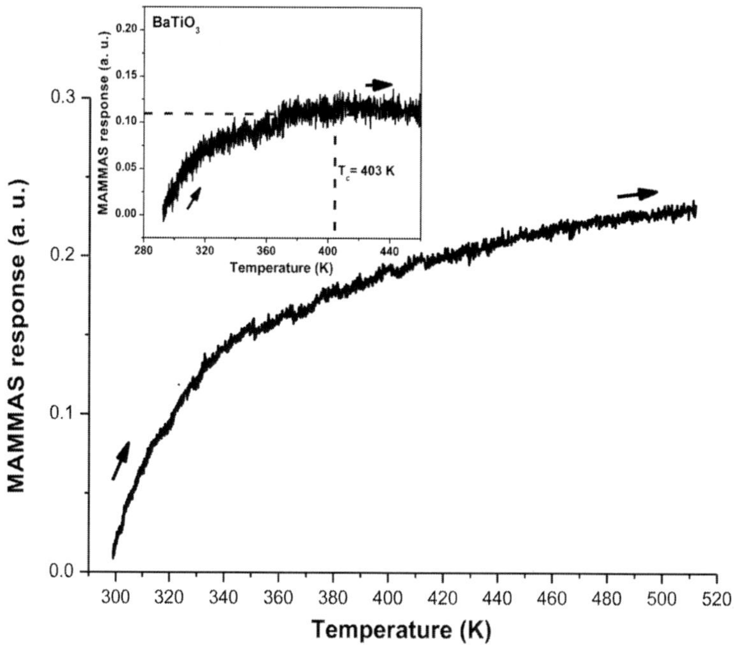

Figure 14. MAMMAS response of the magnetoelectric YM recorded for increasing temperature; the inset shows the MAMMAS response of $BaTiO_3$ powders in the 300-520 K temperature range.

Figure 14 shows the MAMMAS response recorded in the heating run for the magnetoelectric YM. Starting from room temperature (300 K), as the temperature is increased, the microwave absorption increases monotonically and follows an exponential type law. Recently it has been shown that the paramagnetic microwave absorption should follow a continuous decrease when increasing the temperature [34]. Since the observed MAMMAS profile for the multiferroic YM shows a different behavior, we propose that this profile is due also to electric dipoles, and is the continuation of the process A to high temperatures. For a comparison, we also show the MAMMAS response of $BaTiO_3$ powders in the 300-460 K temperature range, see inset of Figure 14. It shows a continuous increase of the microwave absorption when increasing the temperature, following also an exponential type law, and for T > 403 K the microwave absorption reaches a constant value. This behavior is indicative of the gradual vanishing of the absorption process by electric dipoles with long-range interactions, and it is due to the ferro-paraelectric phase transition in $BaTiO_3$; the Curie temperature (T_c= 403 K) obtained by the MAMMAS technique is in a good agreement with the reported values [60,61]. Therefore, by similarity with $BaTiO_3$, we can assign the microwave absorption profile for the magnetoelectric YM in the 300-520 K temperature range as associated with dielectric losses.

3.3.3. LFMA Technique

Figures. 15(a) and 16(a) show the LFMA spectra recorded in the 77-300 K and 300-520 K temperature ranges, respectively. We observe that all the curves exhibit a linear behavior with a positive slope and non-hysteretic traces. The positive slope implies that this microwave absorption is a minimum around zero magnetic field and that it increases with field; in other words, this is a magnetic field-dependent absorption. The LFMA line shows the absence of any irreversible microwave energy-absorption process. This linear behavior of microwave absorption for the magnetoelectric YM has also been observed in other materials [23,34], but this behavior strongly contrasts with that shown by ferrites Ni-Zn [27] and cobalt [62] ferrites. LFMA signal is closely related with the low-field magnetization of the sample in the long-range order regime; i.e., LFMA is lineal in the paramagnetic phase and emerges as the temperature is decreased below magnetic order.

Let us remember that in a dielectric material the Maxwell displacement current, $j=d(\varepsilon\varepsilon_0 E)/dt$, is induced by the microwave electric field E. Since $j=\sigma E$, where σ is the microwave conductivity ($\sigma=\frac{1}{2}\varepsilon_0 \mathrm{Im}(\varepsilon)\omega$), we describe the magnetic field dependence of the microwave absorption, $P=\sigma E_0^2$, in terms of changes in microwave conductivity $\sigma(H)$ with an applied static magnetic field H. Experimentally the observed microwave absorption at high temperatures where the dominant response is due to electric dipoles, is

$$\Delta P= P(H)-P(0)= \Delta\sigma E_0^2 \qquad (6)$$

where $\Delta\sigma= \sigma(H)-\sigma(0)$ is the change in the microwave conductivity induced by the magnetic field H. Since dielectric loss in absence of static magnetic field is given by $P(0)= \sigma(0)E_0^2$, the observed response can be written,

$$\Delta P= [\Delta\sigma/\sigma(0)]\, P(0) \qquad (7)$$

The term $\Delta\sigma/\sigma(0)$ describe relative changes in microwave conductivity induced by the external field, that is, magnetoconductance. The explanation of magnetoresistance in perovskite manganites [63] relies on assumptions about magnetic ordering or exchange coupling between manganese ions with different valency states.

In the present case, the low-field microwave absorption by electric dipoles should be considered as the enhancement of dielectric loss by the external magnetic field, or microwave magnetoconductance, which leads to:

$$dP/dH = [P(0)/\sigma(0)]\, d\sigma/dH \qquad (8)$$

where the factor $P(0)/\sigma(0)$, the square of the microwave electric field E_0^2, is constant.

For low temperatures, where the dominant response is due to magnetic moments, the low-field microwave absorption is given as:

$$\frac{dP}{dH} = \frac{1}{2}\omega\mu_0 H_{mw}^2 \frac{d\,\mathrm{Im}(\mu)}{dH} \qquad (9)$$

Here, the amplitude of the microwave magnetic field, H_{mw}, is constant.

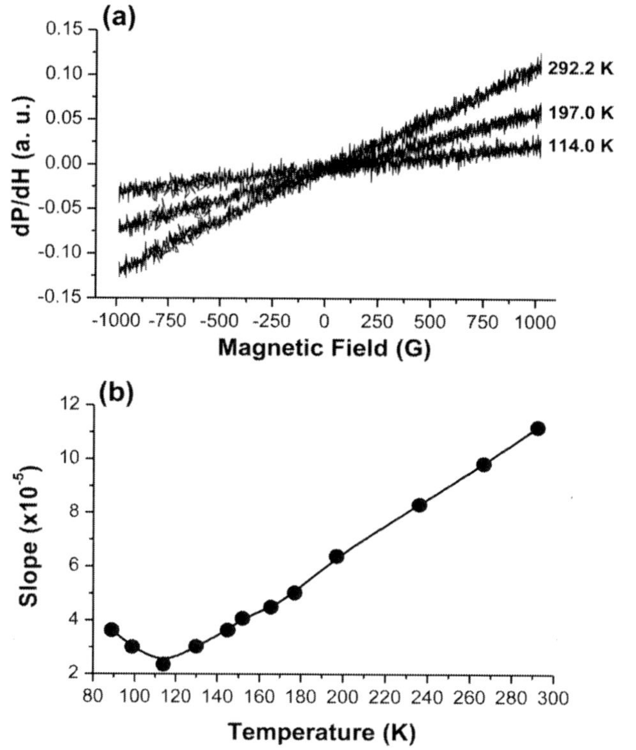

Figure 15. (a) LFMA spectra of the magnetoelectric YM for selected temperatures, with magnetic modulation of 0.63 Oe and microwave power 1 mW; (b) the temperature dependence of the absorption line slope for the 77-300 K temperature range. The solid lines are only guides for the eyes.

Figures 15(b) and 16(b) show the behavior of the slope of the LFMA line in the 77-300 K and 300-520 K temperature ranges, respectively. We observe that the profiles obtained are similar to the shape of the corresponding MAMMAS responses. This clearly points to a common origin for both measurements, giving evidence that LFMA spectra are generated by the same absorption processes above described, namely the absorption dynamics for electric dipoles and magnetic moments in different regions of temperature. Also, as the MAMMAS responses are similar to the plot of the slope vs. temperature, we can establish an experimental correlation between the MAMMAS response and the second derivative microwave power absorption (d^2P/dH^2); i.e., for low magnetic field (≤1000 G) the MAMMAS response is proportional to d^2P/dH^2.

3.4. Magnetoelectric YC

3.4.1. EPR Technique

In Figure 17, we show the EPR spectra recorded in the 85-300 K temperature range for the magnetoelectric YC. It can be observed that resonance spectra exhibit a single broad symmetric Lorentzian line along the entire temperature range, due to the spin of the Cr^{+3} (S=3/2) ions. The temperature dependences of the EPR parameters obtained from these fits are plotted in Figure 18.

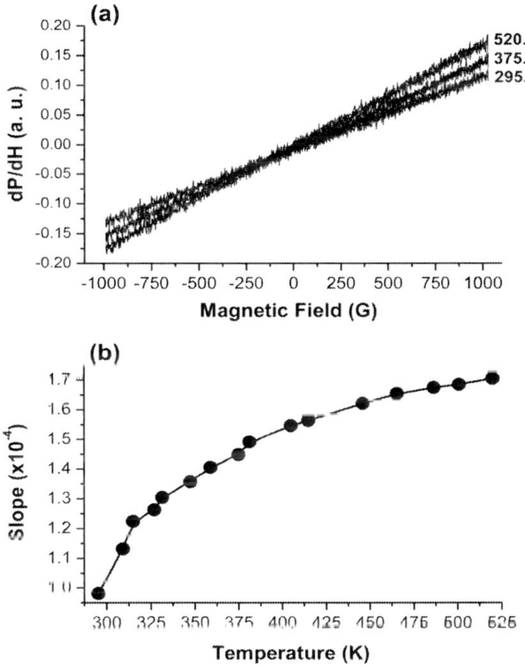

Figure 16. (a) LFMA spectra of the magnetoelectric YM for selected temperatures, with a magnetic modulation of 0.63 Oe and microwave power 2mW; (b) the temperature dependence of the absorption line slope for the 300-520 K temperature range. The solid lines are only guides for the eyes.

The temperature dependence of I_{EPR} is shown in Figure 18(a). Starting from 300 K, I_{EPR} increases continuously with the temperature decrease, exhibiting a maximum at T_{max}= 182 K; this behavior indicates the paramagnetic nature of this signal [25,26]. Additionally, an inflection point at T_{on}= 220 K is also observed, and there are two possible explanations for this feature. The first possibility is that a ferromagnetic impurity in the form of a second phase, with a Curie point close to 220 K and a significant concentration, could explain this feature; but the X-ray diffraction spectrum does not show the presence of a second phase [37]. A more sound explanation of this feature can be due to build-up of magnetic correlations preceding the transition to the long-range antiferromagnetic ordering.

After T_{max}, when the temperature continues to decrease, I_{EPR} rapidly diminishes until T_N=139 K; this rapid decrease can be associated with the decrement of the quantity of absorbing paramagnetic centers, due to the process of antiparallel spin alignment and which is associated with the antiferromagnetic order at T_N. For 85 K$\leq$T<T_N a slow increase in I_{EPR} is observed, where this behavior is a signature of a weak ferromagnetism for this temperature region [28]; the previous behavior can be interpreted in terms of a resulting magnetic moment, i.e., a canting of Cr^{+3} ion sublattices in the antiferromagnetic matrix generates the magnetic moment. The canting of the magnetic moments can be explained by an anisotropic exchange due to Dzyaloshinsky-Moriya interaction [64].

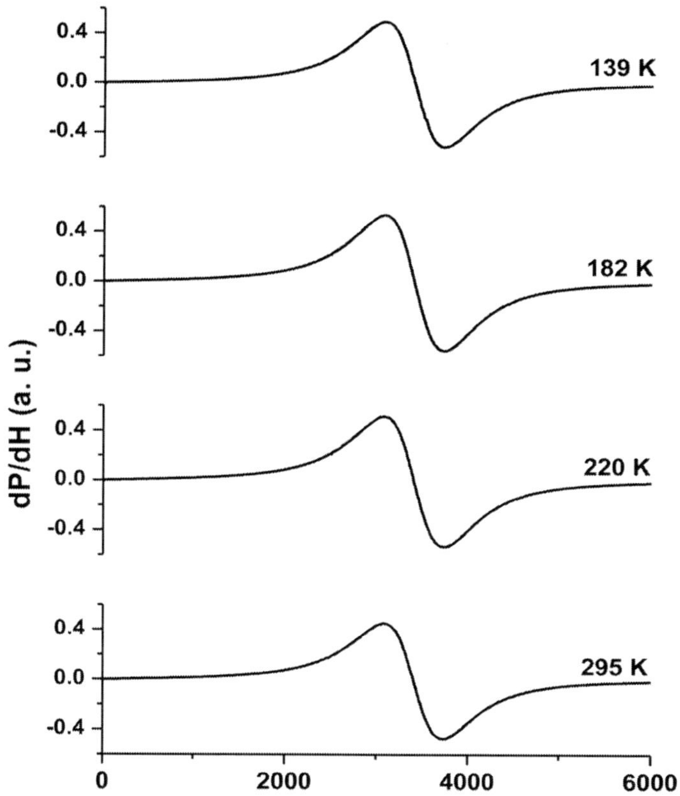

Figure 17. EPR spectra of the magnetoelectric $YCrO_3$ for selected temperatures.

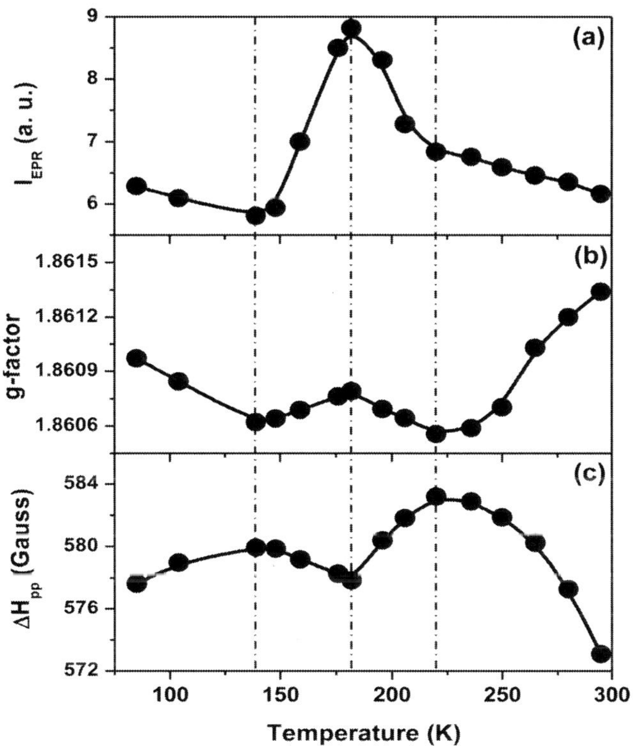

Figure 18. Temperature dependence of (a) the integral intensity -I_{EPR}, (b) the g-factor and (c) the peak-to-peak linewidth -ΔH_{pp}, in the 85-300 K temperature range. Solid lines are guides for the eye only.

Figures 18(b) and 18(c) show the temperature dependence of the g-factor and ΔH_{pp}, respectively. Our experiments, for all temperatures, give a g-factor smaller than that of a free electron (g=2.0023, S=1/2), and it is due to spin value of Cr^{+3} (S= 3/2) ions and to the spin-orbit coupling. The g-factor (ΔH_{pp}) shows a decrease (increase) in the 220-300 K temperature range, and as the temperature decreases further, the g-factor (ΔH_{pp}) increases (decrease), with a maximum (minimum) at 182 K. It is remarkable that the g-factor (ΔH_{pp}) increases (decrease) from 220 K to 182 K. This increase (decrease) can be due to magnetic fluctuations, i.e., fluctuations in the establishment of the long-range order that precedes the transition to the antiferromagnetic order at T_N. For T≤182 K, the g-factor (ΔH_{pp}) decreases (increase) until 139 K, this feature can be attributed to the long-range antiferromagnetic ordering at T_N=139 K, as a result of the superexchange interaction between Cr ions. The increase (decrease) of the g-factor (ΔH_{pp}) in the 85-139 K temperature region is an indication that a weak ferromagnetic behavior is present [28,65], and which we confirmed with the presence of magnetic hysteresis at 5 K [37]. Additionally, the EPR spectra for 300-510 K temperature range also exhibits a single broad symmetric Lorentzian line, due to the spin of the Cr^{+3} ions. For this temperature region, the dependences of the EPR parameters (ΔH_{pp}, I_{EPR}, and the g-factor) are plotted in Figure 19.

Figure 19(a) shows the temperature dependence of ΔH_{pp} for $YCrO_3$ powders, which diminishes monotonically as temperature increases from 300 to 430 K. In paramagnetic materials, the microwave absorption decreases as temperature increases [24-26], and therefore the magnetoelectric $YCrO_3$ has a behavior paramagnetic for this temperature region.

As temperature increases further, T>430 K, ΔH_{pp} decreases continuously but with a lower change rate. This behavior could be due to a ferro-paraelectric transition, as has been observed in other magnetoelectric materials [30,31], and that we confirmed directly by means of dielectric measurements [32]. Then, we associate the ΔH_{pp} behavior to presence of a local non-centrosymmetry, that it accompanies to ferro-paraelectric transition in this material [66,67]; and which produces discreet variations in the vicinity of the Cr^{+3} ions, generating a reorganization of the spin-spin and spin-magnetic field interactions. However, more evidence is gathered below.

The temperature dependence of I_{EPR} is shown in Figure 19(b). Starting from room temperature, I_{EPR} decreases continuously as temperature increases, exhibiting a minimum at 430 K, followed by an increase until 475 K; and for T> 475 K, a slow decrease is observed. A very similar behavior is observed in the magnetoelectric PFN, and that it has been associated with a ferro-paraelectric phase transition [30].

Therefore, the minimum observed is due to onset of the dielectric transition, where the change in electrical ordering caused by the ferro-paraelectric phase transition leads to a redistribution of the electron spins; and which is in concomitance with the local character of the Cr off-centring, affecting the Cr-O-Cr angle that defines these magnetic interactions. The Curie temperature obtained by EPR technique is in a good agreement with our dielectric measurements [32].

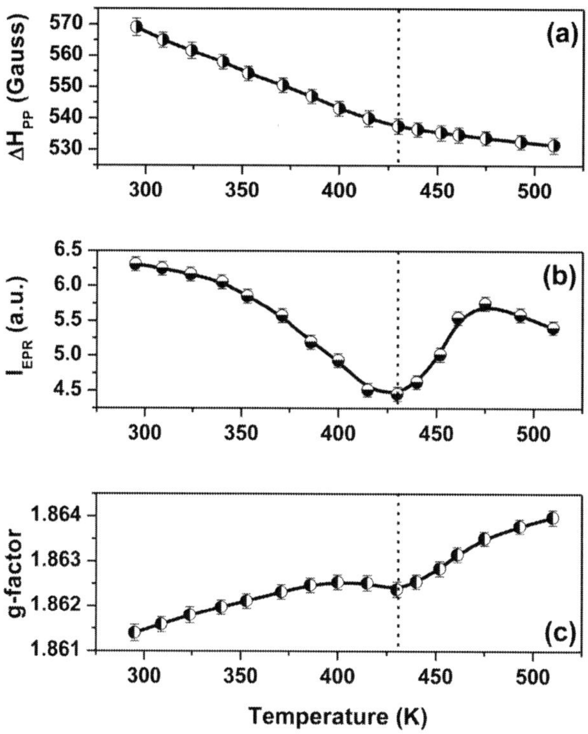

Figure 19. Temperature dependence of (a) the peak-to-peak linewidth -ΔH_{pp}, (b) the integral intensity - I_{EPR} and (c) the g-factor in the 300-510 K temperature range. Solid lines are guides for the eye only.

Figure 19(c) shows the behavior of the g-factor vs. temperature, which is estimated from the resonance field (H_{res}), with g= $h\nu/\mu_B H_{res}$; where h is the Planck constant, ν is the frequency and μ_B is the Bohr magneton. Our experiments give a g-factor smaller than that of a free electron (g= 2.0023, S= 1/2), along the entire temperature range. For a powder sample is expected that the internal field affects are average, and it is believed that the observed variation of the g-factor with temperature is intrinsic to sample. Additionally, this behavior can be explained through the spin value of Cr^{+3} (S=3/2) ions and to changes in the spin-orbit coupling. The effective g-factor (g_{eff}) of a paramagnetic center is given by $g_{eff}=g(1\pm\kappa/\Delta)$, where Δ is the crystal-field splitting and κ is the spin-orbit coupling constant. A gradual decrease of the orbital ordering takes place when the temperature is increased from 300 to 400 K, originating a change in the spin-orbit coupling as well as in the crystal-field splitting, and which can give rise to the observed increase in the g-factor for this temperature range. As the temperature increases further, the g-factor decreases, with a local minimum at 430 K (g_{min}= 1.8624); we associate this feature with the onset of the ferro-paraelectric phase transition. The additional g-factor increase (from 430 K to 510 K; g= 1.8640) is associated with the paramagnetic behavior of the Cr ions [30]. Also, in Figure 19(c), we can establish two behavior regions that are limited to 430 K, and that they can be associated with the diffuse character of the ferro-paraelectric transition [66,68]; which is detected starting from 400 K.

3.4.2. LFMA Technique

We turn now to LFMA results. Figure 20(a) shows LFMA spectra in the 85-295 K temperature range. The curves exhibit a linear behavior with a positive slope and non-hysteretic traces. The positive slope implies that this non-resonant microwave absorption is a minimum around zero magnetic field and that it increases with applied magnetic field. LFMA lines show the absence of any irreversible microwave energy-absorption process in the ordered and disordered phases.

Figure 20(b) shows the behavior of the slope of the LFMA line obtained of the fits (Eq. 2) for the 85-295 K temperature range. The slope increases monotonically when the temperature decreases from 295 K, reaching a maximum value at 183 K. This increasing behavior is characteristic of a paramagnetic phase [34], in a good agreement with the EPR results. Additionally, a change in slope at 220 K is observed, this inflection point is more apparent in the plot d(Slope)/dT vs. T and it is shown as a minimum at 220 K; see the inset of Figure 20(a). This change indicates the presence of an additional absorption process, which we propose as originated by a ferroelectric transition or by magnetic fluctuations at those temperatures.

A ferroelectric transition would originate a change in the relative orientation of magnetic moments due to a structural change, affecting the superexchange interactions between Cr ions, and it would generate a change in the dynamics of microwave absorption [31]; but in this temperature range a ferroelectric transition has not been reported, which is something remote. The other explanation is as follows: this behavior can be due to magnetic fluctuations [34], i.e., fluctuations in the establishment of the long-range order that proceeds to the antiferromagnetic transition; recently, the measurements of specific heat in the magnetoelectric $YCrO_3$ [69] suggests this process, where the magnetic specific heat

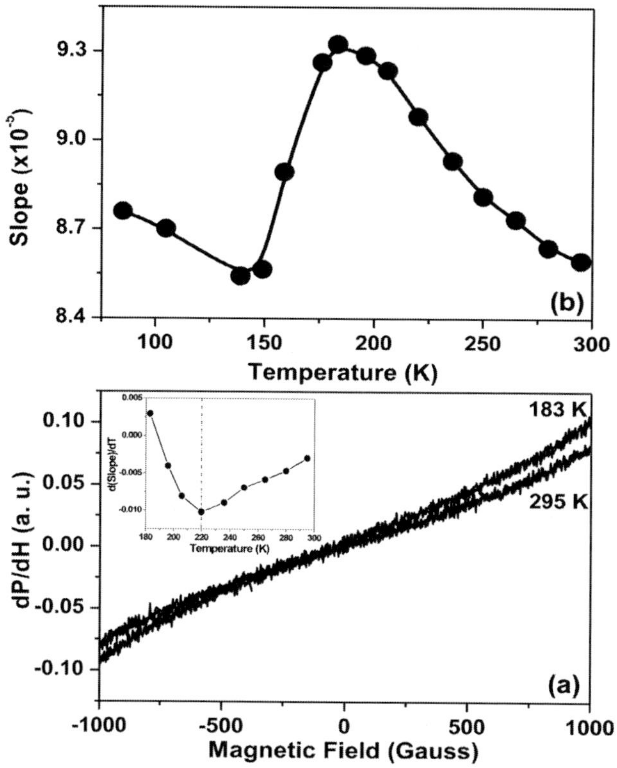

Figure 20. (a) LFMA spectra for 295 K and 183 K, with H_{mod}= 0.63 G and microwave power of 1 mW. (b) The temperature dependence of the absorption line slope for temperature interval of 85-295 K. The inset of Figure 20(a) shows the temperature dependence of d(Slope)/dT in the 183-295 K temperature range. The curves connecting points are only guides for the eye.

contribution begins to dominate on the phonon specific heat contribution at temperatures below ~225 K.

As the temperature is decreased further, T<183 K, the slope decreases very fast with decreasing temperature until T_N= 139 K; in this region the quantity of absorbing centers diminishes considerably due to the process of antiparallel spin alignment. Below T_N, the slope has an approximately lineal increase; this behavior is a signature of weak ferromagnetism in this region of temperature [34], in a good agreement with EPR measurements.

The whole profile of the slope of the LFMA line depend on the thermal dependence of the spin dynamics prior to their antiparallel alignment at T_N; and its intensity follows the variations on the number of absorption centers, which in turn is controlled by the establishment of the presence of a weak ferromagnetism at low temperature.

For the temperature interval of 300-510 K, LFMA spectra also exhibit a linear behavior with a positive slope and non-hysteretic traces, see Figure 21 (a); where LFMA lines show the absence of any irreversible microwave energy-absorption process in the ordered and disordered electric phases. Figure 21(b) shows the behavior of the slope of the LFMA line obtained of the fits, Eq. 2, for the 300-510 K temperature range. Starting from 300 K, as the temperature increased, the slope decreases monotonically; reaching a minimum value at T_{min}= 430 K.

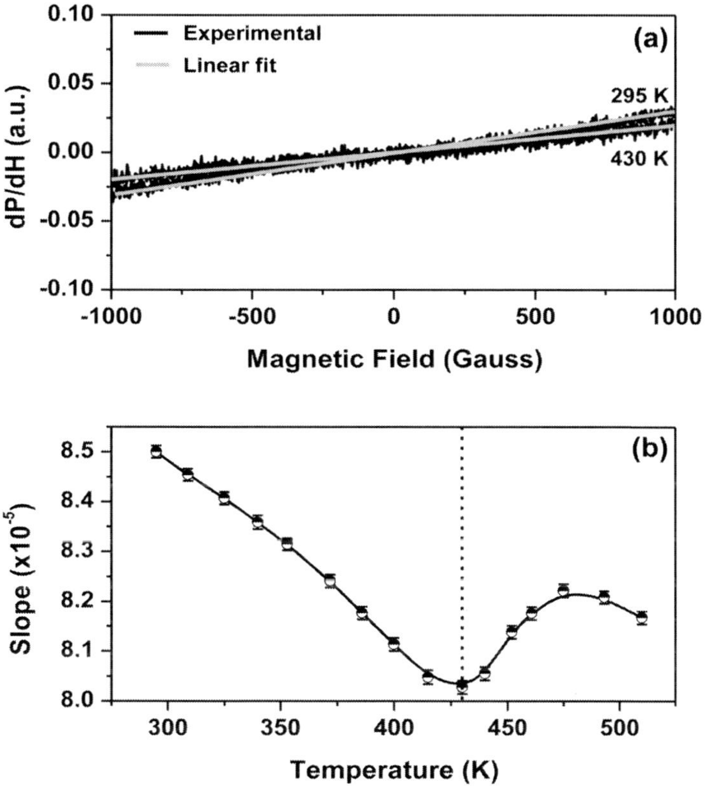

Figure 21. (a) LFMA spectra for 295 K and 430 K, with H_{mod}= 0.63 G and microwave power of 1 mW; the solid lines correspond to the fits obtained of Eq.(2). (b) The temperature dependence of the absorption line slope for temperature interval of 300-510 K; the curves connecting points are only guides for the eye.

This decreasing behavior is characteristic of a paramagnetic phase [34], in a good agreement with the EPR results. After T_{min}, when the temperature continues to increase, the slope increases until 475 K; for T> 475 K a slow decrease in slope is observed. A very similar behavior has been also observed in the magnetoelectric PFN [34]. We associate the minimum (T_{min}) of this profile with the onset of the ferro-paraelectric transition, as it is suggested by dielectric measurements [32]; and which is accompanied for a local non-centrosymmetry [67]. Where it leads to a discreet redistribution of the Cr^{+3} ions, originating a change in the relative orientation of magnetic moments; and it affects the superexchange interactions between Cr ions, generating a change in the microwave absorption dynamics. On the other hand, it is necessary to mention that the microwave absorption is related with the imaginary part of the magnetic permeability [70]. As the material under study is a dielectric, the only current which can flow is the Maxwell displacement current (E_{in}). Since the sample, in the resonant cavity, is located in the place of maximum ac magnetic field [33,34], the microwave absorption must decrease when the dielectric permittivity (ε) increases to satisfy the boundary conditions ($D_{normal}= \varepsilon E_{in}$ and $E_{out}= D_{out} \sim 0$; where D is the displacement). Therefore, the maximum in dielectric constant should coincide with the minimum of the microwave absorption close to the ferro-paraelectric transition. The Curie temperature (T_{min}= 430 K) obtained by the technique LFMA is in a good agreement with our dielectric measurements

[32]. The whole profile of the slope of the LFMA line depends on the thermal dependence of the absorption dynamics of the paramagnetic moments around ferro-paraelectric transition.

Additionally, the microwave absorption dynamics shows characteristic of a diffuse phase transition, i.e., the transition region is extended throughout a wide temperature range. This characteristic can be ascribed to that the relaxor materials are highly inhomogeneous materials in which a diffuse transition is due to a multitude of local phase transitions with different Curie temperatures in a broad range of temperature [71], as it is shown in $YCrO_3$.

Conclusion

In this work is shown that EPR is the most powerful spectroscopic method available for determine the valence state of paramagnetic ions, local structural information, and the nature of magnetic phases at different temperatures. Additionally, this technique has also allowed the study of the ferro-paraelectric transition in magnetoelectric materials.

MAMMAS and LFMA techniques has provided information on the dependences in temperature and magnetic field of the non-resonant microwave absorption in magnetoelectric materials. More important, these techniques can distinguish between different dissipative dynamics of microwave absorbing centers, providing valuable information about the nature of magnetic and/or electric ordering within materials. We have shown that these techniques are powerful tools for the research of magnetoelectric materials at microwave frequencies.

Acknowledgments

G. Alvarez acknowledges a fellowship from CONACyT-Mexico (Retention Program). Support from project IPN-SIP-20110979 is gratefully acknowledged.

References

[1] H. Schmid, *Ferroelectrics* 162 (1994) 317.
[2] C.N.R. Rao and Claudy Rayan Serrao, *J. Mater. Chem.* 17 (2007) 4931.
[3] C.-W. Nan, M.I. Bichurin, S. Dong, D. Viehland and G. Srinivasan, *J. Appl. Phys.* 103 (2008) 031101.
[4] K.F. Wang, J.-M. Liu and Z.F. Ren, *Adv. Phys.* 58 (2009) 321.
[5] M. Fiebig, Th. Lottermoser, D. Fröhlic, A.V. Goldzev and R.V. Pisarev, *Nature (London)* 419 (2002) 818.
[6] D. Khomskii, *Physics* 2 (2009) 20.
[7] D.N. Astrov, *Sov. Phys. JETP* 11 (1966) 708.
[8] G.T. Rado and V.J. Folen, *J. Appl. Phys.* 33S (1962) 1126.
[9] I.E. Chupis, *Low Temp. Phys.* 36 (2010) 477.
[10] T.-H. Wang, C.-S. Tu, H.-Y. Chen, Y. Ding, T.C. Lin, Y.-D. Yao, V.H. Schmidt and K.-T. Wu, *J. Appl. Phys.* 109 (2011) 044101.
[11] J.F. Scott, *Nat. Mater.* 6 (2007) 256.

[12] G. Srinivasan, *Annu. Rev. Mater. Res.* 40 (2010) 153.
[13] Z.J. Huang, Y. Cao, Y.Y. Sun, Y.Y. Xue and C.W. Chu, *Phys. Rev. B* 56 (1997) 2623.
[14] X.S. Gao, X.Y. Chen, J. Yin, J. Wu, Z.G. Liu and M. Wang, *J. Mater. Sci.* 35 (2000) 5421.
[15] N. Hur, I.K. Jeong, M.F. Hundley, S.B. Kim and S.-W. Cheong, *Phys. Rev. B* 79 (2009) 134120.
[16] Y. Du, Z.X. Cheng, X.L. Wang, P. Liu and S.X. Dou, *J. Appl. Phys.* 109 (2011) 07B507.
[17] G. A. Smolenskii and I. E. Chupis, *Sov. Phys. Usp.* 25 (1982) 475.
[18] Y. Yang, J.-M. Liu, H.B. Huang, W.Q. Zou, P. Bao, Z.G. Liu, *Phys. Rev. B* 70 (2004) 132101.
[19] R. Font, G. Alvarez, O. Raymond, J. Portelles and J.M. Siqueiros, *Appl. Phys. Lett.* 93 (2008) 172902.
[20] M. Chandra Sekhar and N. Venkata Prasad, *Ferroelectrics* 345 (2006) 45.
[21] C.H. Townes and A.L. Schawlow, *Microwave Spectroscopy,* Dover Publications, New York, 1996.
[22] D. Ramírez-Rosales, R. Zamorano-Ulloa and O. Pérez-Martinez, *Solid State Commun.* 118 (2001) 371.
[23] G. Alvarez, R. Font, J. Portelles, O. Raymond and R. Zamorano, *Solid State Sci.* 11 (2009) 881.
[24] G. Alvarez, in: Herlinda Montiel, Guillermo Alvarez (Eds.), Advances in Non-crystalline Solids: Metallic Glass Formation, Magnetic Properties and Amorphous Carbon Films, Transworld Research Network, Kerala, India, 2010.
[25] John A. Weil, James R. Bolton and John E. Wertz, Electron Spin Resonance: *Elementary Theory and Practical Applications,* John Wiley and Sons Inc., New York, 1994.
[26] J.R. Pilbrow, *Transition Ion Electron Paramagnetic Resonance,* Clarendon Press, Oxford, 1990.
[27] G. Alvarez, H. Montiel, J.F. Barron, M.P. Gutierrez and R. Zamorano, *J. Magn. Magn. Mater.* 322 (2010) 348.
[28] G. Alvarez, R. Font, J. Portelles, R. Valenzuela and R. Zamorano, *Physica B* 384 (2006) 322.
[29] G. Alvarez, R. Zamorano, J. Heiras, M. Castellanos and R. Valenzuela, *J. Magn. Magn. Mater.* 316 (2007) e695.
[30] G. Alvarez, R. Font, J. Portelles and R. Valenzuela, *Mater. Lett.* 62 (2008) 1737.
[31] G. Alvarez, H. Montiel, J.A. Peña, M.A. Castellanos and R. Zamorano, *J. Alloys Compd.* 508 (2010) 471.
[32] G. Alvarez, H. Montiel, M.P. Cruz, A.C. Duran and R. Zamorano, *J. Alloys Compd.* 509 (2011) L331.
[33] G. Alvarez and R. Zamorano, *J. Alloys Compd.* 369 (2004) 231.
[34] G. Alvarez, R. Font, J. Portelles, R. Zamorano and R. Valenzuela, *J. Phys. Chem. Solids* 68 (2007) 1436.
[35] G. Alvarez, Detection of phase transitions by magnetically modulated microwave absorption spectroscopy, in: Jacob I. Levine (Ed.), Magnetic Materials: Research, Technology and Applications, *Nova Science Publishers,* New York, USA, 2009.

[36] G. Alvarez, H. Montiel, M.A. Castellanos, J. Heiras and R. Valenzuela, *Materials Science and Engineering B* 150 (2008) 175.

[37] G. Alvarez, M.P. Cruz, A.C. Durán, H. Montiel and R. Zamorano, *Solid State Commun.* 150 (2010) 1597.

[38] R. Blinc, P. Cevc, A. Zorko, J. Holc, M. Kosec, Z. Trontelj, J. Pirnat, N. Dalal, V. Ramachandran and J. Krzystek, *J. Appl. Phys.* 101 (2007) 033901.

[39] T. Okamura, Y. Torizuka and Y. Kojima, *Phys. Rev.* 82 (1951) 285.

[40] G.A. Smolenskii, A.I. Agranovskia, S.N. Popov and V.A. Isupov, *Sov. Phys-Teach. Phys.* 3 (1958) 1981.

[41] V.A. Bokov, I.E. Mylnikova and G.A. Smolenskii, *Sov. Phys. JEPT.* 15 (1962) 447.

[42] O. Raymond, R. Font, J. Portelles, N. Suárez-Almodovar and J. M. Siqueiros, *J. Appl. Phys.* 97 (2005) 084107.

[43] O. Raymond, R. Font, N. Suárez-Almodovar, J. Portelles and J. M. Siqueiros, *J. Appl. Phys.* 99 (2006) 124101.

[44] L.E. Cross, S. Jang, R.E. Newnham, S. Nmura and K. Uchino, *Ferroelectrics* 23 (1980) 18.

[45] Chintamani N. R. Rao, *Phase transitions in solids: an approach to the study of the chemistry and physics of solids,* McGraw-Hill Inc., USA, 1978.

[46] Chonggui Zhong and Qing Jiang, *Solid State Commun.* 122 (2002) 601.

[47] Qing Jiang and Chong-Gui Zhong, *Phys. Lett. A* 306 (2002) 166.

[48] M.P. Gutiérrez, G. Alvarez, H. Montiel, R. Zamorano and R. Valenzuela, *J. Magn. Magn. Mater.* 316 (2007) e738.

[49] F.J. Owens, *J. Phys. Chem. Solids* 58 (1997) 1311.

[50] A.I. Shames, E. Rozenberg, G. Gorodetsky, J. Pelleg and B.K. Chandhuri, *Solid State Commun.* 107 (1998) 91.

[51] P. Aleshkevych, M. Baran, S.N. Barilo, J. Fink-Finowicki and H. Szymczak, *J. Phys.: Condens. Matter.* 16 (2004) L179.

[52] Frank J. Owens, *J. Phys. Chem. Solids.* 66 (2005) 793.

[53] H. Montiel, G. Alvarez, I. Betancourt, R. Zamorano and R.Valenzuela, *Appl. Phys. Lett.* 86 (2005) 072503.

[54] J.P. Joshi, R. Gupta, A.K. Sood, S.V. Bhat, A.R. Raju and C.N.R. Rao, *Phys. Rev. B* 65 (2002) 024410.

[55] V.A. Ivanshin, J. Deisenhofer, H.A. Krug von Nidda, A. Loidl, A. Mukhin, J. Balbashov and M.V. Eremin, *Phys. Rev. B* 61 (2000) 6213.

[56] A. Falqui, N. Lampis, A. Geddo-Lehmann and G. Pinna, *J. Phys. Chem. B* 109 (2005) 22967.

[57] N. Lampis, C. Franchini, G. Satta, A.G. Lehmann and S. Massidad, *Phys. Rev. B* 69 (2004) 064412.

[58] W.Z. Zhu, A. Kholkin, P.Q. Mantas and J.L. Baptista, *J. Eur. Ceram. Soc.* 20 (2000) 2029.

[59] E. Granado, N.O. Moreno, H. Martinho, A. Garcia, J.A. Sanjurjo, I. Torriani, C. Rettori, J.J. Neumeier and S.B. Oseroff, *Phys. Rev. Lett.* 86 (2001) 5385.

[60] M. Deri, Ferroelectric Ceramics, Gordon and Breach, New York, 1969.

[61] Y. Xu, *Ferroelectric Materials and their Applications,* North-Holland, Amsterdam, 1991.

[62] M.E. Mata-Zamora, H. Montiel, G. Alvarez, J.M. Saniger, R. Zamorano and R. Valenzuela, *J. Magn. Magn. Mater.* 316 (2007) e532.

[63] H.Y. Hwang, S.-W. Cheong, N.P. Ong and B. Batloog, *Phys. Rev. Lett.* 77 (1996) 2041.

[64] W.C. Koehler, E.O. Wollan and M.K. Wilkinson, *Phys. Rev.* 118 (1960) 58.

[65] F.J. Owens, *J. Phys. Chem. Solids* 66 (2005) 793.

[66] C.R. Serrao, A.K. Kundu, S.B. Krupanidhi, U.V. Waghmare and C.N.R. Rao, *Phys. Rev. B* 72 (2005) 220101.

[67] K. Ramesha, A. Llobet, T. Proffen, C.R. Serrao and C.N.R. Rao, *J. Phys.: Condens. Matter* 19 (2007) 102202.

[68] A. Durán, A.M. Arévalo-López, E. Castillo-Martínez, M. García-Guaderrama, E. Moran, M.P. Cruz, F. Fernández and M.A. Alario-Franco, *J. Solid State Chem.* 183 (2010) 1863.

[69] J.W. Kim, Y.S. Oh, K.S. Suh, Y.D. Park and Kee Hoon Kim, *Thermochim. Acta* 455 (2007) 2.

[70] C.A. Balanis, Advanced Engineering Electromagnetics, first ed., John Wiley and Sons, USA, 1989.

[71] D. Viehland, S.J. Jang and L.E. Cross, *J. Appl. Phys.* 68 (1990) 2916.

In: Ferroelectrics: New Research
Editors: Xui Li Huang and Su Lyn Ma

ISBN: 978-1-61942-282-7

Chapter 3

PREPARATION AND PIEZOELECTRIC PROPERTIES OF $K_{0.5}NA_{0.5}NBO_3$-BASED LEAD-FREE PIEZOELECTRIC CERAMICS

K. W. Kwok and Dumin Lin
Department of Applied Physics and Materials Research Centre,
The Hong Kong Polytechnic University, Kowloon, Hong Kong, China

ABSTRACT

For environmental protection reasons, potassium sodium niobate ($(K_{0.5}Na_{0.5})NbO_3$)-based ceramics have been considered as promising alternatives to lead-based materials which are currently used in majority of ultrasonic applications. However, it is very difficult to process the ceramics to full density by ordinary sintering under atmospheric conditions, which deteriorates their properties and restricts their uses in industrial productions. In this chapter, the preparation of KNN-based ceramics with well-densified structure, good piezoelectric and dielectric properties are reported. Novel sintering aids CuO and MnO are used to improve the densification and hence the electrical properties of the ceramics. The additional hardening effects of the sintering aids are also discussed. The piezoelectric properties of the ceramics are further enhanced via the formation of new solid solutions with other ferroelectrics, such as $LiSbO_3$, $LiTaO_3$, $BaTiO_3$ and $Ba(Zr_{0.05}Ti_{0.95})O_3$. After the formation of new solid solutions, the ferroelectric tetragonal-ferroelectric orthorhombic phase transition is shifted to lower temperatures, such that the ceramics contain both the tetragonal and orthorhombic structures at room temperature. Owing to the more possible polarization states arising from the coexistence of the two phases, the piezoelectric and dielectric properties of the ceramics are enhanced significantly.

INTRODUCTION

Piezoelectric ceramics based on lead zirconate titanate solid solution, $Pb(Zr,Ti)O_3$ (PZT), have been widely used in actuators and sensors as well as microelectronic devices because of their excellent ferroelectric and piezoelectric properties. However, as lead is highly toxic and will vaporize during processing of the ceramics, there is a rising concern about manufacture as well as disposal of products containing PZT. A number of lead-free piezoelectric ceramics such as $BaTiO_3$-based ceramics [1], $Bi_{0.5}Na_{0.5}TiO_3$-based ceramics [2,3], Bi-layered structure ceramics [4,5], tungsten bronze-type ceramics [6] and alkaline niobate-based ceramics [7-19] have then been studied extensively for replacing the lead-containing ceramics in various applications. Owing to the low Curie temperatures for $BaTiO_3$-based ceramics, the high coercive fields and low depolarization temperatures for $Bi_{0.5}Na_{0.5}TiO_3$-based ceramics, the large crystal anisotropy and poor piezoelectricity for Bi-layered structure ceramics, and the difficulty in densification by ordinary processing and low piezoelectricity for tungsten bronze-type ceramics, the use of these materials in practical applications have been restricted.

On the other hand, because of the high Curie temperature (above 400°C) and good piezoelectric properties, $K_{0.5}Na_{0.5}NbO_3$ (KNN) has been considered a promising alternative to lead-based piezoelectric ceramics. It is a solid solution of ferroelectric $KNbO_3$ and antiferroelectric $NaNbO_3$ with the composition K/Na = 50/50 close to the morphotropic phase boundary. A dense and well-sintered KNN ceramic (e.g., prepared by the hotpressing technique) possesses a high density (ρ = 4.46 g/cm^3) and good piezoelectric properties (piezoelectric coefficient d_{33} = 160 pC/N and planar-mode electromechanical coupling coefficient k_p = 0.45) [7,8]. However, because of the high volatility of alkaline elements at high temperatures, it is very difficult to obtain dense and well-sintered KNN ceramics using a conventional sintering process, and the ceramics usually exhibit worse piezoelectric properties (d_{33} = 80 pC/N, k_p = 0.36, and ρ = 4.25 g/cm^3) [8,20]. A number of studied have been carried out to improve the sinterability as well as the piezoelectric properties of the KNN ceramics; these include the use of sintering aids and the formation of solid solutions of KNN with other ABO_3 perovskite oxides.

In this chapter, the preparation of KNN-based ceramics with well-densified structure and good piezoelectric properties using a conventional sintering technique is presented. Novel sintering aids CuO and MnO_2 are used to improve the densification of the ceramics [21-23]. The piezoelectric properties of the ceramics are further enhanced via the formation of new solid solutions with other ferroelectrics, such as $LiSbO_3$, $LiTaO_3$, $BaTiO_3$ and $Ba(Zr_{0.05}Ti_{0.95})O_3$ [24-27].

KNN-Based Ceramics with Sintering Aids

CuO and MnO_2 are typical sintering aids widely used for dielectric ceramics owing to their low melting temperatures and formation of liquid phases during sintering [28-31]. They may also enter the lattices, producing the hardening effect on the ceramics. In this section, the preparation and electrical properties of KNN-based ceramics with sintering aids CuO and MnO_2 are presented, and the resulting effects on the physical properties of the ceramics are discussed [21-23]. $K_{0.5}Na_{0.5}NbO_3$ + x mol% CuO (KNN-Cu-x) and $K_{0.5}Na_{0.5}Nb_{0.94}Sb_{0.06}O_3$ + x mol% MnO_2 (KNNS-Mn-x) ceramics were prepared by a conventional ceramic fabrication technique. The experimental details of the fabrication process, including oxide-mixing, calcination and sintering at atmospheric pressure, have been reported and can be found elsewhere.

Densification of the Ceramics

Figures 1 and 2 show the SEM micrographs of the KNN-Cu-x and KNNS-Mn-x ceramics, respectively. It can be seen that the KNN-Cu-0 (i.e. KNN) ceramic (sintered at 1100°C for 4 h) is not well-densified and contains a small amount of pores (Figure 1a). The relative density (determined by the Archimedes method) of the ceramic is low, about 92%. The grains are large, having a diameter in the range of 2 to 2.5 μm. After the addition of CuO, the ceramics become denser and the grains are slightly smaller (Figures 1b - 1d). The relative density of the ceramics increases to around 95%. These clearly show that the addition of CuO is effective in promoting the densification of KNN ceramics. In general, CuO is used to improve the sintering performance of microwave ceramics [32] and lead-based ceramics [33-36]. Owing to the formation of a liquid phase, the sintering temperature is reduced (about 100 to 250°C) and the densification is improved considerably for those ceramics. For the KNN-Cu-x ceramics, the sintering temperature decreases from 1100 to 1060°C as x increases from 0 to 2.0. As shown in Figures 1c - 1d, a liquid phase is formed in the ceramics with large amount of CuO. It is hence suggested that the reduction in sintering temperature as well as the improvement in densification should be attributed to the formation of liquid phase at the grain boundary.

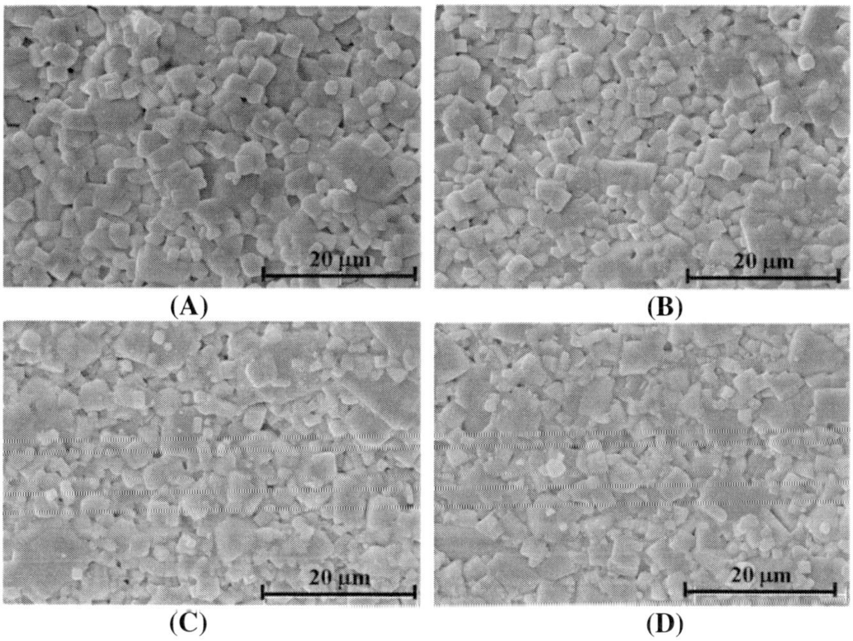

Figure 1. SEM micrographs of the KNN-Cu-x ceramics: (a) x = 0.0, (b) x = 1.0, (c) x = 1.5, and (d) x = 20.

Figure 2. SEM micrographs of the KNNS-Mn-x ceramics: (a) x = 0.0, and (b) x = 1.50.

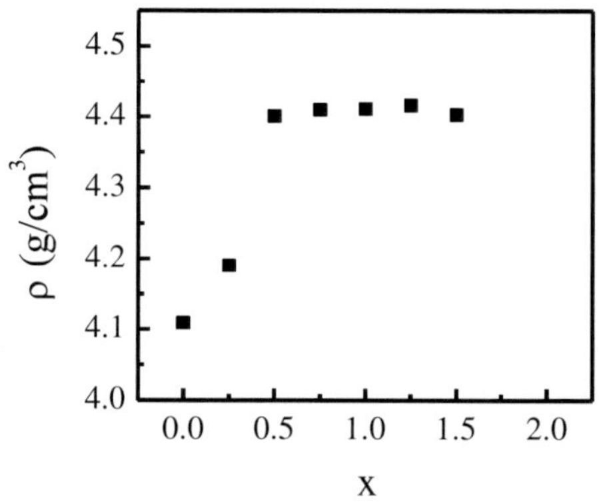

Figure 3. Variation of the relative density ρ with x for the KNNS-Mn-x ceramics.

Similar to the KNN-Cu-x ceramics, the KNNS-Mn-x ceramics become denser as compared to the KNNS-Mn-0 ceramics (Figure 2). The variation of the density ρ with x for the ceramics is shown in Figure 3. It can be seen that the observed ρ increases with increasing x, and becomes saturated at a small value of x (i.e., 0.50). The relative density of the ceramics with x ≥ 0.50 is about 97%. These indicate that MnO_2 is an effective sintering aid and a small amount is enough to improve the densification of the ceramics [28-29,37]. However, unlike CuO, no liquid phase is observed in the KNNS-Mn-x ceramics although the sintering temperature also decreases after the addition of MnO_2, e.g., from 1130 to 1100°C as x increases from 0 to 1.50.

All the KNN-Cu-x ceramics possess a pure perovskite structure with orthorhombic symmetry (the XRD patterns are indexed by JCPDS card 32-0822), and no secondary phase is observed (Figure 4). There are also no significant differences between the diffraction patterns. Probably due to the small amount, the liquid phase of CuO cannot be detected in the XRD. The single-phase structure may also suggest that CuO has diffused in to the KNN lattices, with Cu^{2+} (0.73 Å) entering the Nb^{5+} (0.68 Å) sites. Similar results of a single-phase structure have also been observed for the KNNS-Mn-x ceramics. It has been known that Mn^{4+} (MnO_2) is stable at low temperatures (<1020°C), while Mn^{3+} (0.66 Å) can stably exist at high temperatures such as 1100-1200°C [38]. Accordingly, part of Nb^{5+} at the B sites of the KNN lattices may be replaced by Mn^{3+}. In both series of ceramics, the substitution of Cu^{2+} and

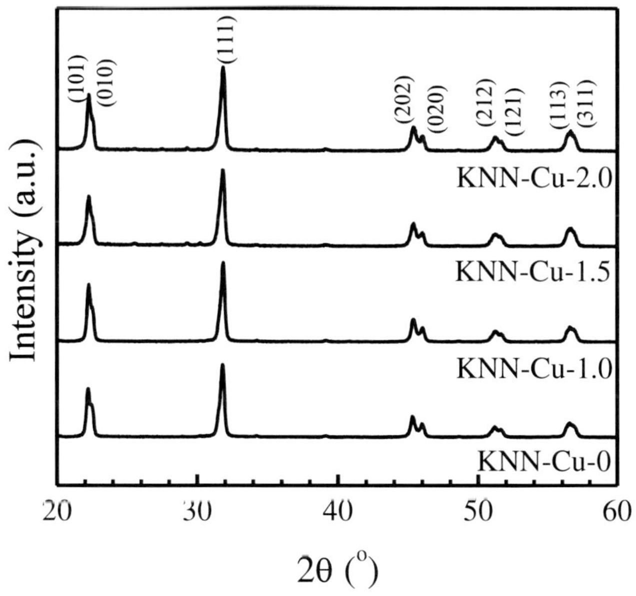

Figure 4. XRD patterns of the KNN-Cu-x ceramics with x = 0.0, 1.0, 1.5 and 2.0.

Mn^{3+} ions for Nb^{5+} would create oxygen vacancies, which promote the lattice diffusion and then the grain growth.

Dielectric, Ferroelectric and Piezoelectric Properties of the Ceramics

As shown in Figure 5, all the KNN-Cu-x ceramics undergo the cubic-tetragonal phase transition at T_C (~410°C) and the tetragonal-orthorhombic phase transition at $T_{O\text{-}T}$ (~200°C). Although the observed T_C decreases slightly with increasing x, the transition peaks are still sharp, indicating that the KNN-Cu-x ceramics retain the classic ferroelectric characteristics after the addition of CuO. It is also seen that the observed $T_{O\text{-}T}$ decreases about 25°C after the addition of CuO, and then remains almost unchanged with increasing x. Similar to CuO, the addition of MnO_2 does not introduce significant effects on the phase transitions of the KNNS-Mn-x ceramics. However, owing to the substitution of Sb^{5+} for Nb^{5+}, the KNNS-Mn-x ceramics exhibit a lower T_C (~310°C) and $T_{O\text{-}T}$ (~130°C).

The relative permittivity ε_r and loss tangent tan δ measured at room temperature and 1 kHz using an impedance analyzer (HP 4192A) for the KNN-Cu-x and KNNS-Mn-x ceramics are shown in Figure 6. Because of the shift of the tetragonal-orthorhombic phase transition towards room temperature, the observed ε_r of the KNNS-Mn-0 ceramic becomes higher than that of the KNN-Cu-0 ceramic (840 vs 400). As shown in Figure 6, CuO and MnO_2 introduce similar effects on the dielectric properties of the ceramics, causing the observed ε_r and tanδ to decrease with increasing x and then to remain almost unchanged at x > 0.5. This indicates that, owing to the lower valence than Nb^{5+}, both Cu^{+} and Mn^{2+} act also as an acceptor dopant causing the ceramics to become "hardened". It should be noted that the observed tanδ for the KNN-Cu-x and KNNS-Mn-x ceramics with x ≥ 0.5 is very low, about 0.4% and 1.6%, respectively, suggesting that the ceramics are promising candidates for practical applications.

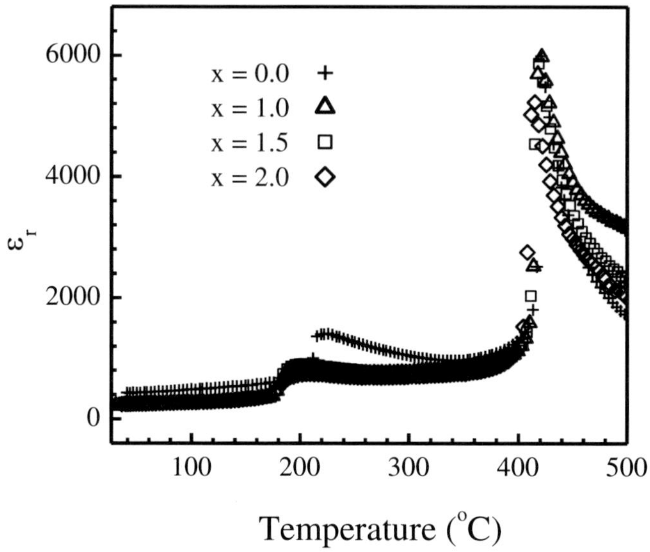

Figure 5. Temperature dependences of the relative permittivity ε_r at 1 kHz for the KNN-Cu-x ceramics with x = 0.0, 1.0, 1.5 and 2.0.

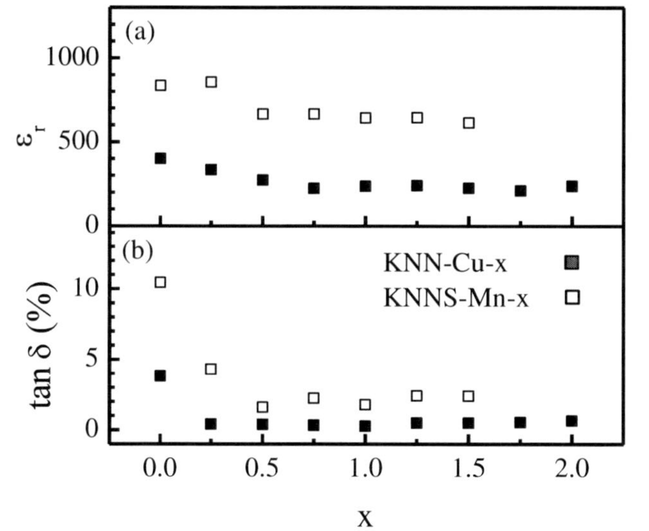

Figure 6. Variations of (a) the relative permittivity ε_r and (b) loss tangent tanδ at 1 kHz for the KNN-Cu-x and KNNS-Mn-x ceramics.

The piezoelectric properties, such as the longitudinal piezoelectric coefficient d_{33}, planar mode electromechanical coupling coefficient k_p and the corresponding quality factor Q_m, of the ceramics are evaluated, giving the results shown in Figure 7. The measurements are carried out one day after a 30-min. poling under a dc field of 5-6 kV/mm at 180-200°C. Similarly, because of the substitution of Sb^{5+}, the KNNS-Mn-0 ceramics exhibits a higher d_{33} than the KNN-Cu-0 ceramic (150 vs 110 pC/N, Figure 7a). Sb has a much larger Pauling electronegativity than Nb (2.05 vs. 1.5). Therefore the partial substitution of Sb^{5+} for Nb^{5+} induces a much stronger covalency in the KNN perovskite which favors the ferroelectricity

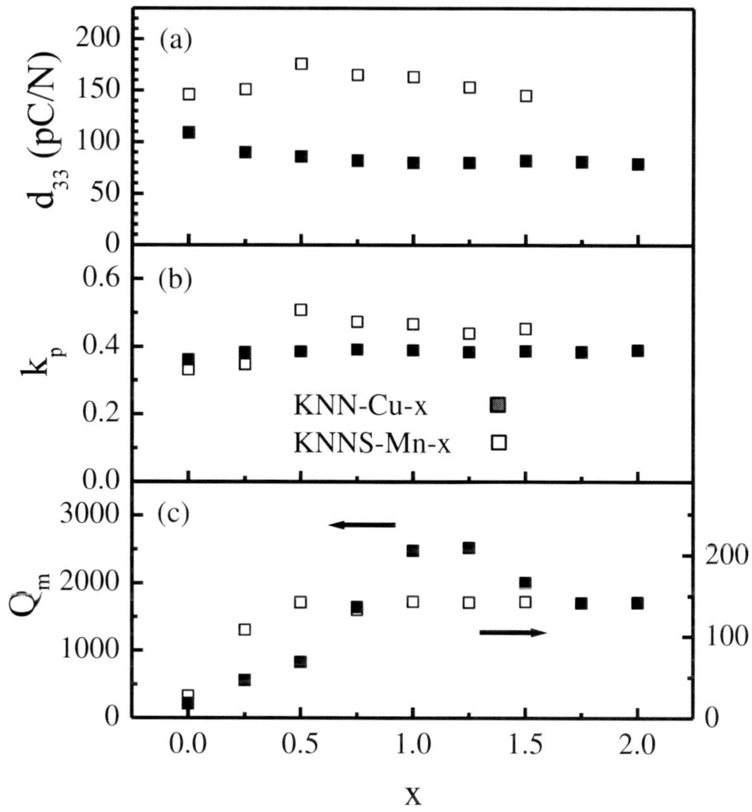

Figure 7. Variations of (a) the longitudinal piezoelectric coefficient d_{33}, (b) planar mode electromechanical coupling coefficient k_p and (c) quality factor Q_m with x for the KNN-Cu-x and KNNS-Mn-x ceramics.

[14,39]. In addition, the ionic radius of Sb (0.62 Å) is smaller than that of Nb (0.69 Å). The substitution of Sb may then give rise to a larger off-centering of the cation, and thus improve the ferroelectricity and hence the piezoelectricity [39]. However, as shown in Figures 7b - 7c, there is no significant difference between their k_p and Q_m, which are determined by the resonance mode and then are not dependent only on the piezoelectric coefficients of the ceramics.

As a result of the acceptor doping effect, the observed d_{33} for the KNN-Cu-x ceramic decreases while Q_m increases with increasing x (Figures 7a & 7c). Probably due to the dependence on other material parameters, such as elastic stiffness, the observed k_p remains almost unchanged with increasing x (Figure 7b). It should be noted that the maximum Q_m achieved in the KNN-Cu-1.25 ceramic is extraordinarily high, about 2500, which is much higher than those of the commercial hard PZT 8 ceramics [40] and comparable to those of some bismuth layer-structured ferroelectrics [41]. Similar results of the enhancement in Q_m resulting from the improved densification and normal acceptor doping effect have been reported for CuO-added lead-based piezoelectric ceramics. However, the improvement is not very large, e.g. only about 25% (from 850 to 1050) for the $Pb((Zn_{1/3}Nb_{2/3})_{0.2}(Zr_{0.50}Ti_{0.50})_{0.8})O_3$ ceramics added with 1.5 wt% CuO [33]. This suggests that there should be other mechanism (which will be discussed in the following section) for the great enhancement of Q_m (from 210 to 2520) for the KNN-Cu-x ceramics. Unlike the KNN-Cu-x ceramics, the observed d_{33}, k_p and Q_m for the KNNS-Mn-x ceramics increase with increasing x. This should be attributed to the improved densification, of which the resulting effect becomes dominant over the normal acceptor doping effect.

Double Hysteresis Loops

Double polarization hysteresis (P-E) loop is a typical feature exhibited by antiferroelectric materials, such as $PbZrO_3$ and high-Zr-content $Pb(Zr,Ti)O_3$ ceramics [42,43]. It is caused by the antiferroelectric-ferroelectric phase transformation. However, it has also been known that after aging, a number of acceptor-doped ferroelectric titanates also exhibit a double P-E loop [44-47]. In general, it is believed that the double P-E loops are caused by a constriction of the polarization. For the rhombohedral-structured PZT ceramics, the polarization is constricted by the random octahedral strains induced by local oxygen octahedral rotations [43]. For the acceptor-doped ferroelectric titanates, defect dipoles are the causes for the constriction [44-46]. Indeed, aging is an essential process for the ferroelectric titanates to go through before exhibiting the double-loop characteristics, which involves a gradual stabilization of the existing ferroelectric domain pattern [43-47]. Various theories such as the grain-boundary theory, surface-layer model, domain-wall theory and volume theory have been proposed for the domain stabilization process [47]. Recently, based on the domain switching studies, it has been shown that aging in acceptor-doped $BaTiO_3$ ferroelectrics is a volume effect [44-46,48]. By extending the symmetry conforming principle of point defects firstly developed for ferroelastic/martensite, Ren [44] suggested that the symmetry of short-range order distribution of point defects (dopant/impurity ions and vacancies) generated by acceptor doping will follow the crystal symmetry of the ferroelectrics when in equilibrium. As a result, because of the non-centric distribution of charged defects, defect dipoles will be formed (after some time, i.e. aging) and provide a restoring force to reverse the switched polarization (i.e. constriction of the polarization). In the following, the double-loop-like characteristics of the KNN-Cu-x ceramics are presented. The underlying mechanism for the double PE loops and the resulting effects on the electromechanical properties, in particular the extraordinarily high Q_m, of the ceramics are discussed.

The polarization hysteresis (P-E) loops measured at 100 Hz using a conventional Sawyer-Tower circuit for the KNN-Cu-x ceramics are shown in Figure 8. As usual, a well-saturated and square-like single P-E loop, with a saturation (P_s) and remanent (P_r) polarization of 28 and 20 $\mu C/cm^2$, respectively, is observed for the KNN-Cu-0 ceramic (i.e. a pure KNN ceramic), However, the P-E loop for the KNN-Cu-0.25 ceramic becomes slightly flattened and constricted, giving $P_s = 23.6\ \mu C/cm^2$ and $P_r = 15.9\ \mu C/cm^2$. As x increases to 0.50, a double P-E loop is observed. As shown in Figure 8, the double-loop-like characteristics become more pronounced with increasing x, and a completely constricted P-E loop is observed for the KNN-Cu-x ceramics with $x \geq 1.0$.

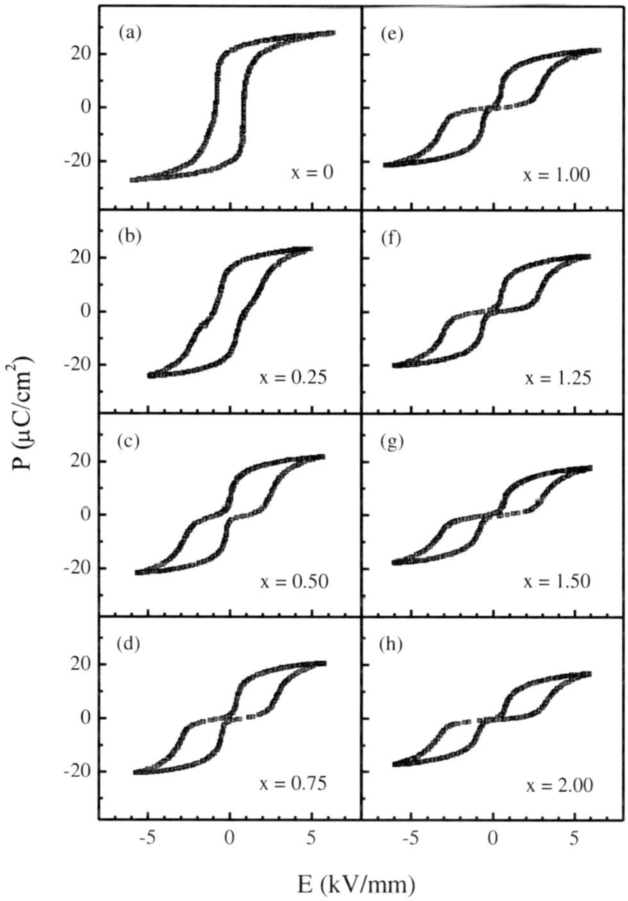

Figure 8. P-E loops of the KNN-Cu-x ceramics measured at 100 Hz and room temperature: (a) x = 0, (b) x = 0.25, (c) x = 0.50, (d) x = 0.75, (e) x = 1.00, (f) x = 1.25, (g) x = 1.50, and (h) x = 2.00.

Although KNN and $BaTiO_3$ have different crystal symmetries (orthorhombic vs tetragonal), they have the same perovskite structures. It is hence expected that the symmetry-conforming principle of point defects proposed for $BaTiO_3$ [44] is also applicable to KNN ceramics. After the doping of Cu^{2+} (for the B-site ions Nb^{5+}), O^{2-} vacancies are formed (for satisfying the requirement of charge neutrality). In the ferroelectric phase, the B-site ion is not located in the center of the oxygen octahedral. As a result, the (statistical) distribution of O^{2-} vacancies around a dopant ion Cu^{2+} is not symmetric and affected greatly by the lattice symmetry. At equilibrium, the symmetry of short-range order distribution of defects (or defect symmetry) follows the crystal symmetry of KNN. The resulting non-centric distribution of defects (positively charged O^{2-} vacancies and negatively charged dopant ions Cu^{2+}) forms defect dipoles P_d along the spontaneous polarization P_s direction. In the P-E loop measurements, the polarization P_s is switched by the external field (abruptly) at 100 Hz. Due to insufficient time for the ions/vacancies to migrate, the defect dipoles P_d remain in the original orientation, thus providing a restoring force to reverse the switched polarization upon removal of the external field. As a result of the restoration of the initial multi-domain pattern (which has zero macroscopic/averaged polarization), a double loop for the KNN-Cu-x ceramics (x ≥ 0.5) is observed (Figures 8c - 8i). For the KNN-Cu-0.25 ceramic, probably due to the low concentration of defects, the restoring force is not large enough to reverse completely the switched polarization and hence a slightly constricted P-E loop is observed (Figure 8b).

Unlike $BaTiO_3$, the double P-E loop is observed in "unaged" KNN-Cu-x ceramics, instead of aged ceramics. This should be attributed to their different phase transitions. $BaTiO_3$ undergoes a cubic-tetragonal phase transition at ~128 °C. The transition is diffusionless and does not involve any exchange of ions [44-46]. As a result, the distribution of the charged defects remains in the cubic symmetry even though the crystal has transformed to a tetragonal symmetry. It takes time (called aging) for the defect symmetry to correct into a polar tetragonal symmetry. The aging time for $BaTiO_3$ is usually long, e.g., 5 days at 80 °C or 28 days at room temperature [44-46]. On the other hand, KNN-Cu-x undergoes a tetragonal-orthorhombic phase transition at ~200°C (Figure 5). It is a diffuse phase transition [49], suggesting that the defects may migrate during the crystal transformation. Moreover, due to the high transition temperature, the defects may have a high migration rate and hence can settles in a distribution with the same symmetry as the crystal after the transformation. Accordingly, aging is not required for the ceramic.

Figure 9 shows, as examples, the P-E loops measured at 100 Hz and different temperatures for the KNN-Cu-1.0 ceramic. The constricted P-E loop starts to open at 130°C (Figure 9c) and becomes fully open at 190°C (Figure 9e). This suggests that the defect dipoles can response to the external field at high temperatures (≥ 130°C) and switch even along with the polarization at temperatures ≥ 190°C. This also provides evidence that the defects can migrate during the diffuse phase transition (Figure 5). It should be noted that at temperatures ≥ 190°C, the distribution of defects may be in a tetragonal symmetry.

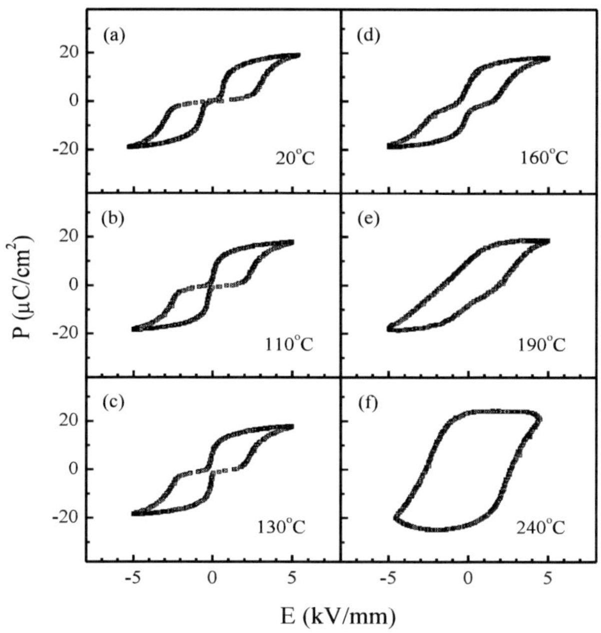

Figure 9. P-E loops of the KNN-Cu-1.0 ceramic measured at 100 Hz and different temperatures: (a) 20°C, (b) 110°C, (c) 130°C, (d) 160°C, (e) 190°C, and (f) 240°C.

To provide additional evidence for the defect migration, the KNN-Cu-1.0 ceramic was annealed at 350 and 150 °C, respectively, for 2 h and then quenched to room temperature. It is expected that because of fast cooling, the defects can be frozen in the original distribution. The P-E loops measured at 100 Hz for the quenched sample are shown in Figure 10. There is only a slight constriction in the P-E loop for the sample quenched from 350°C (Figure 10a), suggesting that the restoring force for reversing the switched polarization is weak. As the defect symmetry is frozen in a cubic or tetragonal symmetry, there are few defect dipoles aligned with the orthorhombic crystal symmetry and hence the restoring force is weak. The P-E loop for the sample quenched from 150°C is completely constricted at E = 0 (Figure 10c), suggesting that the defect symmetry is the same as the crystal symmetry. This also implies that the defect symmetry has transformed to the orthorhombic symmetry after the diffuse phase transition of the ceramic at 200°C. The quenched samples were then aged at room temperature for 5 days. As shown in Figures 10b & 10d, all the aged samples exhibit a constricted P-E loop. This clearly shows that after aging, the defect symmetry for the sample quenched from 350°C has corrected to the orthorhombic symmetry.

As demonstrated in Figure 10, the aging time for the KNN-Cu-1.0 ceramics is relatively short, about 5 days (at room temperature). This implies that the migration rate of defects is not too low (at room temperature) and the defect dipoles may response to a slow-switching electric field. Figure 11 shows, as examples, the P-E loops measured at different frequencies for the KNN-Cu-0 and KNN-Cu-1.0 ceramics. It can be seen that the well-saturated and square-like P-E loop for the KNN-Cu-0 ceramic remains almost unchanged as the frequency decreases from 100 to 0.25 Hz (Figures 11a - 11c). At 0.1 Hz, the loop becomes slightly

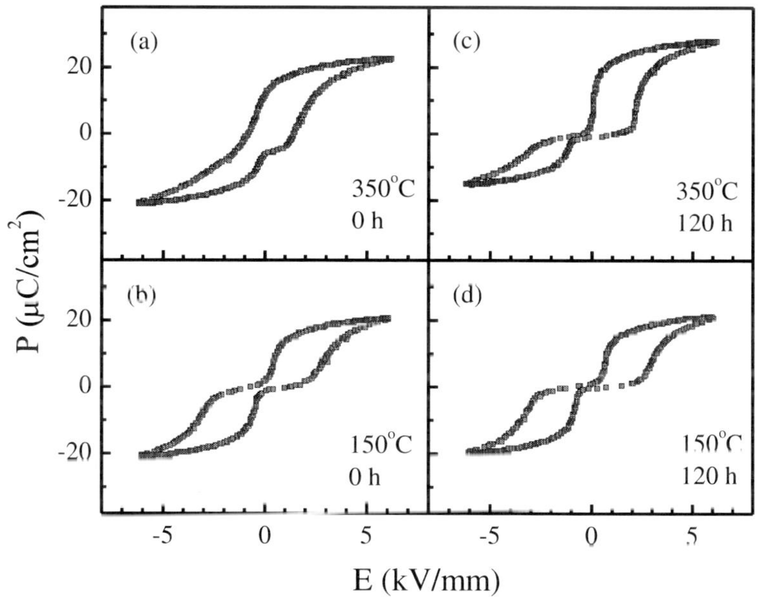

Figure 10. P-E loops of the quenched and aged KNN-Cu-1.0 ceramics measured at 100 Hz and room temperature: (a) as quenched from 350°C, (b) as quenched from 150°C, (c) quenched from 350°C and aged at room temperature for 120 h, and (d) quenched from 150°C and aged at room temperature for 120 h.

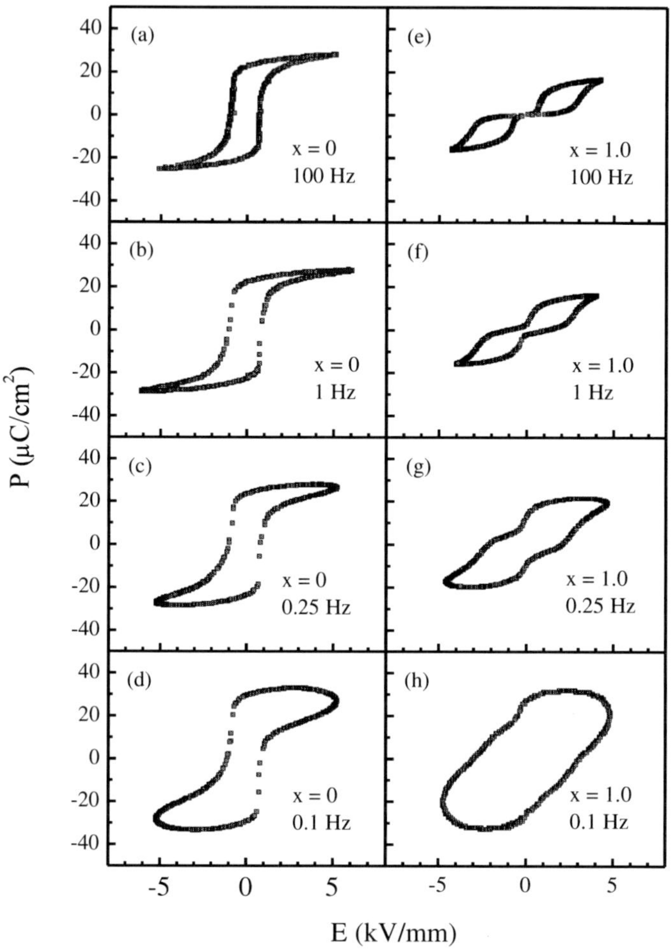

Figure 11. P-E loops of the KNN-Cu-0 and KNN-Cu-1.0 ceramics measured at different frequencies and room temperature: (a) x = 0 at 100 Hz, (b) x = 0 at 1 Hz, (c) x = 0 at 0.25 Hz, (d) x = 0 at 0.1 Hz, (e) x = 1.0 at 100 Hz, (f) x = 1.0 at 1 Hz, (g) x = 1.0 at 0.25 Hz, and (h) x = 1.0 at 0.1 Hz,.

"swelled" (Figure 11d). This should be due to the increase in electrical conduction at low frequencies. Unlike KNN-Cu-0, the double P-E loop for the KNN-Cu-1.0 ceramic "opens" gradually as the frequency decreases, and becomes a normal single loop at 0.1 Hz (Figures 11e - 11h). This clearly shows that under a slow-switching electric field (e.g. ≤ 0.1 Hz), the defects have enough time to migrate and hence the defect dipoles can rotate with the polarization. As a result, the restoring force decreases and the reversal of the switched polarization cannot be completed.

Indeed, electrical conduction at low frequencies would contribute to the opening and swelling of the P-E loops, in particular, at 0.1 Hz (Figure 11). In order to provide additional evidence for the net polarization produced after the P-E loop measurements, the piezoelectric coefficient d_{33} of the ceramics has been measured. Prior to the measurement, the samples have been short-circuited for 24 h to remove the conducting charges. Because of the large remanent polarization (Figure 10), the observed d_{33} of the KNN-Cu-0 ceramic is large (~100 pC/N) and does not change with the frequency (for the P-E measurements). For the KNN-Cu-

1.0 ceramic, the observed d_{33} remains almost zero as the frequency decreases from 100 to 1 Hz. This is in agreement with the completely constricted P-E loop. As the frequency decreases from 1 to 0.1 Hz, d_{33} increases from 0 to ~30 pC/N. This clearly indicates that there is a net polarization after the low-frequency P-E measurement.

As discussed, due to the low migration rates of defects, the defect dipoles cannot response to a fast-switching electric field and hence provide a restoring force to reverse the switched polarization. That is, the defect dipoles will "pin" the motion of ferroelectric dipoles, in particular for those at high frequencies, making them difficult to deviate from their original orientation. This "pinning" effect results in the strong hardening of the piezoelectric properties, in particular Q_m which is evaluated at a resonance frequency of ~300 kHz, as shown in Figure 7. It should be noted that the ceramics are poled under a dc field at 180-200°C for 30 min. So the defects can migrate easily and form defect dipoles along the direction of the aligned polarization.

SOLID SOLUTIONS OF KNN WITH OTHER FERROELECTRICS

It is well known that the morphotropic phase boundary (MPB) plays a very important role in the improvement of piezoelectric properties of lead-based perovskite piezoelectric ceramics, such as $Pb(Zr,Ti)O_3$ [50] and $Pb(Mg_{1/3}Nb_{2/3})O_3$-$PbTiO_3$ [51]. In general, the piezoelectric properties of the ceramics become maximum near the MPB region. MPB is defined as an abrupt structural change for a solid solution with variation in composition [52]. For the lead-based ceramics near the MPB, they contain both the rhombohedral and tetragonal phases and thus more possible polarization states, which consequently lead to the significant enhancement in the piezoelectric properties. On the basis of the reported work and the analogy of PZT-based ceramics, it is believed that the key approach for improving the piezoelectric properties of KNN-based ceramics is, by lowering the ferroelectric tetragonal-ferroelectric orthorhombic phase transition ($T_{O\text{-}T}$), to form coexistence of the tetragonal and orthorhombic phases at room temperature. This could be achieved by the formation of solid solutions of KNN with other ABO_3-type ferroelectrics. In this section, the preparation and electrical properties of novel solid solutions KNN-$LiSbO_3$ [24], KNNS-$LiTaO_3$ [25], KNN-$BaTiO_3$ [26] and KNN-$Ba(Zr_{0.05}Ti_{0.95})O_3$ [27] are presented. The phase transitions of the new solid solutions and the resulting effects on the piezoelectric properties are also discussed.

A conventional ceramic fabrication technique was used to prepare $(1\text{-}x)K_{0.5}Na_{0.5}NbO_3$-$xLiSbO_3$ (KNN-LS-x), $(1\text{-}x)K_{0.5}Na_{0.5}(Nb_{0.94}Sb_{0.06})O_3$–$xLiTaO_3$ (KNNS-LT-x), $(1\text{-}x)K_{0.5}Na_{0.5}NbO_3$-$xBaTiO_3$ (KNN-BT-x) and $(1\text{-}x)(K_{0.5}Na_{0.5})NbO_3$-$xBa(Zr_{0.05}Ti_{0.95})O_3$ (KNN-BZT-x), ceramics. The experimental details of the fabrication process, including oxide-mixing, calcination and sintering at atmospheric pressure, have been reported and can be found elsewhere [24-27]. Owing to the low melting temperature of Li-compounds (e.g., Li_2CO_3), a liquid phase is formed during the sintering and hence the densification of the KNN-LS-x and KNNS-LT-x ceramics is improved. On the other hand, CuO and MnO_2 are used to improve the densification of the KNN-BT-x and KNN-BZT-x ceramics, respectively. In general, all the ceramics are dense and possess a single-phase perovskite structure.

Phase Transitions of the Ceramics

The phase transitions of the ceramics are studied via the measurements of the relative permittivity ε_r as a function of temperature using an impedance analyzer (HP4192A). Figure 12 shows the temperature dependence of ε_r for the KNN-LS-x ceramics. It has been shown that $KNbO_3$ exhibits similar phase transitions as $BaTiO_3$, i.e., transforming from the cubic paraelectric to the tetragonal ferroelectric phase at 435°C (T_C) and from tetragonal ferroelectric to the orthorhombic ferroelectric phase at 225°C ($T_{O\text{-}T}$). The phase transitions for the solid solution of $KNbO_3$ with $NaNbO_3$ ($K_{0.5}Na_{0.5}NbO_3$) is similar, but with transition temperatures changed slightly [52,53]. In agreement with the reported work, the observed temperature dependence of ε_r for the KNN-LS-0 (i.e. KNN) ceramic reveals the two transitions, giving T_C = 421°C and $T_{O\text{-}T}$ = 226°C (Figure 12a). After the addition of $LiSbO_3$, the ceramics retain the ferroelectric characteristic, with T_C (Curie temperature) shifted to lower temperatures (Figures 12b - 12d). It can be seen that the ceramics also exhibit the tetragonal-orthorhombic phase transition. Unlike T_C, $T_{O\text{-}T}$ decreases much faster with increasing x. For the ceramics with $x \geq 0.06$, the tetragonal-orthorhombic phase transition becomes weak and shifted to temperatures below 100°C, which is shown clearly in the insets of Figures 12c - 12d. It has been shown that a partial substitution of A-site ions $(K_{0.5}Na_{0.5})^+$ by Li^+ can cause T_C to increase while $T_{O\text{-}T}$ to decrease [9]. On the other hand, a partial substitution of B-site ions Nb^{5+} by Ta^{5+} can decrease both T_C and $T_{O\text{-}T}$ rapidly [19,54]. Therefore, the co-substitution of A-site and B-site ions by Li^+ and Sb^{5+}, respectively, may result in different rates of decrease in T_C and $T_{O\text{-}T}$, as in our case shown in Figure 12.

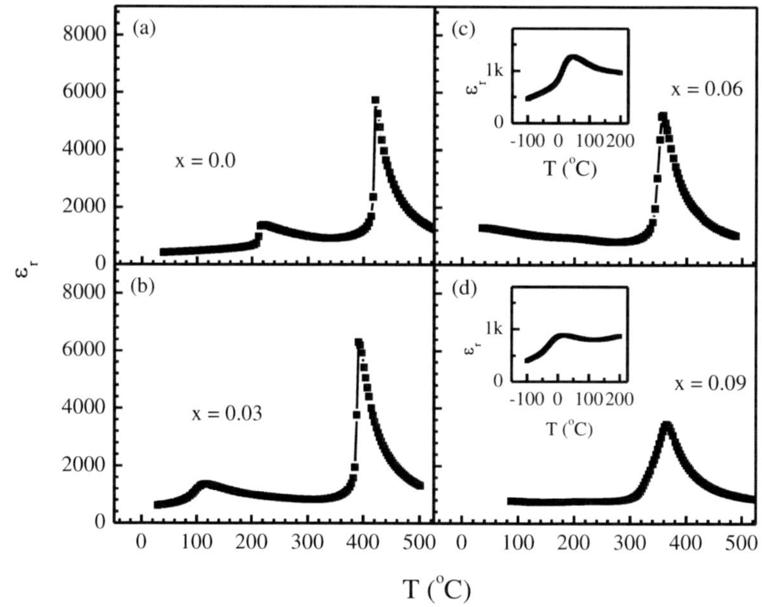

Figure 12. Temperature dependence of the relative permittivity ε_r at 10 kHz for the KNN-LS-x ceramics: (a) x = 0.0, (b) x = 0.03, (c) x = 0.06, and (d) x = 0.09. The insets show the dependences of ε_r on temperature T in the range of temperature from -100 to 200°C.

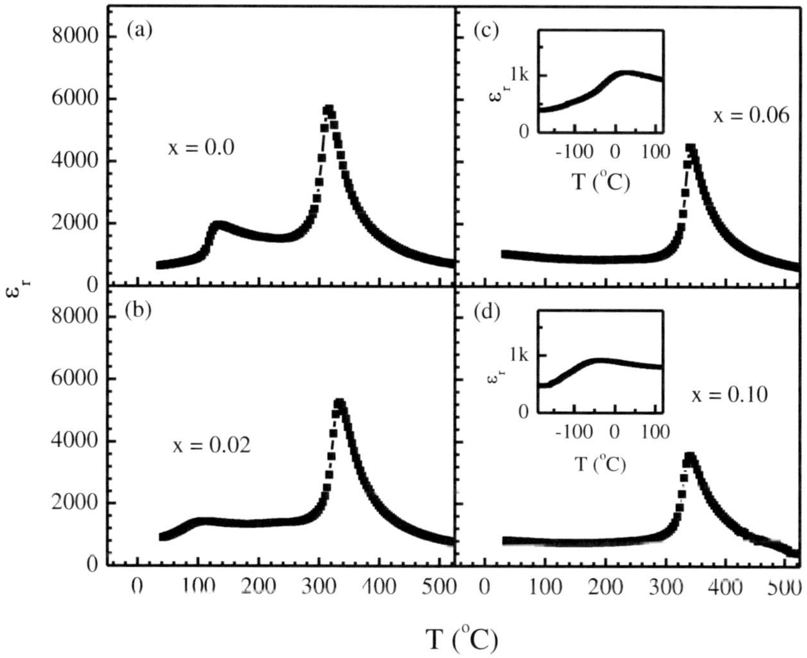

Figure 13. Temperature dependence of the relative permittivity ε_r at 10 kHz for the KNNS-LT-x ceramics: (a) x = 0.0, (b) x = 0.02, (c) x = 0.06, and (d) x = 0.10. The insets show the dependences of ε_r on temperature T in the range of temperature from -190 to 120°C.

Except for the KNN-BT-x ceramics with x > 0.06, all the KNNS-LT-x, KNN-BT-x and KNN-BZT-x ceramics retain the ferroelectric characteristics and undergo the cubic-tetragonal and tetragonal-orthorhombic phase transitions in the temperature range of -200 to 550°C (Figs 13-15). As shown in the phase diagrams (Figure 16), the observed $T_{O\text{-}T}$ for all the solid solutions decreases as x increases, resulting in some compositions with their $T_{O\text{-}T}$ occurred near the room temperature. This suggests that those compositions may contain both the tetragonal and orthorhombic phases at room temperature. It is also noted that the observed T_C for the KNN-LS-x and KNNS-LT-x ceramics remains almost unchanged, while those for the KNN-BT-x and KNN-BZT-x ceramics decrease linearly with increasing x.

The (room-temperature) XRD patterns for the KNN-LS-x ceramics are shown in Figure 17. It can be seen that a single-phase perovskite structure was formed in the ceramics. As the crystal structure of KNN (perovskite structure) is very different from $LiSbO_3$ (ilmenite structure) [55], the results suggest that Li^+ and Sb^{5+} have diffused into the KNN lattices, with Li^+ entering the $(K_{0.5}Na_{0.5})^+$ sites and Sb^{5+} occupying the Nb^{5+} sites, to form a homogeneous solid solution. As shown in the enlarged XRD patterns (Figure 17b), the ceramics with x ≤ 0.04 have an orthorhombic perovskite structure. As x (i.e., the concentration of $LiSbO_3$) increases, a tetragonal phase appears and increases continuously. At x ≥ 0.08, the ceramic becomes possessing the tetragonal phase only. These suggest that the (perovskite) tetragonal and orthorhombic phases coexist in the KNN-LS-x ceramics with 0.05 < x < 0.07, which is in agreement with the observation from the phase transition as shown in Figure 16a. Similar to $BaTiO_3$, the tetragonal phase of the ceramics is a primitive cell, while the orthorhombic phase is nonprimitive. It is also noted that the diffraction peaks shift slightly towards high diffraction angles as x increases. This may be attributed to the smaller ionic radii of Li^+ and Sb^{5+} than those of $(K_{0.5}Na_{0.5})^+$ and Nb^{5+}.

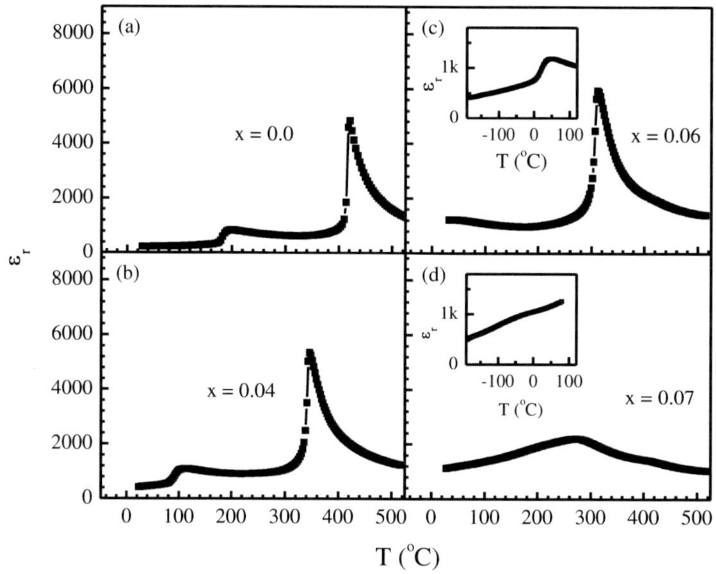

Figure 14. Temperature dependence of the relative permittivity ε_r at 10 kHz for the KNN-BT-x ceramics: (a) x = 0.0, (b) x = 0.04, (c) x = 0.06, and (d) x = 0.07. The insets show the dependences of ε_r on temperature T in the range of temperature from -190 to 120°C.

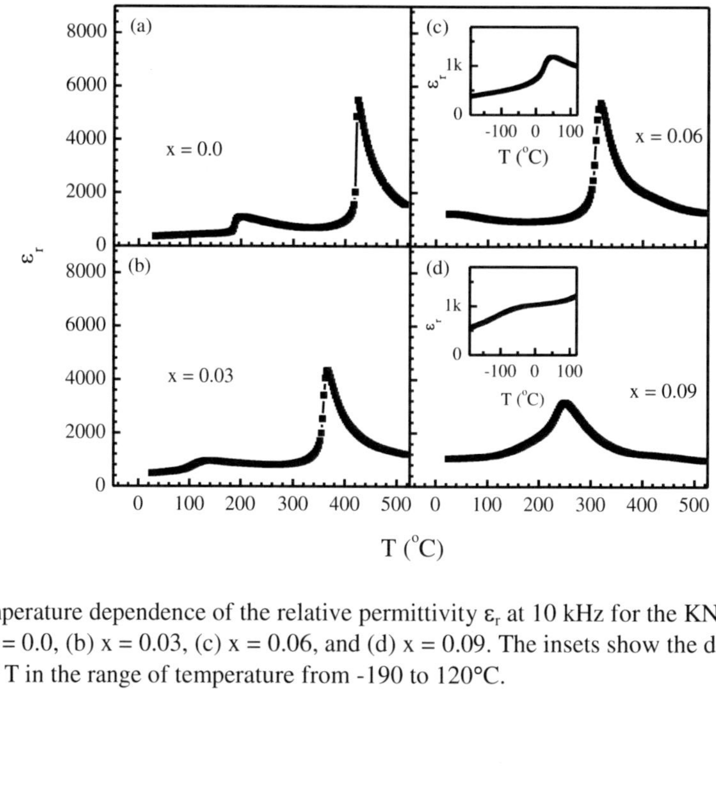

Figure 15. Temperature dependence of the relative permittivity ε_r at 10 kHz for the KNN-BZT-x ceramics: (a) x = 0.0, (b) x = 0.03, (c) x = 0.06, and (d) x = 0.09. The insets show the dependences of ε_r on temperature T in the range of temperature from -190 to 120°C.

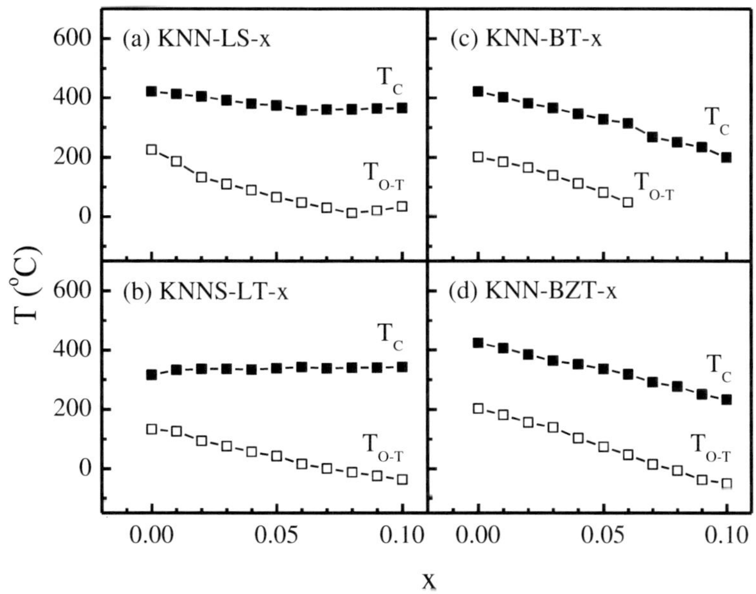

Figure 16. Phase diagram of the ceramics: (a) KNN-LS-x, (b) KNNS-LT-x, (c) KNN-BT-x, and (d) KNN-BZT-x.

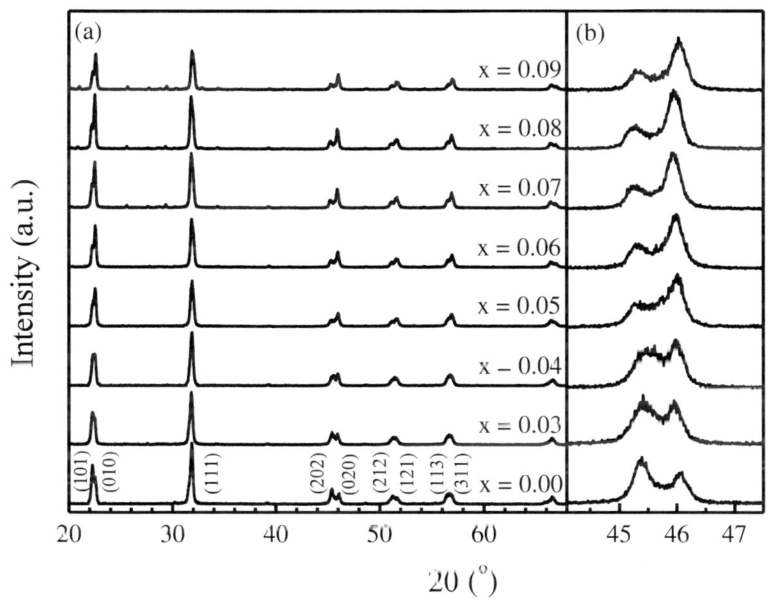

Figure 17. XRD patterns of the KNN-LS-x ceramics with x = 0.0, 0.03, 0.04, 0.05, 0.06, 0.07, 0.08 and 0.09 in the range of 2θ (a) from 20° to 70°, and (b) from 44° to 47.5°.

Similar to KNN-LS-x, the KNNS-LT-x, KNN-BT-x and KNN-BZT-x ceramics possess a single-phase perovskite structure (Figures 18-20), suggesting that homogeneous solid solutions are formed. As also revealed by the enlarged XRD patterns, both the tetragonal and orthorhombic phases exist in the KNNS-LT-x ceramics with $0.02 < x < 0.05$ (Figure 18b),

KNN-BT-x ceramics with 0.04 < x < 0.07 (Figure 19b) and the KNN-BZT-x ceramics with 0.04 < x < 0.07 (Figure 20b). Similarly, these results are in agreement with the observations from the phase diagrams as shown in Figures 16b - 16c.

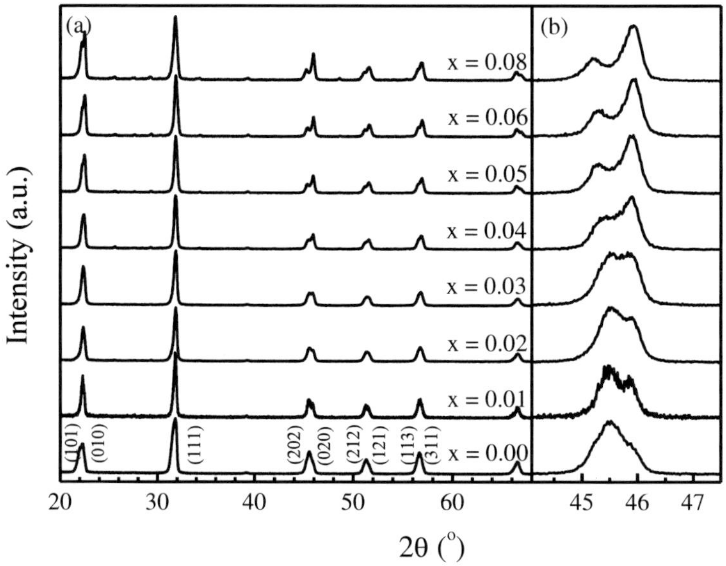

Figure 18. XRD patterns of the KNNS-LT-x ceramics with x = 0.0, 0.01, 0.02, 0.03, 0.04, 0.05, 0.06 and 0.08 in the range of 2θ (a) from 20° to 70°, and (b) from 44° to 47.5°.

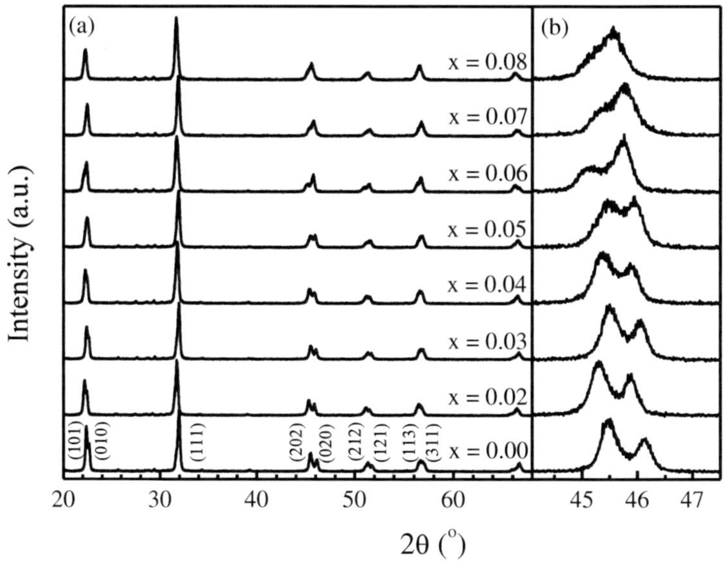

Figure 19. XRD patterns of the KNN-BT-x ceramics with x = 0.0, 0.02, 0.03, 0.04, 0.05, 0.06, 0.07 and 0.08 in the range of 2θ (a) from 20° to 70°, and (b) from 44° to 47.5°.

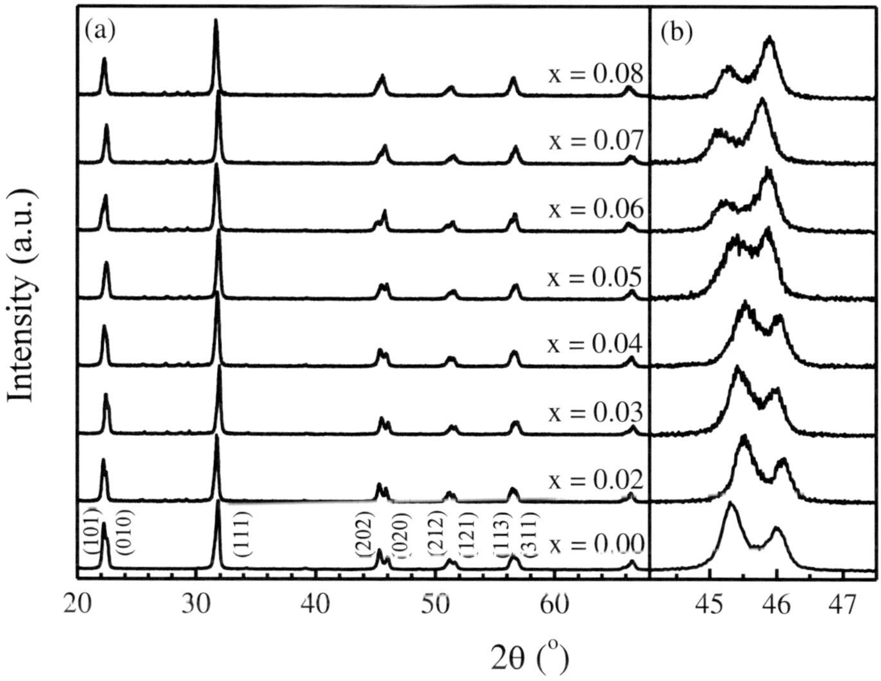

Figure 20. XRD patterns of the KNN-BZT-x ceramics with x = 0.0, 0.02, 0.03, 0.04, 0.05, 0.06, 0.07 and 0.08 in the range of 2θ (a) from 20° to 70°, and (b) from 44° to 47.5°.

Piezoelectric and Dielectric Properties of the Ceramics

The longitudinal piezoelectric coefficient d_{33} of all the ceramics are evaluated, giving the results shown in Figure 21. The measurements are carried out at room temperature one day after a 30-min. poling under a dc field of 4-5 kV/mm at 180-200°C. For each solid solution, the observed d_{33} increases with increasing x and then decreases, giving a maximum value which is more than 100% larger than that of a KNN ceramic. For example, the observed d_{33} for the KNN-BZT-0.06 and KNN-BZT-0 are 234 and 105 pC/N, respectively (Figure 21d). It can also be seen that, together with the results of XRD and phase diagram (Figures 16-20), the composition with the maximum d_{33} for each solid solution is the one containing both the tetragonal and orthorhombic phases. However, it has been noted that the observed remanent polarization P_r decreases with increasing x, suggesting that the enhancement in d_{33} should not be attributed to the change in the net polarization. P_r is proportional to the net polarization which is directly related to the piezoelectric properties of a polarized ceramic. Figure 22 shows, as examples, the variations of P_r with x for the KNN-LS-x and KNNS-LT-x ceramics, while Figure 23 shows the polarization hysteresis loops for the KNN-BT ceramics.

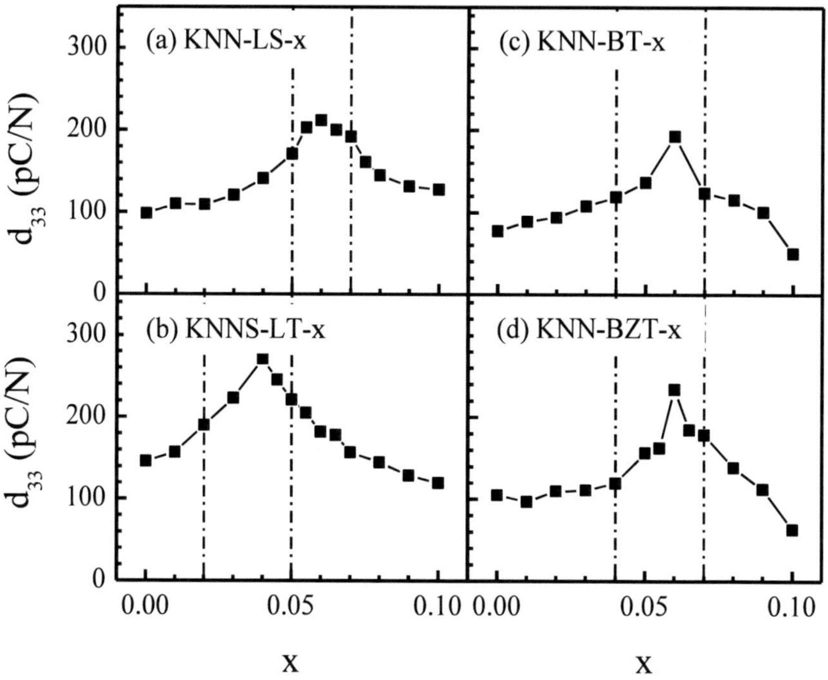

Figure 21. Variations of the longitudinal piezoelectric coefficient d_{33} with x for the ceramics: (a) KNN-LS-x, (b) KNNS-LT-x, (c) KNN-BT-x and (d) KNN-BZT-x. The two dot lines enclose the compositions containing both the tetragonal and orthorhombic phases.

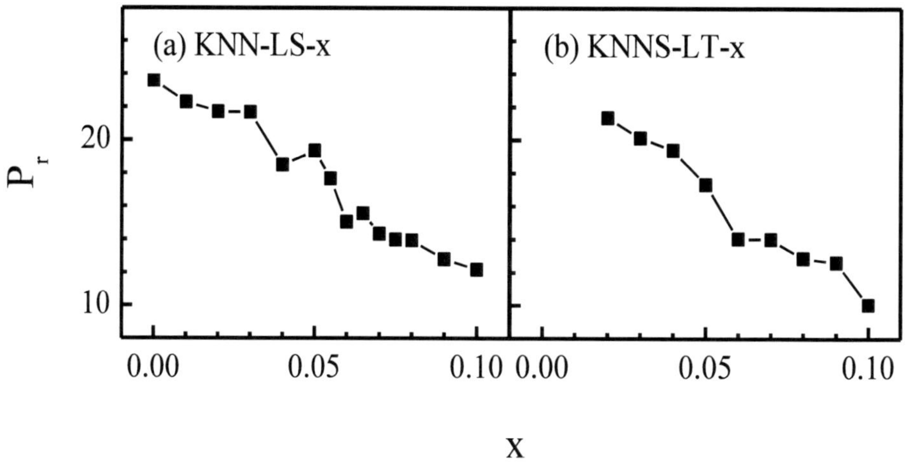

Figure 22. Variations of the remanent polarization P_r with x for the ceramics: (a) KNN-LS-x and (b) KNNS-LT-x.

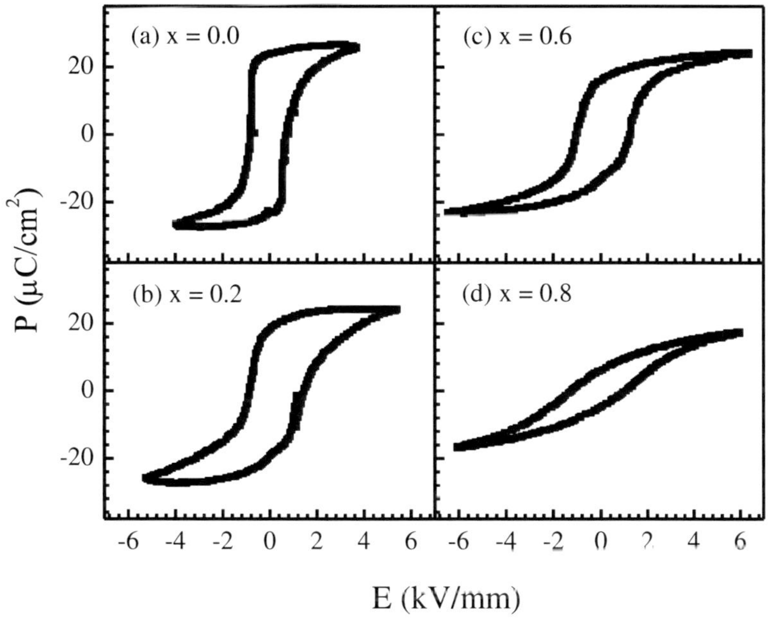

Figure 23. P-E loops of the KNN-BT-x ceramics with x = 0.0, 0.2, 0.06 and 0.08.

Besides P_r, d_{33} is also strongly dependently on ε_r. According to the thermodynamic theory [56], d_{33} can generally be expressed as a function of ε_r, the spontaneous polarization P_S (which can be approximated by P_r), and the electrostrictive coefficient Q_{11}:

$$d_{33} = 2Q_{11}\varepsilon_o\varepsilon_r P_S \qquad (1)$$

where ε_o is the permittivity of free space. Q_{11} is related to the domain structure and should not change significantly by doping. The variations of ε_r with x for all the ceramics are shown in Figure 24. It can be seen that the observed ε_r for the KNN-LS-x cand KNNS-LT-x ceramics exhibits similar dependence on x as d_{33}; i.e., it first increases and then decreases with increasing x, giving a maximum value at the composition containing both the tetragonal and orthorhombic phases (Figures 24a - 24b). Moreover, as shown in Figure 25, the variation of $\varepsilon_r P_r$ with x for these ceramics is very similar to that of d_{33}. Although the observed ε_r for the KNN-BT-x and KNN-BZT-x ceramics does not exhibit similar dependence with x, it increases continuously with increasing x and exhibits a local maximum at the composition with the tetragonal and orthorhombic phases coexisted (Figures 24c - 24d). It has been known that the piezoelectric properties of the PZT ceramics become maximum near the MPB (i.e., the Zr/Ti ratio = 53/47). A MPB is defined as an abrupt structural change for a solid solution with variation in composition [52]. Typically, the change is nearly independent of temperature, giving a nearly vertical line in the phase diagram such as those for the PZT and $Pb(Mg_{1/3}Nb_{2/3})O_3$-$PbTiO_3$ systems [50]. It is generally believed that the enhancement in piezoelectric properties of the ceramics near the MPB is mainly attributed to the more possible polarization states resulting from the coexistence of the two phases. Unlike PZT and other systems, the phase transitions for the KNN-based ceramics depend not only on the compositions but also on the temperature (Figure 16). Although the phase boundary between the orthorhombic and tetragonal phases may not be a MPB, it is believed to be the major origin for the enhancement in piezoelectric properties of the ceramics. Accordingly, together with our results (Figures 16 - 25), the great enhancement in d_{33} for the ceramics should be attributed to the coexistence of the orthorhombic and tetragonal phases, which produces more possible polarization states.

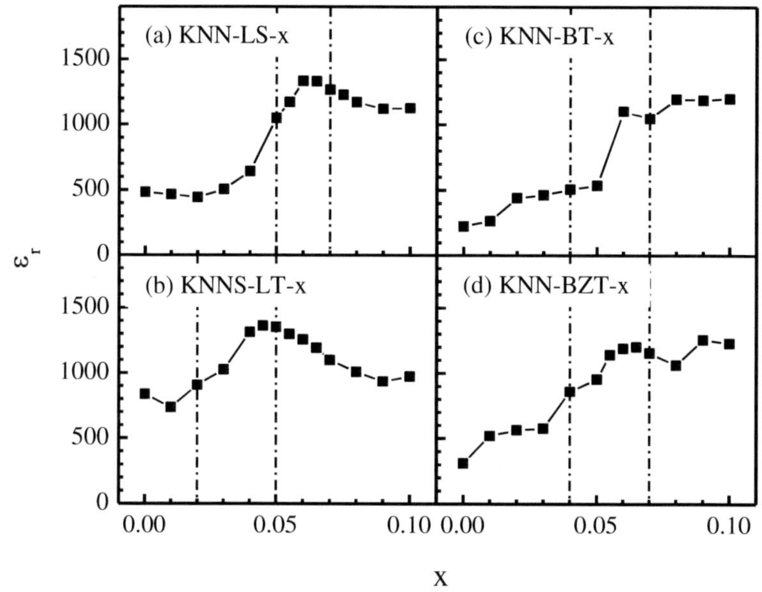

Figure 24. Variations of the relative permittivity ε_r with x for the ceramics: (a) KNN-LS-x, (b) KNNS-LT-x, (c) KNN-BT-x and (d) KNN-BZT-x. The two dot lines enclose the compositions containing both the tetragonal and orthorhombic phases.

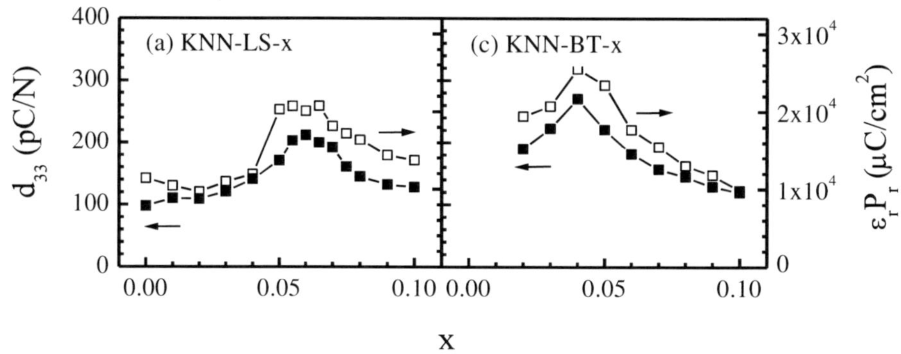

Figure 25. Variations of the longitudinal piezoelectric coefficient d_{33} and the product of the relative permittivity ε_r and remanent polarization P_r with x for the ceramics: (a) KNN-LS-x and (b) KNNS-LT-x.

CONCLUSION

KNN-based ceramics with well-densified structure, good piezoelectric and dielectric properties have been successfully prepared by a conventional sintering technique. Our results reveal that the addition of sintering aids and the formation of new solid solutions are effective approaches to improve the sintering performance as well as piezoelectric properties of KNN-based ceramics. CuO and MnO are effective sintering aids, a small amount of which can promote the densification of the ceramics to a high theoretical density of above 95%. Owing to the formation of defect dipoles, the KNN ceramics added with CuO become "hardened" considerably, and exhibit a completely constricted P-E loop and an extraordinarily high Q_m. The formation of new solid solutions with other ferroelectrics, such as $LiSbO_3$, $LiTaO_3$, $BaTiO_3$ and $Ba(Zr_{0.05}Ti_{0.95})O_3$, would shift the ferroelectric tetragonal-ferroelectric orthorhombic phase transition to low temperatures. As a result, the ceramics contain both the tetragonal and orthorhombic structures at room temperature. Owing to the more possible polarization states arising from the coexistence of the two phases, the piezoelectric and dielectric properties of the ceramics are enhanced significantly.

ACKNOWLEDGMENTS

The authors would like to acknowledge supports from the Postdoctoral Fellowship Grant (G-YX98), Research Committee (A-PK78) and the Centre for Smart Materials of The Hong Kong Polytechnic University.

REFERENCES

[1] Yu, Z.; Ang, C.; Guo, R; Bhalla, A. S. *J. Appl. Phys.* 2002, 92, 1489-1493.
[2] Takenaka, T.; Maruyama, K.; Sakata, K. *Jpn. J. Appl. Phys.* 1991, 30, 2236-2239.
[3] Lin, D.; Xiao, D.; Zhu, J.; Yu, P. *Appl. Phys. Lett.* 2006, 88, 062901.
[4] Jain, R.; Chauhan, A. K. S.; Gupta, V.; Sreenivas, K. *J. Appl. Phys.* 2005, 97, 124101.
[5] Karthik, C.; Ravishankar, N.; Varma, K. B. R. *Appl. Phys. Lett.* 2006, 89, 042905.
[6] Xie, R.; Akimune, Y.; Matsuo, K.; Sugiyama, T.; Hirosaki, N.; Sekiya, T. *Appl. Phys. Lett.* 2002, 80, 835-837.
[7] Egerton, L.; Dillon, D. M. *J. Am. Ceram. Soc.* 1959, 42, 438-422.
[8] Jaeger, R. E.; Egerton, L. *J. Am. Ceram. Soc.* 1962, 45, 209-213.
[9] Guo, Y.; Kakimoto, K.; Ohsato, H. *Appl. Phys. Lett.* 2004, 85, 4121-4123.
[10] Park, H. Y.; Ahn, C. W.; Song, H. C.; Lee, J. H.; Nahm, S.; Uchino, K.; Lee, H. G.; Lee, H. *J. Appl. Phys. Lett.* 2006, 89, 062906.
[11] Wang, R.; Xie, R.; Hanada, K.; Matsusak, K.; Bando, H.; Itoh, M. *Phys. Stat. Sol.* A 2005, 202, R57-R59.
[12] Hollenstein, E.; Davis, M.; Damjanovic, D.; Setter, N. *Appl. Phys.* Lett. 2005, 87, 182905.
[13] Dai, Y.; Zhang, X.; Zhou, G. *Appl. Phys. Lett.* 2007, 90, 262903.

[14] Saito, Y.; Takao, H.; Tani, T.; Nonoyama, T.; Takatori, K.; Homma, T.; Nagaya, T.; Nakamur, M. *Nature* 2004, 432, 84-87.

[15] Zang, G.; Wang, J.; Chen, H.; Su, W.; Wang, C.; Qi, P.; Ming, B.; Du, J.; Zheng, L.; Zhang, S.; Shrout, R. T. *Appl. Phys. Lett.* 2006, 88, 212908.

[16] Zhang, S.; Xia, R.; Shrout, T. R. *J. Appl. Phys.* 2006, 100, 104108.

[17] Zhang, S.; Xia, R.; Shrout T. R.; Zang, G.; Wang, *J. Sol. Stat. Communi.* 2007, 141, 675-679.

[18] Zuo, R.; Fang, X.; Ye, C. *Appl. Phys. Lett.* 2007, 90, 092904.

[19] Matsubara, M.; Kikuta, K.; Hirano, S. *J. Appl. Phys.* 2005, 97, 114105.

[20] Ahn, Z.S.; Schulze, W.A. *J. Am. Ceram. Soc.* 1987, 70, C18-C21.

[21] Lin, D.; Kwok, K.W.; Chan, H.L.W. J. *Phys. D: Appl. Phys.* 2008, 41, 045401.

[22] Lin, D.; Kwok, K.W.; Chan, H.L.W. *Appl. Phys. Lett.* 2007, 90, 232903.

[23] Lin, D.; Kwok, K.W.; Tian, H.Y.; Chan, H.L.W. *J. Am. Ceram. Soc.* 2007, 90, 1458–1462.

[24] Lin, D.; Kwok, K.W.; Lam, K.H.; Chan, H.L.W. *J. Appl. Phys.* 2007, 101, 074111.

[25] Lin, D.; Kwok, K.W.; Lam, K.H.; Chan, H.L.W. *J. Phys. D: Appl. Phys.* 2008, 41, 052002.

[26] Lin, D.; Kwok, K.W.; Chan, H.L.W. *J. Appl. Phys.* 2007, 102, 074113.

[27] Lin, D.; Kwok, K.W.; Chan, H.L.W. *Appl. Phys. Lett.* 2007, 91, 143513.

[28] Galassi, C.; Roncari, E.; Capiani, C.; Craciun, F. *J. Eur. Ceram. Soc.* 1999, 19, 1237-1241.

[29] Yu, C.S.; Hsieh, H.L. *J. Eur. Ceram. Soc.* 2005, 25, 2425-2427.

[30] Yang, C.F.; Wu, L.; Wu, T.S. *J. Mater. Sci.* 1992, 27, 6573-6578.

[31] Kagata, H.; Inoue, T.; Kato, J.; Kameyama, I. *Jpn. J. Appl. Phys.* 1992, 31, 3152-3155.

[32] Kim, D.W.; Ko, K.H.; Hong, K.S. *J. Am. Ceram. Soc.* 2001, 84, 1286-1290.

[33] Hou, Y.D.; Zhu, M.K.; Wang, H.; Wang, B.; Yan, H.; Tian, C.S. *Mater. Sci. Eng. B* 2004, 110, 27-31.

[34] Ahn, C.W.; Nahm, S.; Ryu, J.; Uchino, K.; Yoon, S.J.; Jung, S.J.; Song, J.S. *Jpn. J. Appl. Phys.* 2004, 43, 205-210.

[35] Chung, K.; Lee, D.; Yoo, J.; Jeong, Y.; Lee, H.; Kang, H. *Sensor Actuat. A* 2005, 121, 142-147.

[36] Ahn, C.W.; Song, H.C.; Nahm, S.; Park, S.H.; Uchino, K.; Lee, H.G.; Lee, H.J. *J. Am. Ceram. Soc.* 2006, 89, 921-925.

[37] Ahn, C.W.; Song, H.C.; Nahm, S.; Park, S.H.; Uchino, K.; Priya, S.; Lee, H.G.; Kang, N.K. *Jpn. J. Appl. Phys.* 2005, 44, L1361-L1364.

[38] Li, G.; Zheng, L.; Yin, R. *J. Appl. Phys.* 2005, 98, 064108.

[39] Tennery, V.J.; Hang, K.W. *J. Appl. Phys.* 1968, 39, 4749-4753.

[40] http://www.piezo-kinetics.com/navy_type_III.php

[41] Megriche, A.; Lebrun, L,; Troccaz, M, *Sensor Actuat. A* 1999, 78, 88-91.

[42] Haertling, G.H. *J. Am. Soc. Ceram.* 1999, 82, 797-818.

[43] Dai, X.; Xu, Z.; Viehlaud, D. *J. Am. Ceram. Soc.* 1995, 78, 2815-2827.

[44] Ren, X. Nat. Mater. 2004, 3, 91-94.

[45] Zhang, L.X.; Chen, W.; Ren, X. *Appl. Phys. Lett.* 2004, 85, 5658-5660.

[46] Liu, W.; Chen, W.; Yang, L.; Zheng, L.; Wang, Y.; Zhou, C.; Li, S.; Ren, X. *Appl. Phys. Lett.* 2006, 89, 172908.

[47] Lambeck, P.V.; Jonker, G.H. J. Phys. Chem. Solids 1986, 47, 453-461.

[48] Zhang, L.; Ren, X. *Phys. Rev. B* 2006, 73, 094121.
[49] Cho, D. H.; Ryu, M. K.; Park, S. S.; Cho, S. Y.; Choi, J. G.; Jang, M. S.; Kim, J. P.; Cho, C. R. *J. Korean Phys. Soc.* 2005, 46, 151-154.
[50] Noheda, B.; Cox, D. E.; Shirane, G.; Guo, R.; Jones, R.; Cross, L. E. *Phys. Rev. B* 2000, 63, 014103.
[51] Noheda, B.; Cox, D. E.; Shirane, G.; Gao, J.; Ye, Z. G. *Phys. Rev. B* 2002, 66, 054104.
[52] Jaffe, B.; Cook, W. R.; Jaffe, H. Piezoelectric Ceramics; Academic: London, 1971, pp. 53–64, 135–183, and 185-194.
[53] Maeder, M. D.; Damjanovic, D.; Setter, N. J. *Electroceram.* 2004, 13, 385-392.
[54] Matsubara, M.; Yamaguchi, T.; Sakamoto, W.; Kikuta, K.; Yogo, T.; Hirano, S. *J. Am. Ceram. Soc.* 2005, 88, 1190-1196.
[55] Kumada, N.; Takahashi, N.; Kinomura, N. J. *Solid State Chem.* 1996, 126, 121-126.
[56] Damjanovic, D. *Rep. Prog. Phys.* 1998, 61, 1267-1324.

In: Ferroelectrics: New Research
Editors: Xui Li Huang and Su Lyn Ma
ISBN: 978-1-61942-282-7

Chapter 4

RECENT PROGRESS ON LEAD-FREE THIN FILMS

D. Y. Wang and S. Li
School of Materials Science and Engineering,
The University of New South Wales, Sydney, Australia

1. INTRODUCTION

In the past, innovations towards the design of actuators, sensors and ultrasonic transducers were the driving force for the rapid development of piezoelectric ceramics. The high performance piezoelectric activity remains a property confined to lead-based systems, with $Pb(Zr_xTi_{1-x})O_3$ [PZT] as the nominal industry standard. [1] At present, the industry faces the obstacle of environmental regulations, as the high toxicity of lead present in the widely used PZT family (~60%-lead content) of piezoelectrics poses a severe threat to human health. The commercial electronics industry is currently in the process of a transition to lead-free products. Legislations regarding the waste containing lead pollutants have been in force in all member states of European Unions and Japan, as detailed by the directives on waste from electrical and electronic equipment (WEEE), restriction of hazardous substances (RoHS) and end-of life vehicles (ELV). [2] Some states of the United States and China have also followed the trend of lead-free transition and started exploring legislation to reduce the use of lead in electronic industries. It is expected that similar regulations like the EU-RoHS will be established soon in Australia, New Zealand, and Thailand etc. [3]

In order to circumvent the legislated restrictions on the use of toxic lead for civil applications in the near future, bulk lead-free piezoelectric materials have been extensively investigated to explore the potential candidates with electrical properties comparable to their lead-based counterparts. It is no surprise that the search for lead-free piezoelectric ceramics is described as one of the pillars in today's roadmap for ceramics. [3] To date, the reported lead-free piezoelectric ceramics are of perovskite-type, tungsten bronze-type and bismuth layer-structured-type. Perovskite-type lead-free ceramics with an ABO_3 crystal structure seem to be good candidates to replace PZT and have been actively researched in recent years. Two perovskite families of lead-free materials have currently emerged as the most promising candidates to replace PZT: (1) $(Bi_xNa_{1-x})TiO_3$ [BNT] and (2)$(K_xNa_{1-x})NbO_3$ [KNN]. [3-6]

Although no lead-free materials with an overall piezoelectric performance comparable to PZT systems have been found so far, recent studies have shown that both BNT and KNN-based systems do exhibit some advantages over lead-based materials for specific applications. [7]

From the scientific point of view, the intrinsic nature of the high piezoelectric and ferroelectric properties in PZT system are presented in relation to their respective Curie temperatures (T_C) and the existence of a morphotropic phase boundary (MPB). [8] This implication, to some extent, provides a perspective and introduces guidelines on the development of novel lead-free piezoelectrics. In fact, there were very early considerations in the 1970s to replace lead in piezoceramics. Nevertheless, the blossom of searching for lead-free piezoceramics just started in the early 2000's, in particular, the BNT and KNN-based systems, as shown in Figure 1, providing statistics for refereed publications on BNT and KNN -based piezoceramics from 2004 up to now. Current efforts in pursuing and designing lead-free piezoelectric ceramics involve three different approaches: (1) Searching for new lead-free piezoelectric materials empirically, using phenomenological models (structure-property relations) and first principle calculations; (2) Improving the performance of existing lead-free materials by elaborating substitutions and dopants or introducing MPB; (2) Perfecting the existing lead-free materials with textured growth (controlling structural instabilities and extrinsic contributions). Analogous to PZT, enhanced properties are noted for MPB compositions in the BNT binary and ternary systems. The consequence of a ferroelectric to anti-ferroelectric transition well below T_C, however, limits their usefulness. Though comparable with respect to T_C, the high levels of piezoelectricity reported in the KNN family are the result of enhanced polarizability associated with the orthorhombic-tetragonal polymorphic phase transition (PPT) being compositionally shifted downward. Although one can argue that there is room for improvements in bulk, the fact that so far no lead-free ceramics possess similar properties as their lead containing counterparts have stimulated new strategies in the development of novel lead-free piezoelectric materials.

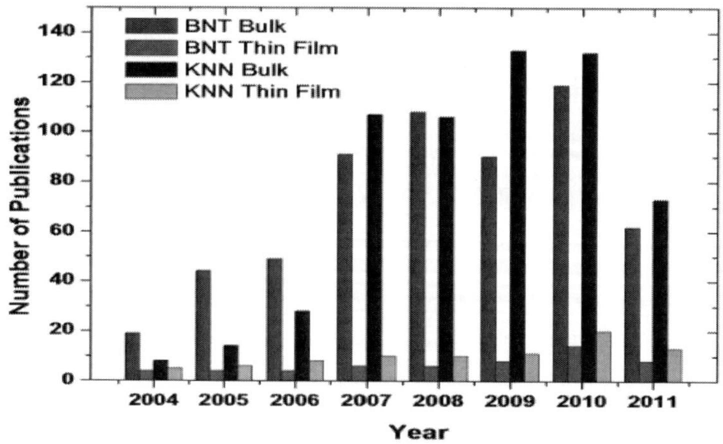

Figure 1. Number of publications on lead-free BNT and KNN-based piezoelectric materials in refereed journals for the time range from 2004 to July 2011. The statistic was created by counting the relevant papers citing Saito et al [9], Takenaka et al [10-11], Wang et al [12-13] and Sasaki et al [14]. Beyond that, a search was conducted in "ISI Web of Science" with keywords such as "lead-free thin film", "potassium sodium titanate", "bismuth sodium titanate" and similar. The collected papers were again searched for relevant references.

In contrast to the numerous studies on lead-free piezoelectric single crystals and ceramics, literatures tackling with lead-free thin films have been lagging far behind, as demonstrated in Figure 1. The lack of progress in lead-free thin films may be attributed to the challenges associated with preserving volatile A-site cations such as bismuth, sodium and potassium in the films and avoiding significant concentrations of A-site and oxygen vacancies to achieve high-quality thin films. [15] Progress on the growth and characterization of BNT-based and KNN-based thin films, still remains at an infancy stage with rather scattered physical properties. [16-48] In this sense, the development of high-performance lead-free piezoelectric thin films for MEMS, micro-actuators and transducers applications has proven to be an attractive, albeit challenging, endeavour. A list of BNT and KNN-based thin films in open literatures and their properties is presented in Table 1 and 2, respectively.

Table 1. Properties of some BNT-based thin films

Thin Film Materials	Remnant Polarization P_r ($\mu C/cm^2$)	Dielectric Permittivity ε_{33} @10/100kHz	Piezoelectric Coefficient d_{33} (pm/V)	Piezoelectric Coefficient e_{31} (C/m^2)	Reference
$(Bi_{0.5}Na_{0.5})TiO_3$	20.9	171	--	--	[16]
$(Bi_{0.5}Na_{0.5})TiO_3$	8.3	270	24.8	--	[17]
$0.93(Bi_{0.5}Na_{0.5})TiO_3$-$0.07BaTiO_3$	15	--	--	-14.4	[18]
$0.94(Bi_{0.5}Na_{0.5})TiO_3$-$0.06BaTiO_3$	12	105	94	--	[19]
$0.945(Bi_{0.5}Na_{0.5})TiO_3$-$0.055BaTiO_3$	12	400	--	--	[20]
$0.85(Bi_{0.5}Na_{0.5})TiO_3$ -$0.15(Bi_{0.5}K_{0.5})TiO_3$	11.3	238	95	--	[21]
$Bi_{0.5}(Na_{0.85}K_{0.15})Ti_{0.75}Sc_{0.25}O_3$	18.6	--	67	--	[22]
$0.88(Bi_{0.5}Na_{0.5})TiO_3$ -$0.08(Bi_{0.5}K_{0.5})TiO_3$ -$0.04BaTiO_3$	30	530	--	-2.2	[23]

Table 2. Properties of some KNN-based thin films

Thin Film Materials	Remnant Polarization Pr ($\mu C/cm^2$)	Dielectric Permittivity ε_{33} @10/100kHz	Piezoelectric Coefficient d_{33} (pm/V)	Piezoelectric Coefficient e_{31} (C/m^2)	Reference
$(K_{0.16}Na_{0.84})NbO_3$	22	320	--	-2.4	[32]
$K_{0.48}Na_{0.52}NbO_3$	22.5	380	45	--	[33]
$K_{0.5}Na_{0.5}NbO_3$	24	185	--	-3.6	[34]
$K_{0.5}Na_{0.5}NbO_3$	21.1	234.7	64.5	--	[35]
$K_{0.5}Na_{0.5}NbO_3$-$0.05BaTiO_3$	--	--	40	--	[36]
$0.94K_{0.52}Na_{0.58}NbO_3$ - $0.06LiNbO_3$	9.7	--	192	--	[37]
$(K_{0.44}Na_{0.52}Li_{0.04})$-$(Nb_{0.84}Ta_{0.10}Sb_{0.06})O_3$	10	325	53	--	[38]

It remains a mystery why lead-free piezoelectric materials do not exhibit very high piezoelectricity. [49] Low piezoelectric coefficient, high leakage current/conductivity as well as relatively low ferroelectric to non-ferroelectric phase transition temperature are the major obstacles to restrict the use of existing lead-free piezoelectrics in practical applications. The advance in thin film technology provides an opportunity to significantly enhance the physical properties of the existing materials through i.e. epitaxial growth. The major challenge in the development of high-performance lead-free piezoelectric thin films lies in the production of stoichiometric thin films along with good crystallinity, as the processing of lead-free piezoelectric thin films is not well established as compared to PZT systems due to the fact of limited understanding of the thermal stability and growth mechanism. It is expected this problem could be surmounted by the optimization of deposition parameters and application of site engineering. Site engineering by cation substitution has been demonstrated to be a feasible means to significantly reduce the leakage current and coercivity in many oxide perovskite ferroelectrics by suppressing the total number of free carriers and facilitating the movement of the domain wall. [27, 31, 46] Following this logic, site engineering, in principle, holds the promise to tune the various remarkable physical properties of lead-free thin films.

This paper reviews our recent efforts on developing lead-free piezoelectric thin films with specific focus on the BNT and KNN-based systems. Physical properties of cation-modified BNT and KNN thin film materials with optimal composition in the vicinity of morphotropic phase boundary are presented.

2. BNT-Based Thin Films

Being one of the most successful lead free piezoelectrics, $(Bi_{0.5}Na_{0.5})TiO_3$ has been highlighted by several research reviews. [1, 3, 4, 5, 15, 50] BNT was found to be an excellent ferroelectric/piezoelectric material with high remnant polarization P_r of 38 $\mu C/cm^2$ and piezoelectric coefficient d_{33} of 73 pC/N. [51-52] Another great advantage of BNT compositions lies in its high solubility for various elements, thus offering the possibility of the formation of modified compositions with enhanced properties to surmount its intrinsic high conductivity and coercive field. Binary and ternary derivatives of BNT with a composition of morphotropic phase boundary have attracted particular attention in the past two decades. [10, 14, 53]

In recent years, great efforts have been made in the growth and characterization of BNT-based thin films. Zhou et al [25] and Tang et al [16] reported the dielectric, ferroelectric and leakage current characteristics of sol-gel-derived BNT thin films grown on platinized silicon substrates. Yu et al [17] and Remondière et al [30] studied the piezoelectric properties of BNT thin films. Yu et al. first quantitatively evaluated the piezoelectric properties of their sol-gel derived BNT thin films and a moderate d_{33} coefficient of 24.8 pm/V was revealed. It should be noted that chemical approaches were adopted in a majority of reported works to prepare the BNT-based thin films due to the good chemical homogeneity and the ease of stoichiometric control. Improvement in the sol-gel processing techniques resulted in dramatically enhanced piezoelectric response with d_{33} coefficients of 94 pm/V and 95 pm/V in $(Bi_{0.5}Na_{0.5})_{0.94}Ba_{0.06}TiO_3$ and $Bi_{0.5}(Na_{0.85}K_{0.15})TiO_3$ thin films, respectively. [19, 21] Physical approaches, such as pulsed laser deposition (PLD) and RF magnetron sputtering have been applied to deposit epitaxial BNT-based thin films. Recently, Abazari et al [23] has reported [001]-oriented MPB $(Bi_{0.5}Na_{0.5})TiO_3$-$(Bi_{0.5}K_{0.5})TiO_3$-$BaTiO_3$ (BNT-BKT-BT) thin films epitaxially grown on $SrRuO_3$ buffered $SrTiO_3$ substrates by PLD. Pronounced ferroelectric and piezoelectric properties have been achieved in BNK-BKT-BT thin films. More promising physical properties were reported by Harigai et al [18] and Adachi et al [28] in [001] and [110] oriented epitaxial $(Bi_{0.5}Na_{0.5})TiO_3$-$BaTiO_3$ thin films produced by RF magnetron sputtering, with surprisingly high transverse piezoelectric coefficients e_{31} of -14.4 C/m^2 and d_{31}^* of -221 pC/N, respectively. Growth and characterization of BNT-based thin film so far have predominantly been examined in the form of polycrystalline and a complete picture is not yet available mainly due to the lack of systematical investigations and comprehensive understandings on the structural, ferroelectric and piezoelectric properties of these thin films. The following is a brief review of what we have achieved in the development of BNT-based thin film materials, namely $Bi_{0.5}(Na_{0.7}K_{0.2}Li_{0.1})_{0.5}TiO_3$ and cation-modified $(Bi_{0.5}Na_{0.5})_{0.94}Ba_{0.06}TiO_3$, by pulsed laser deposition. [26-27]

2.1. BNKLT Thin Films

It has been shown that the A-site substituted BNT ceramic $Bi_{0.5}(Na_{1-x-y}K_xLi_y)_{0.5}TiO_3$ [abbreviated as BNKLT] possesses excellent ferroelectric and piezoelectric properties. [8] Among these systems, $Bi_{0.5}(Na_{0.7}K_{0.2}Li_{0.1})_{0.5}TiO_3$ exhibited the best piezoelectric properties with d_{33} = 231 pC/N and k_t = 0.505 at room temperature, [54] which are certainly desirable for piezoelectric device applications. In this section, we reported the deposition, ferroelectric properties and piezoresponse of BNKLT thin films.

Polycrystalline $Bi_{0.5}(Na_{0.7}K_{0.2}Li_{0.1})_{0.5}TiO_3$ thin films were grown on $Pt/Ti/SiO_2/Si$ substrates using pulsed laser deposition. The crystallographic characterization by x-ray diffraction (XRD) confirmed that the films exhibited a single-phase perovskite with no preferred orientations. The ferroelectric characterization was performed using a metal-ferroelectric-metal (MFM) configuration. Top electrodes with a diameter of 200 μm and a thickness of 200 nm were prepared by RF magnetron sputtering, followed by the standard photolithography process. Figure 2 shows the polarization hysteresis (P-E) loops of the BNKLT thin film measured under different electric fields. Well-defined P-E hysteresis loops are observed. Probably due to the size effect and multi-domain state, [55] the observed remnant polarization P_r is much smaller than that of a bulk BNBKL ceramic (13.9 vs 37.5 $\mu C/cm^2$, inset of Figure 2), while the observed coercive field E_c is much higher (10.2 vs 2.5 MV/m). Similar results have generally been observed in PZT films. As shown in the inset of Figure 2, the P-E loop of the bulk ceramic becomes saturated at a relatively low electric field of 8 MV/m. However, for the BNKLT thin film, the P-E loop does not saturate even at a high electric of 30 MV/m. This should be arisen from the high electric field required for switching the polarization and the large leakage current at such a high electric field. (e.g. 8×10^{-4} A/cm^2 under a dc bias field of 30 MV/m). The large leakage current should be due to the high mobility and electropositivity of the oxygen vacancies, Bi^{3+} vacancies and alkaline ions. [25] It should be noted that the leakage current (for the film) under a dc bias field of 8 MV/m is much smaller, only about 8×10^{-6} A/cm^2. The piezoelectric coefficient d is related to the (net) polarization P (which may be approximated by P_r), the permittivity ε and the electrostriction

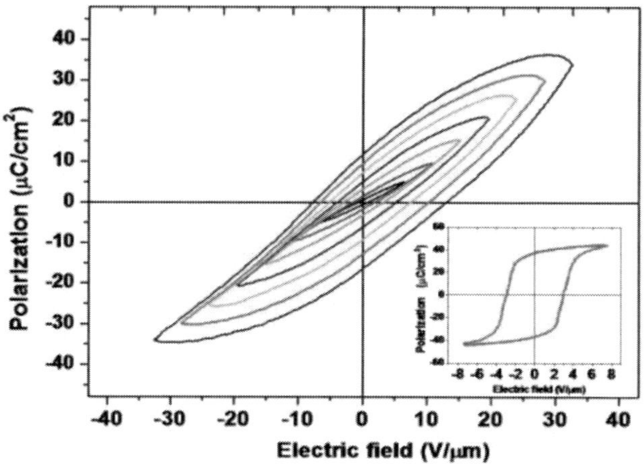

Figure 2. P-E hysteresis loops of the BNKLT thin film grown on $Pt/Ti/SiO_2/Si$ substrate. Inset is the *P-E* hysteresis loop of the BNKLT bulk ceramic. [Reproduced with permission from Ref. [26], Copyright 2008, AIP]

coefficient Q via a general equation [56] $d = 2Q\varepsilon P$. Therefore, because of the large P_r value, it is expected that our BNKLT thin films will exhibit a relatively high piezoelectric coefficient.

Analysis of the surface morphology and the evolution of domain structure under different temperatures were carried out by simultaneous acquisition of photographic views and domain imaging using an atomic force microscope equipped with a Pt/Ir coated Si tip in PFM mode.

For the PFM imaging, an ac modulating voltage of 10 V (peak-to-peak) at 12 kHz was applied to the bottom electrode of the thin film, while the tip was electrically grounded. The BNKLT thin film possesses a relatively smooth surface with average grain size of 100-120 nm as shown in Figure 3 (a). Figures 3 (b)-(d) show the local piezoresponse phase image of the BNKLT thin film at 25, 100 and 180°C, respectively. All of the images were taken at the same region as in Figure 3 (a). At room temperature, the contrast of the piezoresponse image is weak and most of the domains appear yellow in color (Figure 3(b)). Most of the grains are of a single domain structure and only a few of them contain a multi-domain structure. No obvious lamellar domains, as reported in PZT or PT thin films, [57-58] were observed. As the temperature increases to 100°C, which is slightly below the depolarization temperature T_d of BNKLT, the domain structure shows a great change (Figure 3(c)). Some of the domains became partially depolarized (the circled regions), showing a blurry contrast. Nevertheless, the domain structures were maintained at the outer part of some grains, especially near the grain boundary regions (the white patterns), which could be explained by the domain pinning near the grain boundaries. Moreover, the contrast between opposite domains becomes sharper than that at room temperature. This indicates that the domain switching becomes much easier at high temperatures, which is consistent with the macroscopic measurement of the coercive field in BNKLT bulk ceramics: the higher the temperature, the smaller the coercive field. [54] At 180°C, which is well above T_d, no obvious domain structure can be identified in the piezoresponse image (Figure 3(d)), implying that the film is almost completely depolarized. The image shows mainly a grey contrast, except for a few small white regions which may correspond to the normal ferroelectric domains or defective pinned domains. In general, BNT-based ceramics undergo two phase transitions: one is the transition from a rhombohedral phase to a tetragonal phase at T_d, and the other is the transition from a tetragonal phase to a cubic phase at T_m. Although a number of researchers have suggested that the ceramics are of the "anti-ferroelectric phase" at temperatures between T_d and T_m [10], the nature of this phase is still a matter of debate. Nevertheless, the BNKLT ceramics will lose most of their ferroelectric activities (non-ferroelectric state) when the temperature exceeds T_d (~130 °C), [54] which is consistent with the domain structure evolution observed in our BNKLT thin film.

Figure 4 shows the variations of the static piezoresponse (amplitude) and effective $d_{33,f}$ with the bias dc electric field for a grain of the BNKLT thin film measured by PFM incorporated with lock-in technique [59-60]. It can be seen that the film exhibits a typical well-shaped "butterfly" loop. The observed E_c from the piezoresponse hysteresis loop is slightly smaller than that from the *P-E* hysteresis loop (10.2 MV/m, refer to Figure 2). Such a discrepancy in E_c is often observed in ferroelectric thin films. [61-62] This may be due to the difference in the switching mechanism between the microscopic PFM and macroscopic *P-E* measurement. [61] As shown in Figure 4, the film exhibits a maximum $d_{33,f}$ value of ~ 64 pm/V under a bias field of ~ 34 MV/m, which is about one third of the value for a bulk sample.

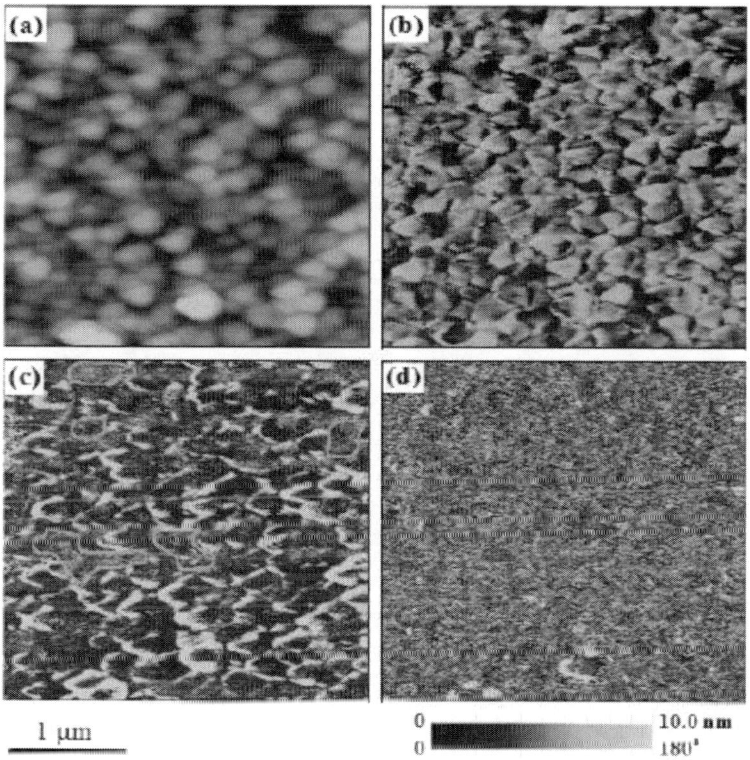

Figure 3. (a) Surface morphology of the BNKLT thin film; and the piezoresponse phase images of the BNKLT thin film at different temperatures: (b) 25°C, (c) 100°C and (d) 180°C. [Reproduced with permission from Ref. [26], Copyright 2008, AIP]

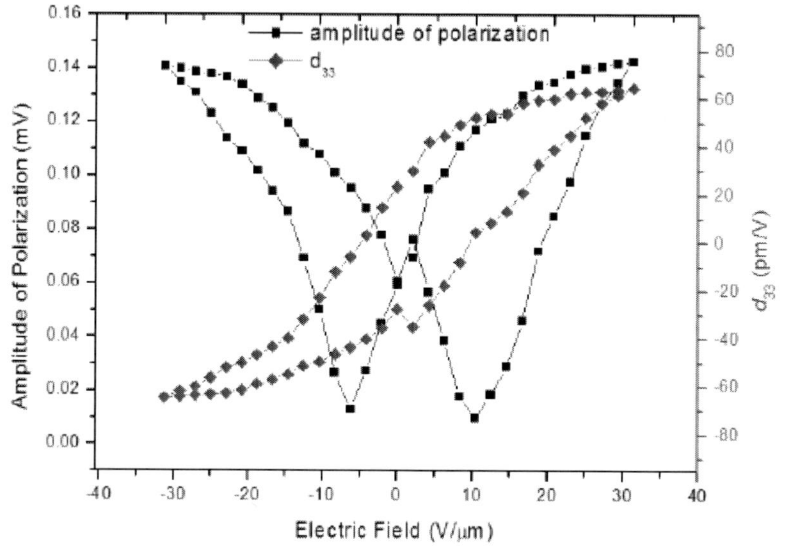

Figure 4. Variations of the amplitude of the static piezoresponse and the effective $d_{33,f}$ with the bias dc field for a grain of the BNKLT thin film measured at room temperature. [Reproduced with permission from Ref. [26], Copyright 2008, AIP].

Owing to the clamping effect imposed by the substrate or the bottom electrode, a piezoelectric thin film usually exhibits a reduced mechanical strain and thus a smaller piezoelectric coefficient as compared to a bulk counterpart. [63] Nevertheless, the observed $d_{33,f}$ value for the BNKLT film is relatively large, showing its potential for replacing the toxic piezoelectric PZT films in future environment-friendly piezoelectric devices.

2.2. BNBT-Based Thin Films

The pioneer work by Takenaka *et al* [10] demonstrated that rhombohedral (F_α)-tetragonal (F_β) morphotropic phase boundary (MPB) in $(Bi_{0.5}Na_{0.5})TiO_3$-$BaTiO_3$ (BNBT) binary system is remarkably effective in promoting piezoelectric properties (d_{33} = 125 pC/N). Subsequent studies revealed the piezoelectric performance of BNBT ceramics with MPB composition could be further enhanced up to d_{33} = ~160-180 pC/N by appropriate dopants, such as potassium, cerium, lanthanum and manganese etc. [53, 64-65] In this section, thin films of La+Ce co-doped and Mn-doped BNBT with composition in the vicinity of MPB are epitaxially deposited on $CaRuO_3$ (CRO) electroded $(LaAlO_3)_{0.3}(Sr_2AlTaO_6)_{0.35}$ [LSAT] (001) single-crystal substrates by PLD. Doping effects with respect to the electrical properties of the BNBT-based thin films were investigated.

The targets of $(Bi_{0.5}Na_{0.5})_{0.94}Ba_{0.06}TiO_3$+ 0.5 mol% CeO_2 + 0.5 mol% La_2O_3 (BNBTLaCe), $(Bi_{0.5}Na_{0.5})_{0.94}Ba_{0.06}TiO_3$ + 0.5 mol% MnO_2 (BNBTMn) used for laser ablation were fabricated by the conventional solid state reaction route. The substrates were cubic (001)-oriented LSAT (a = 3.868 Å), which were first coated with a layer of $CaRuO_3$ (~ 30 nm) by PLD. CRO was chosen as bottom electrode material not only because of its high electrical conductivity in comparison with widely used $SrRuO_3$, [66] but also because its high lattice similarity with LSAT and BNBT (rhombohedral, a = 3.893 Å), as CRO can be

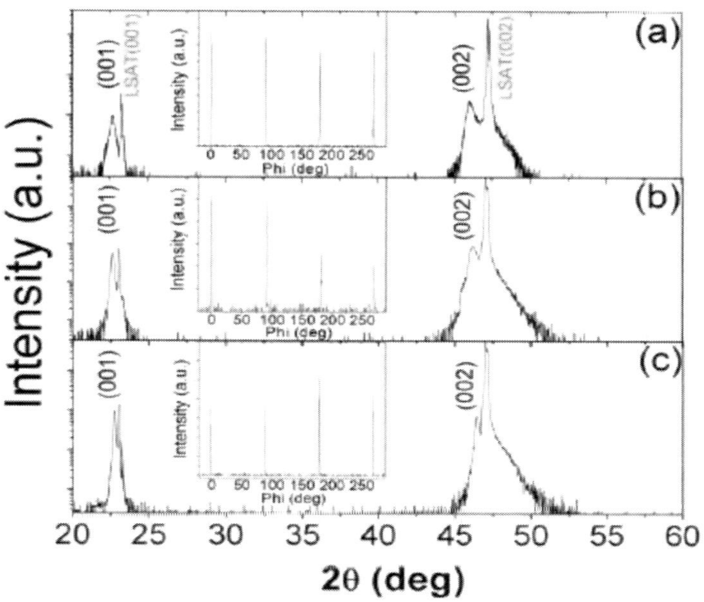

Figure 5. XRD θ-2θ scans of (a) BNBTLaCe (b) BNBTMn and (c) BNBT thin film epitaxially grown on CRO(001)$_c$/LSAT(001) substrates. Insets are the off-axis φ scans of the (202) reflections of corresponding films. [Reproduced with permission from Ref. [27], Copyright 2010, AIP].

interpreted as a pseudo-cubic perovskite structure with $a \approx 3.85$ Å under a constrained environment of epitaxial growth on a cubic substrate. [67] An undoped BNBT thin film was also deposited on CRO(001)$_c$/LSAT(001) under the same conditions for comparison purposes. The films were grown to a thickness of ~ 380 nm to minimize the strain effects.

Figure 5 shows the XRD θ-2θ spectra of BNBTLaCe and BNBTMn thin films as well as an undoped BNBT thin film. Only strong (00*l*) peaks appear in the spectra, indicating the appearance of a pure perovskite phase. The in-plane alignment of the thin films with respect to the major axes of the (001) substrates was also confirmed by the XRD off-axis φ scan of the (202) reflections, as shown in the insets of Figure 5. The appearance of four peaks with a 90° interval indicates a cube-on-cube epitaxial growth relationship between the films and substrates. Lattice parameters (out-of-plane c, in-plane a) and c/a ratios of the thin films were determined from the XRD results. The c/a ratios were found to be 1.0066±0.0002, 1.0019±0.0003 and 1.0001±0.0002 for BNBTLaCe, BNBTMn and BNBT thin films, respectively. This indicates the lattice of doped BNBT thin films are tetragonally distorted, while the undoped BNBT thin film remains a rhombohedral structure, which is in good agreement with previously reported results by other groups. [24]

Figure 6(a) shows the *P-E* loops of BNBT-based thin films measured under different electric fields at a frequency of 1 kHz. Well-defined *P-E* hysteresis loops are observed for all films. The remnant polarization P_r and coercive field E_c are 29.5 μC/cm^2 and 7.4 kV/mm, 24.5 μC/cm^2 and 11.3 kV/mm, and 14.5 μC/cm^2 and 10.2 kV/mm for BNBTLaCe, BNBTMn and BNBT thin films, respectively. It should be noted that P_r of doped films are comparable to bulk values, whereas the E_c values are considerable higher, which may be attributed to the size effect. [61] In general, tetragonality (c/a) is strongly correlated with ferroelectric properties in perovskite oxides. It is believed that the increased c/a ratio in doped thin films contributes to their enhanced ferroelectricity. Figure 6(b) illustrates the C-V characteristics of the BNBT-based films measured at 1 MHz. The observed remarkable non-linear dependence of ε on the dc bias field demonstrates that the ferroelectric properties of the films were originated from the domain switching. In addition, the dielectric permittivity **ε** at zero electric field has been significantly enhanced by the dopants, from 340 for undoped film to 395 and 480 for BNBTMn and BNBTLaCe thin films, respectively. The simultaneous addition of CeO_2 and La_2O_3 has a complex effect of modification, e.g. Ce^{3+} and La^{3+} go into A-site to create A-site vacancies, which facilitate the movement of domain wall, so as to increase the ferroelectric and dielectric properties; Nevertheless, multi-valent Mn (Mn^{2+}, Mn^{3+} and Mn^{4+}) may coexist in perovskite oxide thin films [68] and possibly occupy either A- or B-sites. [46] The simultaneous enhancement of polarization, dielectric permittivity and coercive field caused by Mn doping implies that Mn may act as both "softening" and "hardening" dopant.

The leakage currents through the BNBT-based thin film capacitors were measured at room temperature. Figure 7(a) gives the typical *I-V* curves of the Au/BNBT-based films/CRO heterostructures. All of the films show diode-like *I-V* characteristics with hysteresis and the leakage current is limited at the BNBT-based film/CRO interface.

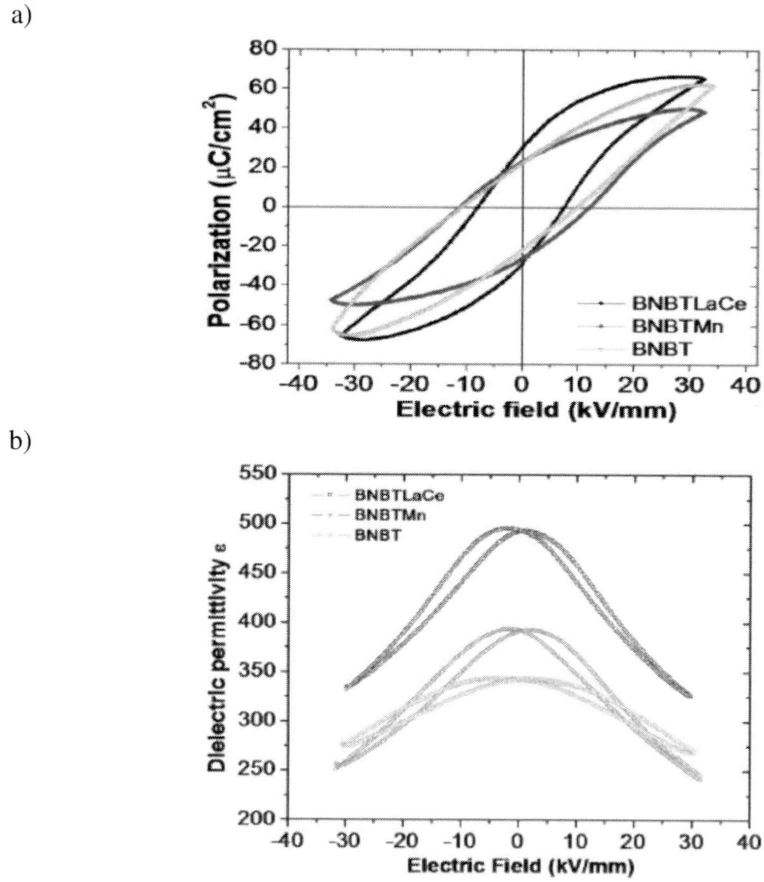

Figure 6. (a) P-E hysteresis loops and (b) Electric field dependence of the epitaxial BNBT-based thin films. [Reproduced with permission from Ref. [27], Copyright 2010, AIP].

A sharp increase of *I* occurs at 6.9 kV/mm, 9.9 kV/mm and 9.3 kV/mm for BNBTLaCe, BNBTMn and BNBT thin films, respectively, which approximately correspond to their coercive fields. The loss of A-site volatile elements, e.g. Na^+ and Bi^{3+}, implies that BNBT-based thin films exhibit *p*-type conductivity, while CRO is *n*-type. Similar *I-V* characteristics subject to resistance switching has been also observed in other perovskite oxide *p-n* junctions, such as (Pb,La)(Zr, Ti)O_3/Nb-SrTiO_3. [69] It is worth to note: unlike the hysteresis due to the dielectric relaxation or trapped carriers at low field as shown in Figure 6 (b), the current flows more during decreasing |V| rather than increasing |V|, which is very much similar to that found in the contaminated Schottky barriers on Si. [70] Another interesting feature in Figure 7(a) is the strong dopant dependence of *I-V* characteristics. Current density of Mn-doped film is dramatically reduced by two orders of magnitude, while current density of BNBTLaCe film just reduces slightly compared with that of undoped BNBT thin films. It has been reported that the Mn doping is effective to lower the leakage current in $BiFeO_3$ and (K, Na)NbO_3 by suppressing the intrinsic *p*-type conductivity of undoped films. [45, 71] This is associated with decreasing the total number of free carriers (holes) that correlated with the A-site vacancies. In our Mn-doped BNBT thin films, similar phenomena could be achieved, as Mn^{2+} ($r \sim 0.90$Å) possesses close ionic radius to Na^+ ($r \sim 1.02$ Å) and Bi^{3+} ($r \sim 1.03$Å), thus very likely occupies the A-site vacancies. In the La+Ce co-doped film, however, Ce^{3+} ($r \sim 1.02$Å) and La^{3+} ($r \sim 1.03$Å) may not only compensate the existing A-site vacancies but also possibly substitute Ba^{2+} ($r \sim 1.35$Å) to create extra A-site vacancies, resulting in the leakage current is insignificantly reduced as compared with the undoped film.

Figure 8(a) and (b) display the variations of the static piezoresponse (PR) amplitude (A) and phase (Φ), respectively, with the bias dc electric field for a grain of the BNBT-based thin films at room temperature. The effective $d_{33,f}$ were derived from $d_{33,f} = A \cdot \cos\Phi$, as shown in Figure 8(c). The amplitude of static PR [Figure 8(a)] exhibits a typical well-shaped "butterfly" loop, while the d_{33} hysteresis loops are distorted and offset.

The offset of d_{33} hysteresis loops could be attributed to the presence of non-switchable domains pinned near the electrode-film interface and built-in electric field (imprint), [72-73] which results from the difference nature of the top and bottom electrodes: the Pt/Ir tip serves as the top electrode whereas the bottom electrode consists of epitaxial CRO.

Taking the average value of maximum $d_{33,f}$ for positive and negative sides of d_{33} loops, the values of $d_{33,f}$ were found to be 70 pm/V, 48 pm/V and 26 pm/V for BNBTLaCe, BNBTMn and pure BNBT thin films, respectively. It can be seen that the dopants have remarkably enhanced the piezoelectric properties of BNBT thin films. Consistent with its good ferroelectricity, the promising piezoelectric performance in BNBTLaCe thin films is attributed to the ease of domain wall movement induced by the A-site substitutions of La^{3+} and Ce^{3+}. Although Mn may also act as an A-site dopant, its "hardening effect" by compensating A-site vacancies and substituting B-site ions partially smears its contribution on facilitation of movement of domain walls, resulting in a relatively lower apparent piezoelectric coefficient. Owing to the clamping effect imposed by the substrates, the piezoelectric coefficients in our doped BNBT thin films are much lower than their bulk values (~160 pC/N). However, the observed $d_{33,f}$ value for our La+Ce co-doped BNBT film is very much comparable to those of the lead-based counterparts ($d_{33,f}$= 10 - 110 pm/V for typical PZT thin films [74]) and one of the highest reported values for lead-free piezoelectric thin films up to this date. In addition, we believe the ferroelectric and piezoelectric properties of these BNBT-based thin films can be further enhanced through judicious control of doping level, as the MPB and optimal doping concentration in thin film forms might be slightly different from those of bulks.

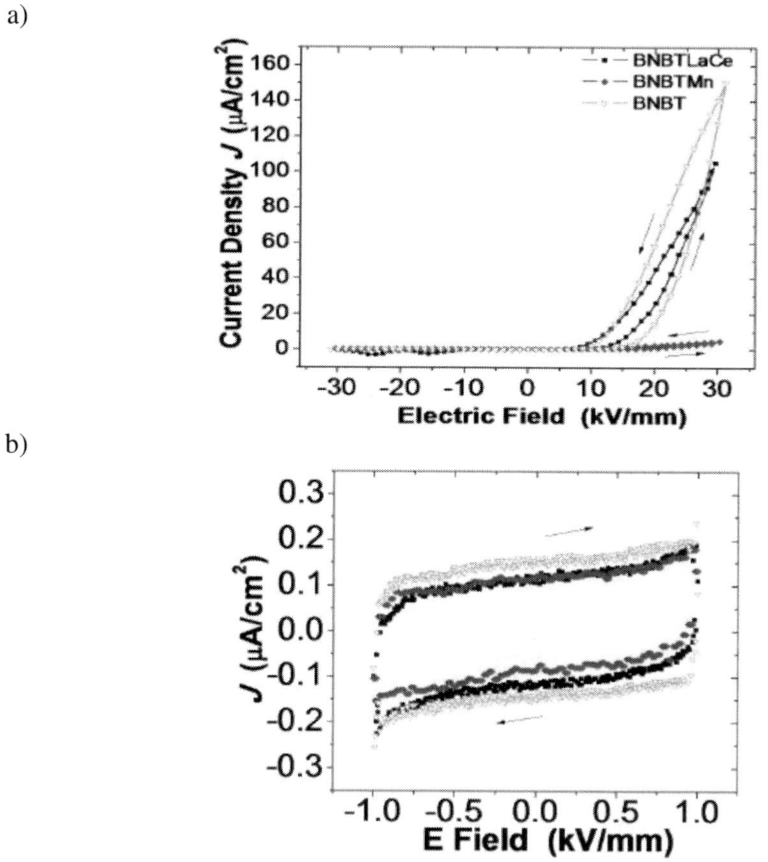

Figure 7. (a) Leakage current density *J* as a function of electric field *E* for the Au/BNBT-based thin film/CRO capacitors at room temperature; (b) the *J*-*E* characteristics at low electric field. [Reproduced with permission from Ref. [27], Copyright 2010, AIP].

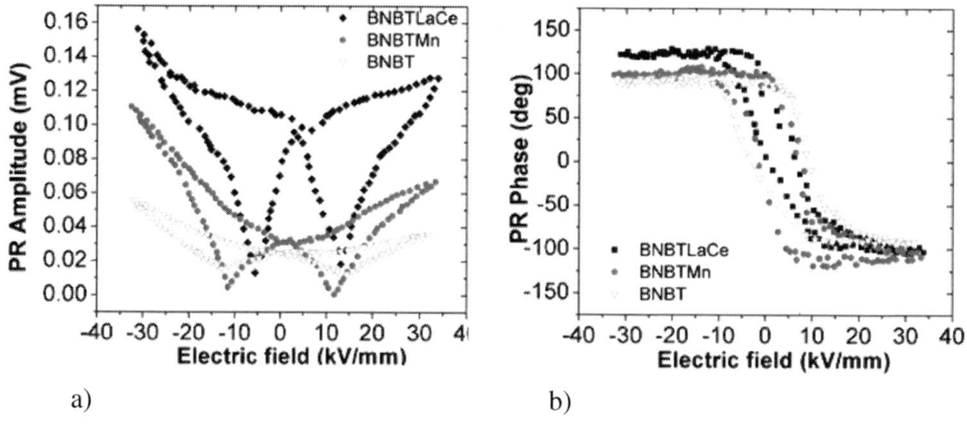

Figure 8. (Continued).

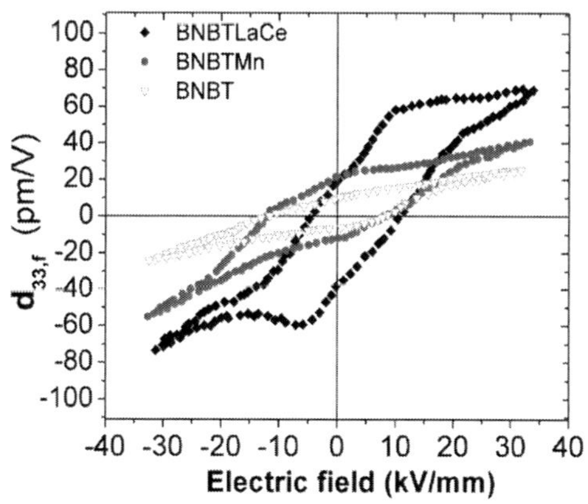

c)

Figure 8. Local piezoresponse (a) amplitude, (b) phase and (c) effective $d_{33,f}$ hysteresis loops of the epitaxial BNBT-based thin films measured at room temperature. [Reproduced with permission from Ref. [27], Copyright 2010, AIP].

3. KNN-Based Thin Films

$(K_{1-x}Na_x)NbO_3$ is also considered to be a promising candidate owing to its excellent ferroelectric and piezoelectric properties, high Curie temperature and biocompatible nature. KNN thin films have been grown on various substrates by both physical and chemical approaches. [32-48] Pioneer works on thin film KNN materials were mainly focused on undoped compositions in the vicinity of morphotropic phase boundary (MPB). Subsequent investigations have found that Li substitution to K and/or Na sites plays an important role in enhancing the piezoelectric properties of the KNN system. [37, 43] However, there are discrepancies in the literatures concerning the ferroelectric hysteresis and leakage current behaviors of LKNN thin films, although the films were grown under similar conditions. A recent breakthrough in the textured highly-piezoelectric Li, Ta, Sb doped KNN (KNN-LT-LS) ceramics by Saito *et al* [9] has triggered intensive studies on KNN-LT-LS-based thin films. [38, 45-46] In spite of relatively high ferroelectric and piezoelectric properties, the films still suffered from a high leakage current (e.g. >10^{-1}A/cm^2@10kV/mm) and details of conduction mechanisms remain unclear. Indeed, the participation of toxic Sb in KNN-LT-LS system does not meet the ecological strategy of the development of lead-free piezoelectrics.

Lin's and Hollenstein's studies [75-76] have revealed Li^+ and Ta^{5+} dual dopants can dramatically enhance the piezoelectric properties of KNN ceramics. The specific composition $(K_{0.48}Na_{0.48}Li_{0.04})(Nb_{0.775}Ta_{0.225})O_3$ (KNLNT hereafter) possesses optimal piezoelectric properties at room temperature due to the coexistence of tetragonal and orthorhombic phases. This MPB-like behavior, and less processing and compositional complexity, especially the absence of toxic Sb content, as compared with KNN-LT-LS, have made KNLNT a promising candidate for biocompatible piezoelectric device applications. In this section, we reported the good ferroelectric and piezoelectric properties, and low leakage current density in KNLNT thin films grown by pulsed laser deposition. Conduction mechanisms of the KNLNT thin films were investigated in detail.

Pulsed laser deposition of KNN-based thin films was one of the most popular techniques to fabricate KNN-based thin films because of the versatility of the technique and ease of

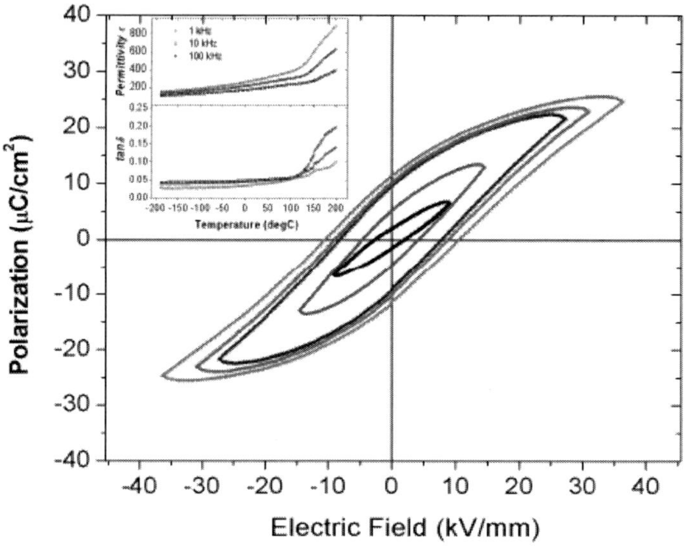

Figure 9. *P-E* hysteresis loops of the KNLNT thin film grown on $Pt(111)/Ti/SiO_2/Si(001)$. Inset is temperature dependence of dielectric permittivity and loss tangent of the KNLNT thin film. [Reproduced with permission from Ref. [39], Copyright 2011, AIP].

processing. [15] In this work, $(K_{0.48}Na_{0.48}Li_{0.04})(Nb_{0.775}Ta_{0.225})O_3$ thin films were deposited on $Pt(111)/Ti/SiO_2/Si(001)$ substrates using pulsed laser deposition. The KNLNT films of thickness ~ 280 nm were prepared by a number of 6000 pulses in an oxygen atmosphere of 27 Pa. The film exhibited a single-phase perovskite with a (110) preferred orientation as determined by XRD.

The fact that (001)/(100) and (002)/(200) doublets split while (110) remains non-splitting suggests the presence of a tetragonal structure in the KNLNT film with the lattice parameters a = 3.9440 Å and c = 3.9548 Å. The KNLNT thin film possesses a very smooth surface with an average grain size of 160-180 nm as observed in AFM (image not shown). Figure 9 shows the *P-E* loops of the KNLNT thin film measured under different electric fields at a frequency of 1 kHz. Well-defined *P-E* hysteresis loops are observed and the saturation of the hysteresis loop as a function of voltage is key evidence of intrinsic ferroelectricity in the KNLNT film. The observed remnant polarization $2P_r$ and coercive field E_c are 22.6 μC/cm^2 and 10.3 kV/mm, respectively.

As compared to a bulk KNLNT ceramic ($2P_r$ = 28.8 μC/cm^2, E_c = 2.4 kV/mm), the observed remanent polarization of the film is a bit lower to that of the bulks, whereas the coercive field of the film is considerably higher. This phenomenon may be attributed to the much smaller grain size of the thin film (160-180 nm vs. 1-2 μm for the bulk). It has been suggested that the fine grain films with film thicknesses less than 2 μm can lead to a limited number of domain variants. [37, 61] The absence of 90° domain walls and the strong pinning of non-180° domain walls lead to a decrease in P_r and an increase in E_c. Dielectric properties of Au/KNLNT/Pt capacitor were measured over a frequency range of 1kHz -100kHz in the temperature range of -200 – 200 ^{o}C with a heating rate of 1 ^{o}C/min. As shown in the inset of Figure 9, both dielectric permittivity and loss tangent of KNLNT thin film increased monotonically with increasing temperature and no apparent phase transition can be observed in the measured temperature range.

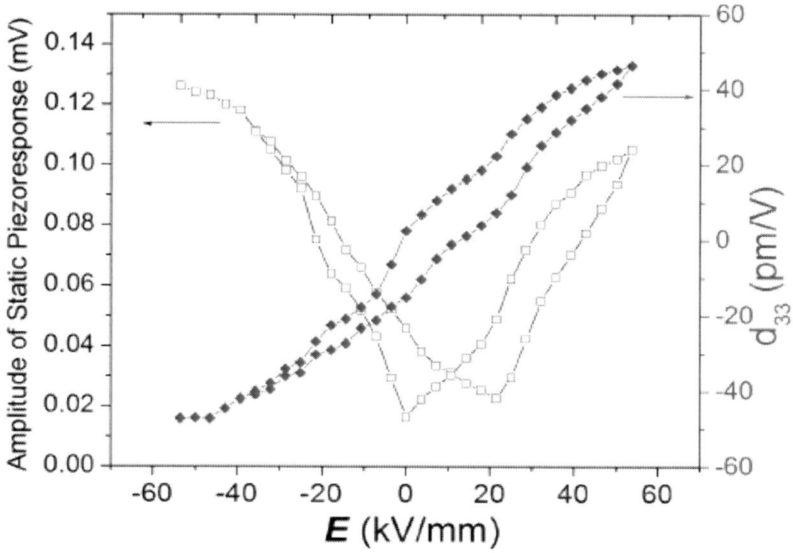

Figure 10. Variations of the amplitude of the static piezoresponse and the effective $d_{33,f}$ with the bias dc field for a grain of the KNLNT thin film measured at room temperature. [Reproduced with permission from Ref. [39], Copyright 2011, AIP].

Figure 10 shows the variations of the static piezoresponse (amplitude) and effective $d_{33,f}$ with the bias dc electric field for a grain of the KNLNT thin film at room temperature.

The amplitude of static piezoresponse exhibits a typical well-shaped "butterfly" loop, while the piezoresponse hysteresis loop is distorted and offset. [72] As shown in Figure 10, the film exhibits a maximum $d_{33,f}$ value of ~ 49 pm/V, which is better than PLD-produced highly *c*-axis oriented KNN ($d_{33,f}$ = 40 pm/V) [77] and Mn-doped KNN-LT-LS ($d_{33,f}$ = 45 pm/V) [38] thin films.

Understanding the mechanisms of the leakage current in complex oxide heterostructures would be a difficult problem. The correlation between the conduction mechanism and electrical properties has been extensively investigated in PZT and multi-ferroic thin films. [78-80] Rigorous understanding of the dominant mechanisms of the leakage current is a very challenging study of KNN based piezoelectric thin films, but also crucial for us to adjust thin film deposition conditions in order to minimize the leakage current density. The leakage currents through the Au/KNLNT/Pt capacitors were measured in a temperature range of 130 - 370 K. Figure 11(a) shows the typical current density J as a function of electric field E and the J-E characteristics were nearly symmetric with respect to voltage polarity at all temperatures. It should be noted that the current density in our KNLNT thin film is remarkably lower than that in the KNN-LT-LS-based films. [38, 45-46] It is speculated that the low current density is attributed to the strong (110) orientation, as the *c*-axis oriented or columnar grained microstructure may offer more leakage current paths through the vertically aligned grain boundaries. [37]

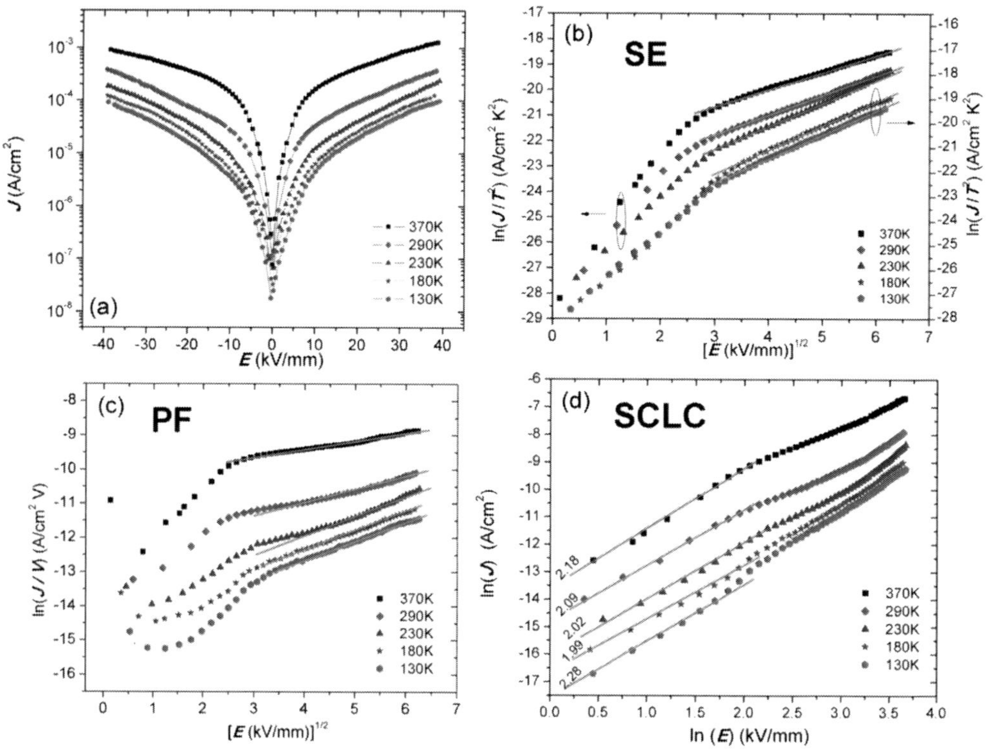

Figure 11. (a) Leakage current density J as a function of electric field E for the Au/KNLNT/Pt thin film capacitor at different temperatures; (b) $\ln(J/T^2)$-$E^{1/2}$ plots with respect to SE; (c) $\ln(J/V)$-$E^{1/2}$ plots with respect to PF emission (d) $\ln(J)$-$\ln(E)$ plots with respect to SCLC. [Reproduced with permission from Ref. [39], Copyright 2011, AIP]

Since no significant asymmetry in J-E behavior upon reversal of the polarity of applied voltage can be observed, we will hereafter study the situation in which Pt electrode is positively biased.

Current transport in dielectric thin films at high electric fields is usually determined to be interface-limited Schottky emission (SE) or bulk-limited Poole-Frenkel (PF) emission. [78] The current density across a Schottky barrier is [79]

$$J_S = A^*T^2\exp -[(\phi_b - \beta_s E^{\frac{1}{2}})/kT] \tag{1}$$

where $\beta_s = (q^3/4\pi\varepsilon_0\varepsilon_r)^{1/2}$, A^* is the Richardson constant, T the temperature, q the electron charge, Φ_b the Schottky barrier height, k the Boltzmann constant, ε_0 the permittivity of free space, ε_r the dielectric constant of the thin film. The Pool-Frenkel conductivity is given by [79]

$$J_{PF} = \sigma_0 E\cdot\exp -[(E_I - \beta_{PF}E^{\frac{1}{2}})/kT] \tag{2}$$

where $\beta_{PF} = (q^3/\pi\varepsilon_0\varepsilon_r)^{1/2}$, σ_0 is a sample-dependent zero-field conductivity and E_I is the trap ionization energy. The functional relationship of SE and PF are similar so that the plots of $\ln(J/T^2)$ vs $E^{1/2}$ and $\ln(J/V)$ vs $E^{1/2}$ for SE and PF, respectively, will yield linear relationships. Figure 11(b) and (c) show that the data satisfies the SE and PF equations irrespective to the measuring temperature, apart from the low E field region (<10kV/mm). Although the fittings are in good agreement with SE and PF, we are also interested in whether sensible physical parameters can be extracted from the data. The relative dielectric constants of the KNLNT film were found to be 1.7 - 4.1 and 5.7 - 14.8 for SE and PF, respectively, as calculated from the slopes of the linear fits. The physical range of ε_r for KNN-based materials is limited from one side by the optical dielectric constant $\varepsilon_\infty \approx 5$ [40] and from the other side by the static dielectric constant $\varepsilon_{dc} \approx 1000$. [75] The calculated values of ε_r in SE are clearly outside this range. This demonstrates that PF emission instead of SE is the dominant conduction mechanism in our KNLNT thin films at high E field strengths.

Another possible leakage current mechanism is the bulk-limited space-charge-limited-current (SCLC), whose current density obeys quadratic electric field dependence in the case of discrete traps: [79]

$$J_{SCLC} = 9\mu\varepsilon_0\varepsilon_r E^2 / 8d \tag{3}$$

where μ is the carrier mobility and d the film thickness. Figure 11 (d) shows the plots of $\ln(J)$-$\ln(E)$ at different temperatures. Regardless of the measuring temperature, slopes of the linear fits were found to be around 2 at low E fields (<10kV/mm), which is compatible with the discrete trap case of SCLC. Thus, we have clarified that leakage current behaviors of KNLNT thin film are dominated by two distinct mechanisms with respect to E field: SCLC at low E field (<10kV/mm) and PF emission at high E field (>10kV/mm). In addition, the J-E characteristics show little dependence on polarity of applied bias, again confirming that the leakage currents are probably bulk limited. [79]

CONCLUSION

In summary, the development of lead-free piezoelectric thin films has made considerable strides in the past few years. In BNT and KNN-based systems, the major difficulty in growing high performance thin films lies in the stoichiometric control in connection to the unavoidable A-site and oxygen vacancies. Our recent efforts on the deposition of cation-modified BNT and KNN-based thin films have led to pronounced ferroelectricity and piezoelectricity and also great potential for improvement. The comprehensive understanding of the thermal-driven evolution of ferroelectric domain structure and the dominant mechanisms of leakage current in lead-free thin films are very crucial and fruitful to adjust thin film deposition conditions in order to enhance their physical properties. Although the lead-free thin films do not match the overall performance of $Pb(Zr,Ti)O_3$ yet, the advances of BNT and KNN-based thin films give well-founded hope that high-quality lead-free thin films can be obtained to serve in environmentally-friendly and biocompatible piezoelectric devices in the near future.

Acknowledgments

This work was supported by the Vice-Chancellor's post-doctoral fellowship program, The University of New South Wales (SIR50/PS16940) and Australian Research Council Discovery Project (Grant No. DP110104629).

References

[1] E. Cross, *Nature* (London) 432, 24 (2004).

[2] M. Pecht, Y. Fukuda and S. Rajagopal, IEEE Trans. Electro. *Packaging Manufac.* 27, 221 (2004).

[3] J. Rödel, W. Jo, K. T. P. Seifert, E. M. Anton, T. Granzow and D. Damjanovic, *J. Am. Ceram. Soc.* 92, 1153 (2009).

[4] T. R. Shrout and S. J. Zhang, *J. Electroceram.* 19, 111, (2007).

[5] P. K. Panda, J. Mater. *Sci.* 44, 5049 (2009)

[6] E. Aksel and J. L. Jones, *Sensors,* 10, 1935 (2010).

[7] D. Damjanovic, N. Klein, J. Li and V. Porokhonskyy, *Fun. Mater. Lett.* 3, 5 (2010).

[8] D. Damjanovic, *Appl. Phys. Lett.* 97, 062906 (2010).

[9] Y. Saito, H. Takao, T. Tani, T. Nonoyama, K. Takatori, T. Homma, T. Nagaya, and M. Nakamur, *Nature* (London) 432, 84 (2004).

[10] T. Takenaka, K. Maruyama and K. Sakata, *J. Jpn. Appl. Phys.* 30, 2236 (1991).

[11] T. Takenaka and H. Nagata, *J. Euro. Ceram. Soc.* 25, 2693 (2005).

[12] X. X. Wang, X. G. Tang and H. L. W. Chan, *Appl. Phys. Lett.* 85, 91 (2004).

[13] X. X. Wang, H. L. W. Chan and C. L. Choy, *Solid State Commun.* 125, 395 (2003).

[14] Sasaki, T. Chiba, Y. Mamiya, and E. Otsuki, *Jpn. J. Appl. Phys.*, Part 1, 38 [9B] 5564 (1999).

[15] Safari and M. Abazari, IEEE Trans. *Ultra. Ferro. Freq. Contr.* 57, 2165 (2010).

[16] X. G. Tang, J. Wang, X. X. Wang and H. L. W. Chan, *Chem. Mater.*, 16, 5293 (2004).

[17] T. Yu, K. W. Kwok and H. L. W. Chan, Proc. IEEE Inter. *Symp. Appl. Ferro.* 168 (2007).

[18] T. Harigai, Y. Tanaka, H. Adachi and E. Fuji, *Appl. Phys. Exp.* 3, 111501 (2010).

[19] D. Z. Zhang, X. J. Zheng, X. Feng, T. Zhang, J. Sun, S. H. Dai, C. J. Gong, Y. Q. Gong, L. He, Z. Zhu, J. Huang and X. Xu, *J. Alloys Compd.* 504, 129 (2010).

[20] D. Alonso-Sanjose, R. Jimenez, I. Bretos and M. L. Calzada, *J. Am. Ceram. Soc.* 92, 2218 (2009).

[21] X. J. Zheng, S. H. Dai, X. Feng, T. Zhang, D. Z. Zhang, Y. Q. Gong, Y. Q. Chen and L. He, *Appl. Surf. Sci.* 256, 3116, (2010).

[22] Y. Y. Wu, X. H. Wang and L. T. Li, *J. Am. Ceram. Soc.* 94, 2518 (2011).

[23] M. Abazari, A. Safari, S. S. N. Bharadwaja and S. Trolier-McKinstry, *Appl. Phys. Lett.* 96, 082903 (2010).

[24] H. W. Cheng, X. J. Zhang, S. T. Zhang, Y. Feng, Y. F. Chen, Z. G. Liu and G. X. Cheng, *Appl. Phys. Lett.* 85, 2319 (2004).

[25] Z. H. Zhou, J. M. Xue, W. Z. Li, J. Wang, H. Zhu and J. M. Miao, *Appl. Phys. Lett.*, 85, 804 (2004).

[26] D. Y. Wang, D. M. Lin, K. S. Wong, K. W. Kwok, J. Y. Dai, and H. L. W. Chan, *Appl. Phys. Lett.* 92, 222909 (2008).

[27] D. Y. Wang, N. Y. Chan, S. Li, S. H. Choy, H. Y. Tian and H. L. W. Chan, *Appl. Phys. Lett.,* 97, 212901 (2010).

[28] H. Adachi, Y. Tanaka, T. Hakakiyo, M. Ueda and E. Fujii, *Appl. Phys. Exp.* 4, 051501 (2011).

[29] M. Bousquet, J. R. Duclere, C. Champeaux, A. Boulle, P. Marchet, A. Catherinot, A. Wu, P. M.Vilarinho, S. Deputier, M. Guilloux-Viry, A. Crunteanu, B. Gautier, D. Albertini and C. Bachelet, *J. Appl. Phys.* 107, 034102 (2010).

[30] F. Rémondière, A. Wu, P .M. Vilarinho and J. P. Mercurio, *Appl. Phys. Lett.* 90, 152905 (2007).

[31] Y. Y. Wu, X. H. Wang and L. T. Li, *J. Am. Ceram. Soc.* 94, 1843 (2011).

[32] Kanno, T. Mino, S. Kuwajima, T. Suzuki, H. Kotera and K. Wasa, IEEE Trans. *Ultra. Ferro. Freq. Contr.* Vol. 54 (12), (2007).

[33] H. J. Lee, I. W. Kim, J. S. Kim, C. W. Ahn, and B. H. Park, *Appl. Phys. Lett.* 94, 092902 (2009).

[34] Shibata, F. Oka, A. Ohishi, T. Mishima and I. Kanno, *Appl. Phys. Exp.* 1, 011501 (2008).

[35] S. Kang, B. Y. Kim, I. T. Seo, T. G. Seong, J. S. Kim, J. W. Sun, D. S. Paik, I. Hwang, B. H. Park and S. Nahm, *J. Am. Ceram. Soc.* 94, 1970 (2011).

[36] J. S. Choi, I. R. Hwang, S. H. Hong, G. T. Oh, J. A. Choi, S. H. Jeon, S. O. Kang, V. Sh. Yalishev, B. H. Park, C. H. Choi, C. W. Ahn, S. Nahm and S. J. Ahn, *Journal of Nanoscience and Nanotechnology,* Vol. 9, 7354, (2009).

[37] F. P. Lai, J. F. Li, Z. X. Zhu,and Y. Xu, *J. Appl. Phys.* 106, 064101 (2009).

[38] Abazari, T. Choi, S. W. Cheong, and A. Safari, *J. Phys. D: Appl. Phys.* 43, 025405 (2010).

[39] D. Y. Wang, D. M. Lin, K. W. Kwok, N. Y. Chan, J. Y. Dai, S. Li and H. L. W. Chan, *Appl. Phys. Lett.* 98, 022902 (2011).

[40] Blomqvist, S. Khartsev, A. Grishin, A. Petraru, and Ch. Buchal, *Appl. Phys. Lett.* 82, 439 (2003).

[41] R. Cho and A. Grishin, *Appl. Phys. Lett.* 75, 268 (1999).

[42] S. Khartsev, A. Grishin, J. Andrèasson, J. H. Koh and J. S. Song, *Inter. Ferro.*, 55, 769 (2003).

[43] W. Ahn, E. D. Jeong, S. Y. Lee, H. J. Lee, S. H. Kang, and I. W. Kim, *Appl. Phys. Lett.* 93, 212905 (2008).

[44] J. G. Wu and J. Wang, *J. Appl. Phys.* 106, 066101 (2009).

[45] Abazari, E. K. Akdoğan, and A. Safari, *Appl. Phys. Lett.* 92, 212903 (2008).

[46] M. Abazari and A. Safari, *J. Appl. Phys.* 105, 094101 (2009).

[47] M. Abazari and A. Safari, *Appl. Phys. Lett.* 97, 262902 (2010).

[48] M. Abazari, E. K. Akdoğan, and A. Safari, *Appl. Phys. Lett.* 93, 192910 (2008).

[49] W. F. Liu and X. B. Ren, *Phys. Rev. Lett.* 103, 257602 (2009).

[50] J. F. Scott, *Chem.Phys.Chem.*, 11, 341 (2010).

[51] G. A. Smolenskii, V. A. Isupov, A. I. Agranovskaya and N. N. Krainic, *Sov. Phys. Solid State* 2, 2651 (1961).

[52] J. Suchanicz and W. S. Ptak, Ferroelectr. *Lett. Sect.* 12, 71 (1990).

[53] Y. Hiruma, H. Nagata and T. Takenaka, *J. Jpn. Appl. Phys.* 45, 7409 (2006).

[54] M. Lin, D. Q. Xiao, J. G. Zhu and P. Yu, *Appl. Phys. Lett.*, 88, 062901 (2006).
[55] Z. Simões, M .A. Ramírez, A. Ries, J. A. Varela, E. Longo and R. Ramesh, *Appl. Phys. Lett.*, 88, 072916 (2006).
[56] Damjanovic, *Rep. Prog. Phys.* 61, 1267 (1998).
[57] X. Zhao, J. Y. Dai, X. G. Tang, J. Wang, H. L. W. Chan and C. L. Choy, *Appl. Phys. A,* 81, 997 (2005).
[58] C. Loppadner, F. Schlaphof, S. Schneider, U. Zerweck, S. Grafdtrom, L. M. Eng, A. Roelofs and R. Waser, *Surf. Sci.* 483, 532 (2003).
[59] Gruverman, O. Auciello, R. Ramesh and H. Tokumoto, *Nanotechnology* 8, A38 (1997).
[60] G.Zavala, J. H. Fendler and S. Trolier-McKinstry, *J. Appl. Phys.* 81, 7480 (1997).
[61] T. Watanabe, H. Funakubo, M. Osada, H. Uchida, I. Okada, B. J. Rodriguez and A. Gruverman, *Appl. Phys. Lett.* 90, 112914 (2007).
[62] V. V. Shvartsman, N. A. Pertsev, J. M. Herrero, C. Zaldo and A. L. Kholkin, *J. Appl. Phys.* 97, 104105 (2005).
[63] Setter, D. Damjanovic, L. Eng, G. Fox, S. Gevorgian, S. Hong, A. Kingon, H. Kohlstedt, N. Y. Park, G. B. Stephenson, I. Stolitchnov, A. K. Taganstev, D. V. Taylor, T. Yamada and S. Streiffer, *J. Appl. Phys.*, 100, 051606 (2006).
[64] X. X. Wang, H. L. W. Chan and C. L. Choy, *Appl. Phys.* A 80, 333 (2005).
[65] X. Y. Zhou, H. S. Gu, Y. Wang, W. Y. Li and T. S. Zhou, *Mater. Lett.* 59, 1649 (2005).
[66] R. J. Kennedy, R. Madden and P. A. Stampe, *J. Phys. D: Appl. Phys.* 34, 1853 (2001).
[67] Ito, H. Masumoto and T. Goto, J. Ceram. *Soc. Jpn.* 115, 683 (2007).
[68] S. Valencia, A. Gaupp and W. Gudat, *Phys. Rev.* B 73, 104402 (2006).
[69] Y. Watanabe, *Phys.Rev.* B 59, 11257 (1999).
[70] Y. Watanabe, *Ferroelectrics* 349, 190 (2007).
[71] S. K. Singh, H. Ishiwara and K. Maruyama, *Appl. Phys. Lett.* 88, 262908 (2006).
[72] L. Kholkin, E. L. Colla, A. K. Tagantsev, D. V. Taylor, and N. Setter, *Appl. Phys. Lett.* 68, 2577 (1996).
[73] S. V. Kalinin, B. J. Rodriguez, S. Jesse, E. Karapetian, B. Mirman, E. A. Eliseev, and A. N. Morozovska, *Annu. Rev. Mater. Res.* 37, 189 (2007).
[74] Muralt, IEEE Trans. Ultrason. *Ferro. Freq. Control*, 47, 903 (2000).
[75] D. M. Lin, K. W. Kwok, and H. W. L. Chan, *J. Appl. Phys.*102, 034102 (2007).
[76] Hollenstein, M. Davis, D. Damjanovic, and N Setter, *Appl. Phys. Lett.* 87, 182905 (2005).
[77] J. S. Kim, I. R. Hwang, S. H. Hong, J. H. Lee, B. H. Park, A. C. Woo and N. Sahn, *J. Korean Phys. Soc.*, 48, 1583 (2006).
[78] B. Nagaraj, S. Aggarwal, T. K. Song, T. Sawhney, and R. Ramesh, *Phys. Rev.,* B 59, 16022 (1999).
[79] Zubko, D. J. Jung, and J. F. Scott, *J. Appl. Phys.* 100, 114113 (2006).
[80] W. Pabst, L. W. Martin, Y. H. Chu and R. Ramesh, *Appl. Phys. Lett.* 90, 072902 (2007).

In: Ferroelectrics: New Research
Editors: Xui Li Huang and Su Lyn Ma

ISBN: 978-1-61942-282-7

Chapter 5

FERROELECTRIC AND MUTIFERROIC STUDY OF BISMUTH FERRITE FROM THE VIEWPOINT OF MATERIALS SCIENCE

Hongyang Zhao,[1] Hideo Kimura,[1,*] Qiwen Yao,[1] Rumi Tanahashi,[1] Zhenxiang Cheng,[2] Xiaolin Wang[2] and Huarong Zeng[3]

[1]National Institute for Materials Science, Tsukuba, Ibaraki, Japan
[2]Institute for Superconducting and Electronic Materials, Faculty of Engineering, University of Wollongong, North Wollongong, NSW, Australia.
[3]Shanghai institute of Ceramics, Chinese Academy of Sciences, Shanghai, China

ABSTRACT

Ferroelectric and mutiferroic materials have been studied from the viewpoint of pure and applied physics. For the applications, a study from the viewpoint of materials science will be important in the future. Examples of materials science study are substitution of elements, selection of substrate in thin films and in-situ observation. Nano-scale domain configurations of multiferroic $BiFeO_3$ thin films and their dynamic response to the externally applied fields have been investigated by piezoresponse force microscopy. High-resolution piezoresponse images reveal that nano-scale domain walls are approximately with ~2 nm in width. In this article, we focus on $BiFeO_3$ as mutiferroic materials and as ferroelectric materials. We will present our results about electric and magnetic properties, and scanning probe microscopy observations. Local polarization reversal and retention behavior will also be presented and discussed.

1. INTRODUCTION

Multiferroic (Magnetoelectric, ME) materials exhibit simultaneous ferroelectric and magnetic orders, which enables the coupling interaction between ferroelectric and magnetic orders [1,2]. This coupling interaction, also named the ME effect, produces various possibilities in the realization of mutual control and detection of electrical polarization and magnetism. As a potential candidate for practical application among the limited choices provided by single-phase multiferroic materials, $BiFeO_3$ has attracted enormous attention, due to its multiferroic properties at room temperature (with T_C = 1103 K, T_N = 643 K) [3-5]. The $BiFeO_3$ is antiferromagnetic with a *G*-type spin configuration along the $[111]_c$ or $[001]_h$ directions in its pseudo-cubic or rhombohedral structures. The $BiFeO_3$ also has a superimposed incommensurate cycloid spin structure, with a periodicity of 62 nm along its $[110]_h$ axis at room temperature, canceling out the macroscopic magnetization [6,7]. However, the long-range ordering of the cycloid spin periodical structure can be broken in thin films, and as a result, the macroscopic magnetic moment can be enhanced instead [8]. Besides the required large magnetic moment, strong ferroelectric performance is also required for $BiFeO_3$ to be used in either magnetoelectric devices or ferroelectric memory chips. Unfortunately, $BiFeO_3$ has poor ferroelectric performance due to severe electric leakage in both thin films and in bulks. It is believed that the electrical leakage is caused by oxygen vacancies and iron ions with different valences via the formation of shallow energy centers [9,10]. In the recent years, we have carried out a series of works aiming to improve the magnetic moment and reduce the electrical leakage, including doping effect in both thin films and bulk samples. In addition, we have also tried to use barrier layer to reduce the electrical leakage. Other effect in doped bismuth ferrite, such as substrate effect and starting materials effect have also been studied.

2. Properties in Multiferroic La and Pb Co-Doping $BiFeO_3$

As a promising multiferroic, many improvements in the ferroelectric and magnetic properties of $BiFeO_3$ have been achieved in recent years. However, there are still some challenges before its application in devices. One of the effective methods for the improvement is to dope a small amount of impurities at A or B sites in this ABO_3 perovskite [3,10-16] structure. This method is simple hence it is good for the industrial technology. It has been proved that doping of La causes structural changes and improves its crystallization [17,18]. As a result, both the ferroelectric and ferromagnetic properties are significantly enhanced. It is regarded that more complicated doped to $BiFeO_3$ based on La doping will be able to further improve the ferroelectric and ferromagnetic properties of $BiFeO_3$.

We tried to dope La and Pb as co-doping elements for the A site of $BiFeO_3$. $La_{0.1}Bi_{0.9}FeO_3$ (La: BFO), $Pb_{0.1}Bi_{0.9}FeO_3$ (Pb: BFO) and $(La_{0.1}Pb_{0.1})Bi_{0.8}FeO_3$ (La, Pb: BFO) thin films on $Pt/Ti/SiO_2/Si$ substrates were fabricated. Thin films were deposited using a pulsed laser deposition (PLD) system. The third harmonic generation of Nd: YAG laser with a wavelength of 355 nm and a repetition rate of 10 Hz was used as the laser source. The ceramic targets of doped $BiFeO_3$ were fabricated by a traditional solid-state reaction method. The starting materials of Bi_2O_3, Fe_3O_4, La_2O_3, PbO were weighed according to the molecular ratio with 10 mol% extra Bi_2O_3. They were mixed, pressed into pellets and sintered at 800 °C for 3 h. Then the ceramics were crushed, ground, pressed into pellets and sintered again at 880 °C for Pb: BFO and 960 °C for La: BFO and La, Pb: BFO for 1h. All of the thin films were fabricated at 550~580 °C and in 20 mTorr dynamic oxygen pressure.

Figure 1 shows the XRD patterns of the targets and thin films of doped $BiFeO_3$. All of them showed the structure of single BFO phase with space group *R*3*c*. However, extra peaks, which were identified as $Bi_2Fe_4O_9$, were observed in the XRD patterns. The impurity phase also included Pb oxide in La, Pb: BFO and Pb: BFO ceramics.

Figure 2 shows the surface morphology of the La, Pb: BFO, La: BFO and Pb: BFO thin films. Although the growth mechanism is not very clear for now, it can be found that root-mean-square roughness (R_{rms}) and grain size (S) are different when doped with different ions: R_{rms} (LaPbBFO) < R_{rms} (PbBFO) < R_{rms} (LaBFO), S (LaPbBFO) < S (PbBFO) < S (LaBFO). But each thin film shows the similar morphology with special tendency that looks like isosceles triangles with the vertex angle pointing to the same direction.

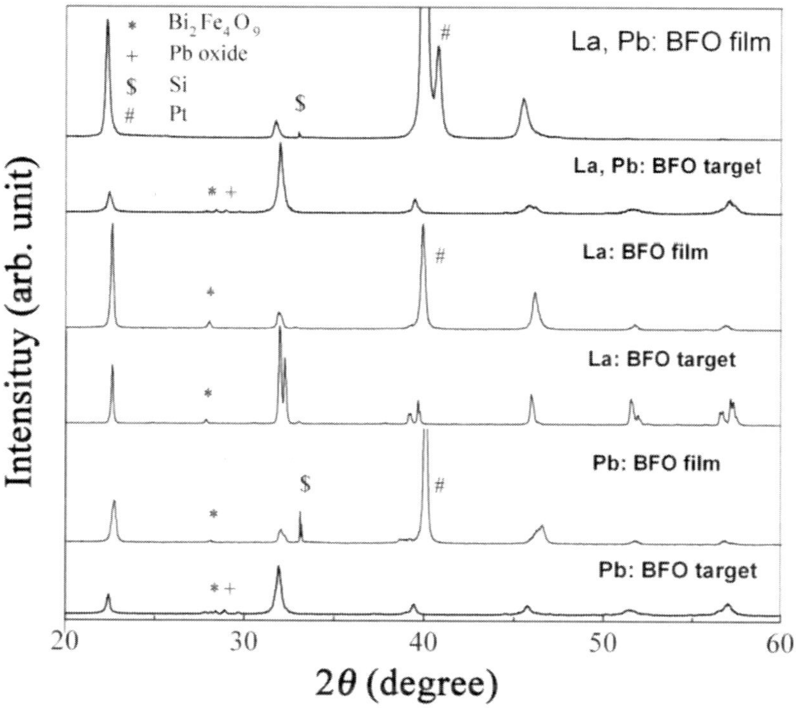

Figure 1. XRD patterns of the ceramics and films of La, Pb: BFO, La: BFO and Pb: BFO.

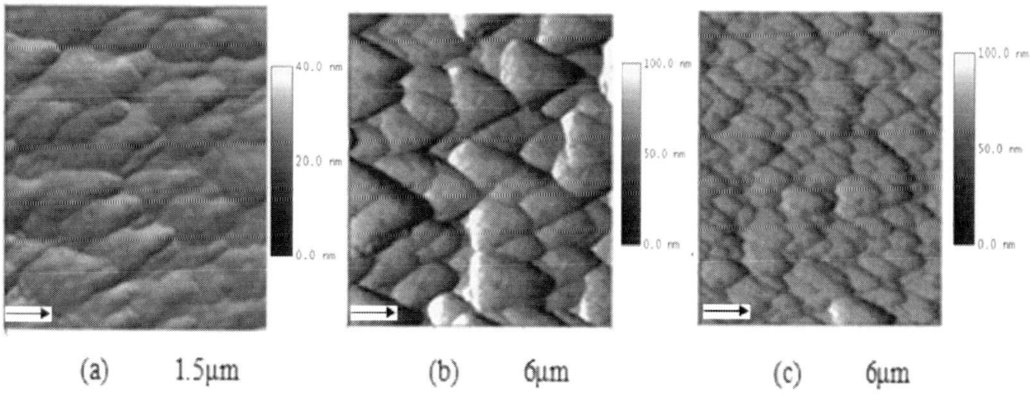

Figure 2. Surface morphology of thin films. (a) La, Pb: BFO, (b) La: BFO and (c) Pb: BFO.

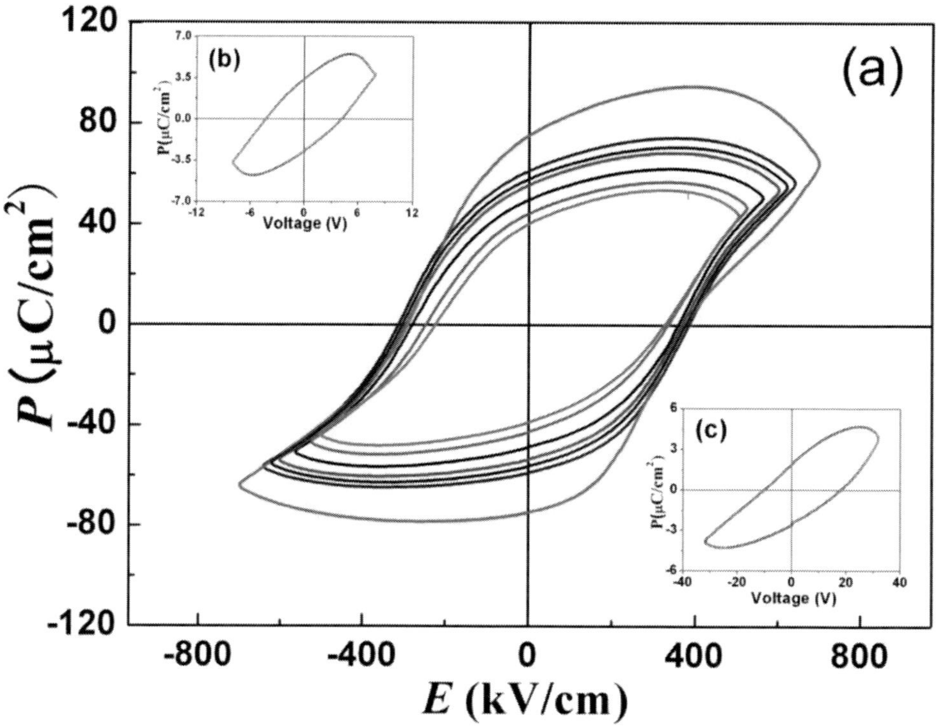

Figure 3. Ferroelectric hysteresis loops (*P-E* loops) of the films on the same Pt/Ti/SiO_2/Si substrate (a) La, Pb: BFO, (b) La: BFO and (c) Pb: BFO.

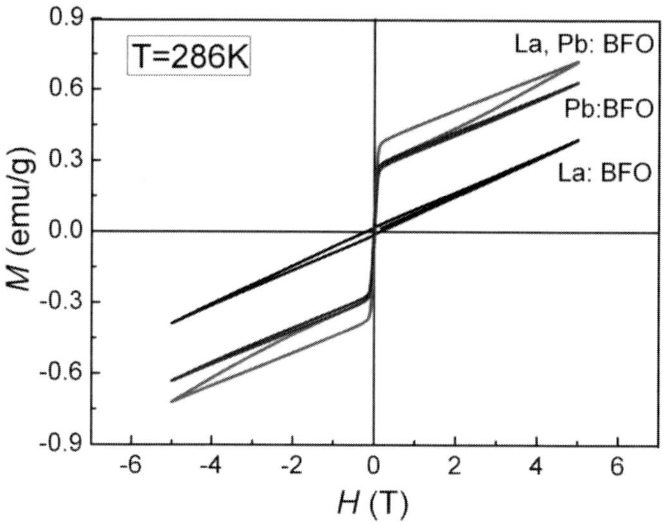

Figure 4. The magnetization curves of the doped $BiFeO_3$ targets at room temperature.

Figure 3 shows the electrical polarization hysteresis loops (*P-E* loops) of the La: BFO, Pb: BFO and La, Pb: BFO films. As shown in this figure, well-saturated hysteresis loops are observed for La, Pb: BFO thin films with the maximum electric field of 700 kV/cm, where P_r and E_c are 150 μC/cm^2 and 291 kV/cm, respectively. As expected the La, Pb: BFO thin film shows a much improved hysteresis loop than La: BFO and Pb: BFO. This may be because that La^{3+} (0.106 nm) and Pb^{2+} (0.119 nm) cations are larger than Bi (0.103 nm), in which the $6s^2$ lone pair cause the off-center distortion in perovskite structure and produces polarization [19]. In perovskite structured ferroelectric materials, there is some relation between the polarization and the size of the A- and the B-site cations. That is to say, the polarization may increase when using a larger A-site cation, which is consistent with other reported results [20].

The magnetization hysteresis (*M-H*) loops of the $BiFeO_3$ thin films could not be observed due to the very strong diamagnetic signal from the substrate [21]. Therefore, the targets of different thin films are measured instead of thin films to compare their magnetic properties.

Figure 4 shows the magnetic properties of the bismuth ferrite bulk samples in mm size. The difference between the three samples is observed. The co-doping of La, Pb: BFO sample shows much improved magnetic properties than the La: BFO and the Pb: BFO. Because of the co-doping of La and Pb, the antiferromagnetic ordered cycloid spin structure of BFO is changed and the magnetization is improved.

The field cooled and zero field cooled (FC-ZFC) magnetization curves were measured below room temperature in a field of 0.5 T (5k Oe), the FC-ZFC measurement revealed that the magnetization decreased linearly as the increase of the temperature below 300 K. The FC curve shows a variation in the temperature range of 30 and 50 K, with a sharp cusp around 40 K. This anomaly is similar with BFO crystal (around 50 K), which reveals the spin-glass behavior in BFO.

3. Enhanced Properties in Multiferroic La and Nb Co-Doping $BiFeO_3$ [11]

Un-doped $BiFeO_3$ film on the $Pt/Ti/SiO_2/Si$ substrate shows a rhombohedral structure with space group *R*3*c*, in apparent contrast to the previous report of a tetragonal *P*4*mm* structure for $BiFeO_3$ polycrystalline thin film on the same substrate [23]. This is based on the character of the single 012 peak around 22° and the splitting of the 104 and 110 peaks around 32°. The $Bi_{0.9}La_{0.1}FeO_3$ film maintains a rhombohedral structure, which is manifested by the splitting of the 006 and 202 peaks around 40°. However, the $Bi_{0.8}La_{0.2}FeO_3$ thin film shows a tetragonal structure, according to the splitting of the 001 and 100 peaks around 22° and the single 111 peak around 40°. The doping of a tiny amount of Nb into La doped $BiFeO_3$ does not change its structure; therefore, the $Bi_{0.9}La_{0.1}Fe_{0.99}Nb_{0.01}O_3$ thin film and the $Bi_{0.8}La_{0.2}Fe_{0.99}Nb_{0.01}O_3$ thin film have the rhombohedral structure and the tetragonal structure, respectively. The La doped $BiFeO_3$ thin films show an evolution of the structure, which shows a similar trend to that of their bulks. Nevertheless, pure and 10 mol% La doped, including Nb co doped, $BiFeO_3$ thin films do not show such strong rhombohedral distortion as their respective bulk samples, so that a tendency towards tetragonal symmetry is apparent. Besides the differences observed above, the texture habits of these thin films on $Pt/Ti/SiO_2/Si$ substrate are quite different. Un-doped $BiFeO_3$ film shows strong epitaxial growth along the $[012]_h$ direction, while the other films prefer to grow along the $[001]_h$ direction for rhombhedral symmetry or the $[111]_t$ direction for tetragonal symmetry, except for the 20 mol% La doped $BiFeO_3$ film, which shows more preferred growth along its $[101]_t/[110]_t$ direction than its $[111]_t$ direction.

Figure 5 shows the ferroelectric *P-E* loops of doped and un-doped $BiFeO_3$ thin film capacitors. The un-doped $BiFeO_3$ film has a sensing margin, $2P_r$ (P_r^+-P_r^-) of 135 μC/cm^2 at an applied electrical field with a maximum value of 547 kV/cm. Note that freely movable charges appear to contribute significantly to the electrical hysteresis loop, as manifested by the rounded features of the loops, which means that the electrical leakage is high in these thin films. The remnant polarization of 10 mol% La doped $BiFeO_3$ thin film has a $2P_r$ value of 137 μC/cm^2 at the breakdown voltage, which is not much different in comparison to the pure $BiFeO_3$. Yet the contribution to the polarization from the freely movable charges is reduced, as evidenced by the less rounded features of the loops. The contribution to the polarization from the freely movable charges is further reduced in the 1 mol% Nb doped $Bi_{0.9}La_{0.1}FeO_3$ thin film, and a $2P_r$ value of 142 μC/cm^2 was observed. In comparison with the pure and 10 % doped $BiFeO_3$ film, rounded feature was not presented, that is, there was no contribution from the freely movable charges in the *P-E* loops of the 20 mol% La doped $BiFeO_3$ thin film. Simultaneously, the value of $2P_r$ was enhanced up to 145 μC/cm^2. Addition of 1 mol% Nb in the 20 mol% doped $BiFeO_3$ film further increases the remnant polarization $2P_r$ value up to 160 μC/cm^2. At the same time, well developed *P-E* loop was observed without any observable features pointing to electrical leakage. Another phenomenon observed from the comparison of the *P-E* loops was the significant increasing of the breakdown voltage by the La and Nb co-doping.

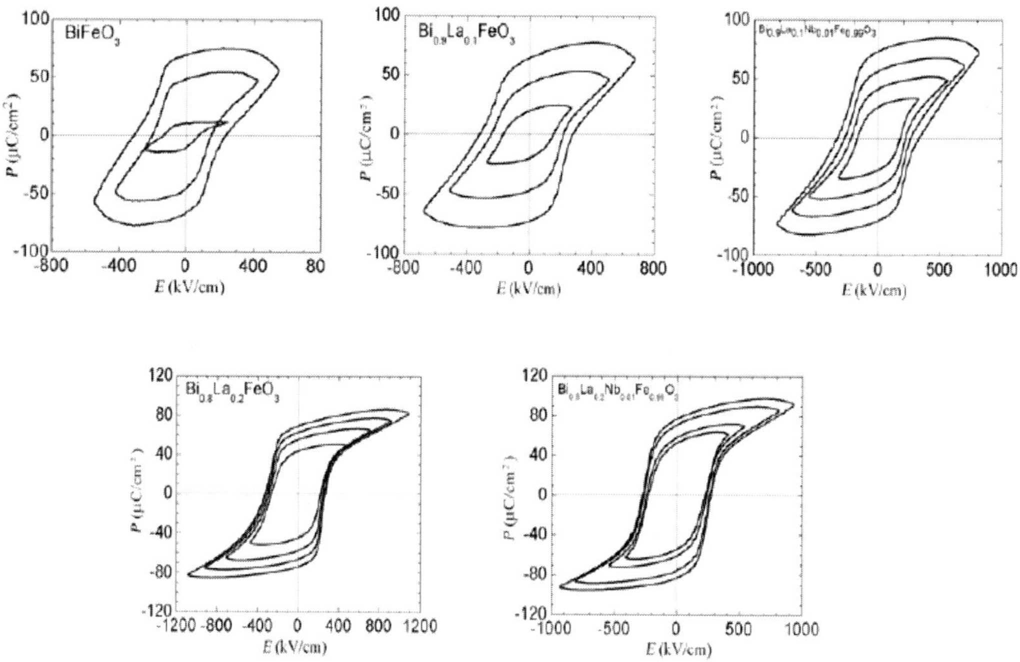

Figure 5. Ferroelectric polarization loops (*P-E* loops) of doped $BiFeO_3$ thin film capacitors on $Pt/Ti/SiO_2/Si$ substrate at a frequency of 100 Hz and room temperature [11]. Significant improvements of the electric polarization due to the reduction in the density of the freely moving charges are observed.

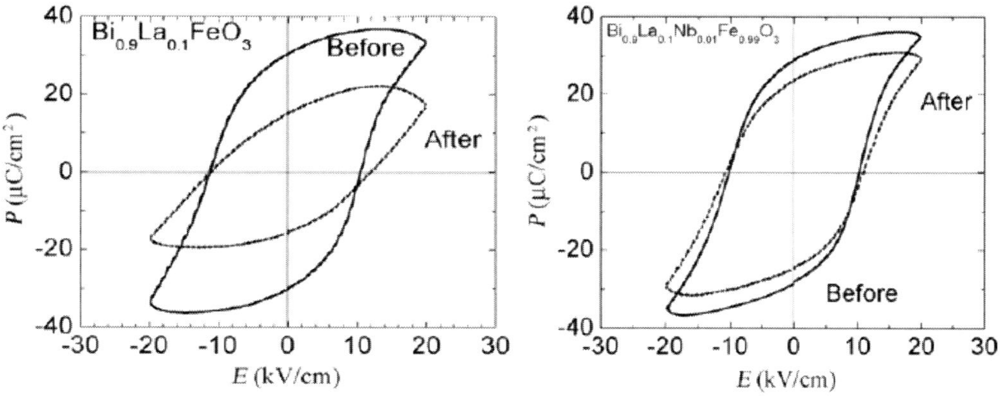

Figure 6. Comparison of fatigue measurements of the $Bi_{0.9}La_{0.1}FeO_3$ and $Bi_{0.9}La_{0.1}Nb_{0.01}Fe_{0.99}O_3$ thin film samples [11]. The *P-E* loops before fatigue testing (marked 'Before') and after the test (marked 'After') are also plotted. The polarization measurement was at 20 V, the switching voltage was 12 V, and the frequency was 5 kHz.

The La doping into $BiFeO_3$ stabilizes the formation of the $BiFeO_3$ phase and reduces the formation of iron rich phases, for example for the $Bi_2Fe_4O_9$ phase on the bulk materials [24,25]. The formation of the iron rich phase will thus change the ratio of elements in the thin film and lead to the production of oxygen vacancies and iron ions with different valences. Oxygen vacancies are one of the main sources of movable charges (free electrons) in $BiFeO_3$. Ions with varying valences will also cause increased conductance, for example, possibly through a double exchange mechanism via Fe^{2+}-O-Fe^{3+} chains. Therefore, the movable charges are significantly reduced by the doping of La. Nb doping further reduces the movable charge density in La doped $BiFeO_3$. It is believed that the mechanism by which the Nb doping works is similar to that for the La doping, through the reduction of oxygen vacancies by a charge neutralization effect due to the higher valence of Nb (+5) than Fe.

A precise comparison of the enhancement in electrical polarization in $BiFeO_3$ by La doping and co-doping of La and Nb is difficult, due to the breaking down of $BiFeO_3$ and 10 mol% La doped $BiFeO_3$ thin films before they were fully switched, but enhancement was observed in the fully switched thin films. It is proposed here that the observed enhancement in remnant polarization is mainly due to the fact that the doped thin films being more insulating, but enhancement from further structural distortion caused by the doping effect cannot be simply ruled out, but such contribution should not be large if any. This is based on a recent observation of a large polarization of about 60 μC/cm^2 in $BiFeO_3$ single crystal, which is very close to the value observed in tetragonally distorted $BiFeO_3$ thin film. This is in contrast to the very small polarization observed in pure $BiFeO_3$ in the early work, proving that tetragonal distortion may not be necessary for large electrical polarization [5,23,26]. Another contribution to the observed enhancement in the polarization in thin film samples is the preferred epitaxial growth direction. It is known that the polarization direction in the rhombohedrally or tetragonally distorted $BiFeO_3$ films is along its $[001]_h$ or the perovskite pseudo-cubic $[111]_C$ direction. Un-doped $BiFeO_3$ thin films on Pt/Ti/SiO_2/Si substrate prefer to grow along the rhombohedral $[012]_h$ direction, which involves great deviation from the polarization maximum direction of $[001]_h$. For the 10 % La doped $BiFeO_3$ thin films, including the 1% Nb co-doped sample, the preferred growth direction is $[001]_h$, which is the polarization maximum direction. Similarly, 20 mol% La doped samples, especially the Nb and 20 mol% La co-doped one, show preferred growth along the $[111]_C$ direction, which is the polarization maximum direction. The preferred growth along the polarization maximum direction in doped and co-doped samples will directly contribute to the observed polarization enhancement.

For the frequency dependence of the real part ε' and the imaginary part ε" of the dielectric permittivity of the doped and un-doped thin film capacitors on $Bi_{0.8}La_{0.2}Nb_{0.01}Fe_{0.99}O_3$, $Bi_{0.9}La_{0.1}Nb_{0.01}Fe_{0.99}O_3$, $Bi_{0.8}La_{0.2}FeO_3$, $Bi_{0.9}La_{0.1}FeO_3$, and $BiFeO_3$, the apparent differences in the dielectric permittivities due to the doping are obvious. Similarly, the dielectric loss is significantly reduced with La and Nb co-doping. At room temperature, the ε" values are 0.038, 0.047, 0.69, 1.78, and 6.47 at 100k Hz respectively, for these five above mentioned films. The five samples can be divided into two groups. One group includes three samples: the solely La doped and un-doped $BiFeO_3$ films, while the other group includes the two of the Nb and La co-doped samples. In contrast to the large loss and small ε' shown by the non-Nb doped samples, the Nb doped samples have large ε' but with small loss. This indicates that Nb doping drastically reduced the free movable charge density in the $BiFeO_3$ thin film, while non-La doping causing the reduction of the density of the freely movable charges in $BiFeO_3$ is apparent. This observation is in accordance with the results from the electrical polarization measurements, that is, due to the significant reduction in the movable charge density, Nb doped $BiFeO_3$ thin films show perfectly saturated *P-E* loops.

Fatigue measurements of the doped $BiFeO_3$ thin film capacitors are shown in Figure 6. The 10 mol% La doped $BiFeO_3$ thin film does not show much improvement in fatigue resistance in comparison to the un-doped thin film. After introducing Nb into La: $BiFeO_3$, the fatigue resistance was improved. The remnant polarization decreased 23 % for the $Bi_{0.9}La_{0.1}Nb_{0.01}Fe_{0.99}O_3$ film in comparison to the 37 % decrease for the $Bi_{0.9}La_{0.1}FeO_3$ film after 10^7 read/write switching cycles. A similar improvement was also observed for the Nb doping into 20 mol% La doped $BiFeO_3$ thin film. The fatigue in perovskite oxides is mainly due to the diffusion and relocation of oxygen vacancies and electronic-charge trapping at the domain wall during the electrical switching. The reduction in the movable charge density through doping is the direct cause for the improvement of the fatigue resistance. It also proves that Nb doping is more effective than La doping in reducing the charge density.

The magnetic hysteresis loops of pure and doped $BiFeO_3$ thin film samples at room temperature show strong diamagnetic signals, it is because of the dominant amount of the substrate in the samples. However, the magnetic field cooled and zero field cooled magnetization measurements clearly show that these thin films have magnetic moment at room temperature. The magnetic hysteresis loops of target bulks show that the doping of Nb and La enhanced the magnetic moment in the bulk. This result is consistent with the observed enhancement of magnetic moment in La doped $BiFeO_3$ ceramics from other reports [24,25].

4. ND: $BiFeO_3/Bi_2FeMnO_6$ BILAYERERD FILMS [27,28]

ABO_3 perovskite structured ferroelectric materials show antiferromagnetic ordering for the same B site magnetic element, except that $BiMnO_3$ is ferromagnet. While for the $A_2BB'O_6$ double perovskite oxides, the combination between B and B' gives rise to a ferromagnetic coupling. They are also expected to be multiferroic materials. The ferroelectric polarization is induced by the distortion, which usually causing a lowering in symmetry. For device application, a large magnetoelectric effect is expected in the $BiFeO_3$ and bismuth-based double perovskite oxides ($BiBB'O_6$), many of which have aroused great interest like Bi_2NiMnO_6 [29,30], $BiFeO_3$-$BiCrO_3$ [29,31]. But few researches have focused on Bi_2FeMnO_6. It is particularly interesting to investigate the multilayered films containing Nd: $BiFeO_3$ and Bi_2FeMnO_6 nanolayers. In the former research, the doping of Nd is proved to further improve the ferroelectricity, so the Nd doping was chosen to study further. According to Goodenough-Kanamori's rules, the 180 degree -Fe^{3+}-O-Mn^{3+}- bonds is quasistatic, partly because the strong Jahn-Teller uniaxial strain in an octahedral site. In the as-grown film, it includes the diamagnetic contribution of the substrate, the paramagnetic contribution of Nd, the antiferromagnetic contribution of $BiFeO_3$ and the couplings of -Fe-O-Mn-. The ferromagnetism would be expected to overcome the other effects in the as-designed films. Here, $Nd_{0.1}Bi_{0.9}FeO_3$ (Nd: $BiFeO_3$), Bi_2FeMnO_6 and bilayered Nd: $BiFeO_3$/Bi_2FeMnO_6 films on Pt/Ti/SiO_2/Si substrate were fabricated. Nd: $BiFeO_3$ films were fabricated at 550 ~ 580 °C with 200 mTorr dynamic oxygen pressure, Bi_2FeMnO_6 films were also fabricated at 550 ~ 580 °C with ~10^{-5} Torr.

Both Nd: $BiFeO_3$ and Bi_2FeMnO_6 phases were observed by the XRD measurements in the bilayered film and there are no other second phases except some peaks from the Pt/Ti/SiO_2/Si substrate. For Bi_2FeMnO_6 films, it shows complicated structure when deposited at different conditions. Bi et al [32] has calculated three structures of Bi_2FeMnO_6 with the space group of $Pm\bar{3}m$, $R3$ and $C2$. It is observed in our experiments for Bi_2FeMnO_6 films, which will be shown elsewhere.

For the surface morphology of the Nd: $BiFeO_3$/Bi_2FeMnO_6 and Nd: $BiFeO_3$ films, Root-mean-square roughness (R_{rms}) and grain size (S) are different: R_{rms} (Nd: $BiFeO_3$) < R_{rms} (Nd: $BiFeO_3$/Bi_2FeMnO_6) < R_{rms} (Bi_2FeMnO_6), S (Nd: $BiFeO_3$) < S (Nd: $BiFeO_3$/Bi_2FeMnO_6) < S (Bi_2FeMnO_6). The morphology of Nd: $BiFeO_3$ film on the Bi_2FeMnO_6 Pt/Ti/SiO_2/Si indicated that Nd: $BiFeO_3$ film has a larger growth rate on Bi_2FeMnO_6 than on the Pt/Ti/SiO_2/Si substrate.

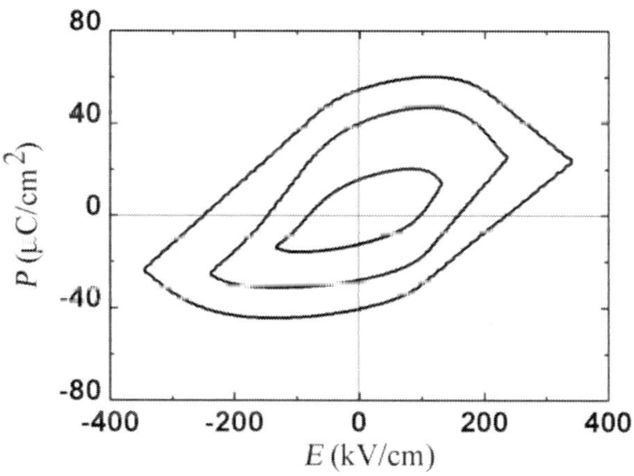

Figure 7. *P-E* hysteresis loops of the Nd: $BiFeO_3$/Bi_2FeMnO_6 film [27].

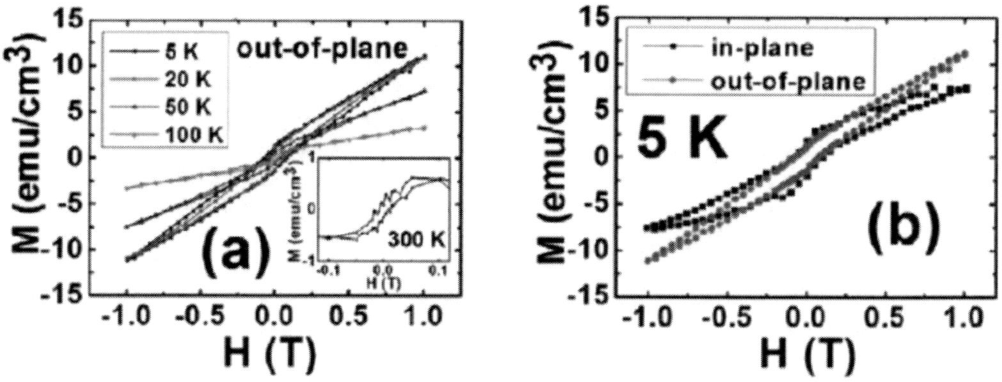

Figure 8. Magnetization hysteresis loops of LBFM film (a) measured at various temperatures and the inset shows a magnification of the hysteresis loops at 300 K; (b) measured with magnetic field applied parallel and perpendicular to the sample plane [28].

Figure 7 shows *P-E* loops of the Nd: $BiFeO_3/Bi_2FeMnO_6$ films. The remnant polarization P_r was 54 μC/cm^2 and E_c was 237 kV/cm. Some anomalies were observed in the *P-E* loop: the loop is asymmetry and the polarization decreased as the increasing of the electric field which was also observed in Ref. [33] in the case of low frequency. This is caused by many reasons but some of them can be neglected like the macroscopic electrode influence and non-uniform polarization on the surface of the film. We consider that there are two main reasons. The film is insulating so there is no movable carriers to balance the bound charge. Therefore, the polarization gradient will be arisen in the film and induced the depolarization field. In addition, there are inhomogeneous domains with different coercivity in the film, some of which are difficult to switch with applied field. Evidence can also be seen in the fatigue results, which showed that the polarization increased with the increasing of the switching cycles. The fatigue can be caused by the domain nucleation, the domain wall pinning due to space charges or oxygen vacancies, interface between the electrode and the film, thermodynamic history of the sample and so on. For the unusual profile of fatigue (polarization increased with the increasing of switching cycles), the different domain walls played important roles during the polarization reversal. Frequency dependence of the real part of the permittivity is measured at room temperature. There is a notable increase at low frequencies. In such bilayered films, there are space charges at the interface between the two layers of Nd: $BiFeO_3$ and Bi_2FeMnO_6 which will affect the ferroelectric properties [34]. As the definition of ferroelectricity is strict, the positive-up-negative-down (PUND) test is necessary [35]. The switching polarization is observed using the triangle waveform as a function of time.

The magnetic moment versus magnetic field loops (*M-H* loops) of Nd: $BiFeO_3/Bi_2FeMnO_6$ film at room temperature (RT) and 5 K were observed. When the room temperature magnetic field dependence of "in-plane" ($H_{\parallel}$) and "out-of-plane" ($H_{\perp}$) magnetization are compared, clear magnetic anisotropy is observed. The hysteresis was also observed both in RT and 5 K loops. The film shows typical ferromagnetic properties and highly saturated magnetization (25 emu/cm^3 at RT and 83 emu/cm^3 at 5 K). The profiles especially the room temperature loop were affected by the substrate which has strong diamagnetic signals. When the saturation magnetization M_s as a function of temperature was measured in below room temperature, the magnetization showed a $T^{1.6}$ dependency below room temperature. The temperature dependence of magnetization in saturation could be described by $M_s(T)=M_s(0)[1\text{-}b(T/T_c)^{1.6}]$ with the best fit for b value of $1.02\pm0.11\times10^{-4}$ $K^{-3/2}$. This fit gives an estimated ferromagnetic transition temperature of T_c=440 K. Based on the hysteresis loop measured at 5 K, the average magnetons per B-site cation are 4.49 μ_B. This value is consistent with those anticipated for high-spin states of Mn^{3+} and Fe^{3+} ions. The observed ferromagnetism may be the result of a 5-degree canted antiferromagnetic ordering that leaves a small net magnetic moment and modest Curie temperature [36,37].

The La: Bi_2FeMnO_6 (LBFM) diffraction peaks of 100, 200 and 300 were observed in the XRD pattern. It indicates the epitaxial growth of the LBFM film on the (100) STO substrate. There is no traceable impurity in the film. But there were unavoidable impurities of bismuth oxides in LBFM ceramics, such impurities also further worsen the crystalline quality in the ceramic samples.

As shown in Figure 8(a), the *M-H* loops of the LBFM film were measured at 5, 20, 50, 100 and 300 K. The inset shows the magnification of the hysteresis loops at 300 K. There was no change of loop width below 20 K, but at higher temperature the loop width decreases rapidly as the temperature increase. The hysteresis doesn't disappear at room temperature. Figure 8 (b) shows the in-plane and out-of-plane magnetic field dependence of magnetization measured at 5 K. The film shows anisotropy, and the out-of-plane was the easy magnetization direction [38,39]. The out-of-plane zero-field-cooled (ZFC) and field-cooled (FC) magnetization curves measured between 5 and 350 K are shown in Figure 9. The *M-H* loops and the ZFC-FC results indicate that the LBFM film has a Néel temperature (T_N) above room temperature. A spin-glass-like behavior was occurred at 18 K at 0.1 T (1k Oe) applied field. This behavior is similar to the reported Bi_2FeMnO_6 film [38]. It is not overwhelmed at 1 T (10k Oe) but the peak temperature changes from 18 to 10 K. As we know, the AC magnetization measurement is very powerful for the investigation of spin glass behavior. We have performed the frequency dependent measurement of magnetization but there are many noises because of the influences of the substrate. The results are not shown because they are useless for the analysis of spin-glass. In such a case, whether the phenomena in the LBFM film can be classified, as a classical spin-glass still remains unknown to us.

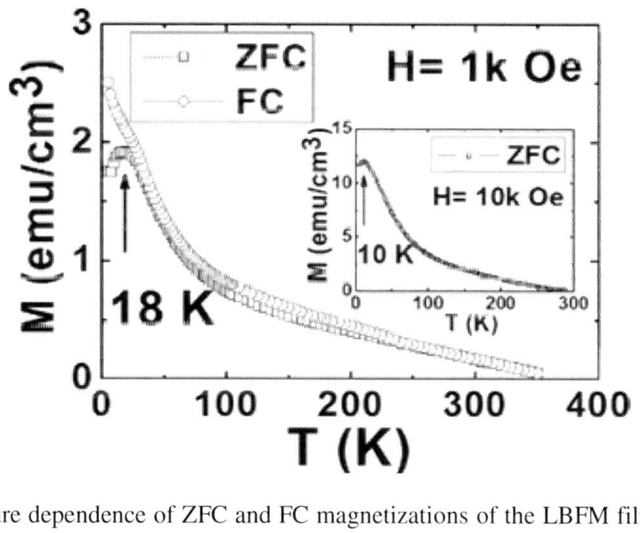

Figure 9. Temperature dependence of ZFC and FC magnetizations of the LBFM film recorded at 0.1 T (1k Oe) [28]. The inset shows the ZFC results at 1 10k Oe.

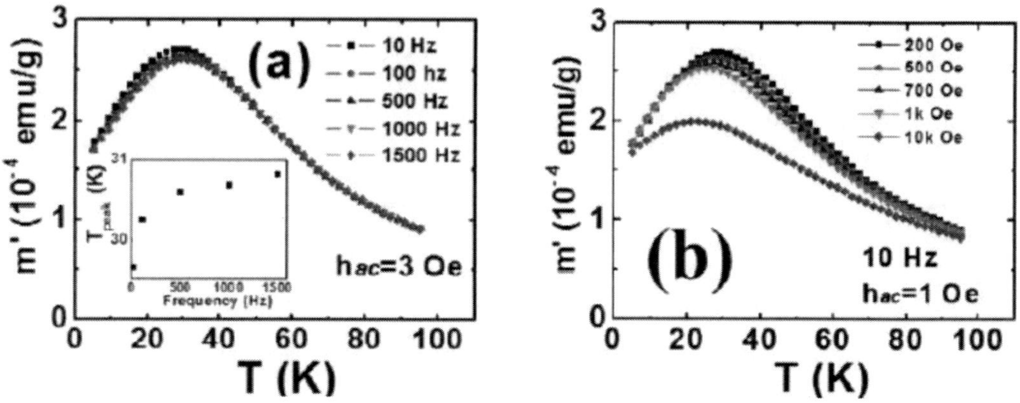

Figure 10. (a) Temperature dependence of the ac magnetization obtained with different frequencies, inset shows the shift of the peak temperature (T_{peak}) with frequency; (b) The ac magnetization [28].

For the magnetization hysteresis loops of the LBFM ceramic measured at various temperatures, the sample shows unsaturated magnetization for an applied magnetic field up to 1 T (10k Oe). The magnetizations at 10k Oe for LBFM ceramic were 2.88, 1.94, 0.46 and 0.26 emu/g, corresponding to the temperature of 5, 50, 160 and 300 K, respectively. The inset shows the amplification of the loop at 160 K, a very small magnetic hysteresis was observed. On the other hand, the hysteresis disappears at 300 K, which is different from the LBFM film (see Figure 8 (a) inset). Because the structure and properties of the LBFM films will depend on the substrate type, film thickness and processing conditions during the deposition. As shown in Figure 8 (b), the stress induced anisotropy from LBFM/STO lattice mismatches was observed.

ZFC and FC curves of the LBFM ceramic were recorded at different applied field of 500, 1k, 5k and 10k Oe. Similar to the Bi_2FeMnO_6 film [38] and the LBFM film, a spin-glass-like behavior was also observed. The ZFC-peak was not suppressed as the field reached 10k Oe and its position shows field dependence. It is difficult to obtain the exact transition temperature. However, it is clear that the ZFC-peak temperature gradually decreases as the applied field increases. Above the spin-glass behavior, the antiferromagnetic transition was observed at 250 K. Bi et al [32] has calculated three structures of Bi_2FeMnO_6. For LBFM, we believe that it exhibits phase transition but the exact phases were unclear. The data from 350 to 250 K were well fitted to the Curie-Weiss law (curve not shown here). The parameters of Curie constant (C), effective paramagnetic moment (μ_{eff}) and Weiss constant (θ) were determined according to the Curie-Weiss fits to be 1.92 emu K mol^{-1} f.u., 3.91 μ_B and -9 K respectively.

As shown in Figure 10 (a), from the real part of the ac susceptibility measured at different frequencies, we can see a peak, which changes very little with different frequency (inset figure). The possible shift to high frequency is not obvious, that is partially due to the less points measured around the peak. However, it can be found that all the curves on the left side shift clearly to high frequencies, while the curves on the right side of the peak remain the same as that of the lowest frequency. Figure 10 (b) shows the real part of the ac magnetization versus temperature at 10 Hz fixed frequency and h_{ac}= 1 Oe. Different static fields of 200, 500, 700, 1k and 10k Oe were applied. It clearly shows that the values of m' decreases and the peak moved to lower temperatures as the field increases. This indicates that the possible spin glass state takes place at lower temperature with increasing magnetic field. In another word, the magnetic field seems to push the spin glass to occur at lower temperatures.

There are hysteresis loops below the spin glass transition temperature and a small negative θ value was obtained through the Curie-Weiss fitting. Therefore, similar to the $BiFeO_3$ [40], there are at least three clear evidences for spin-glass behavior in the LBFM ceramic: first, there is a large difference between FC and ZFC at low temperatures; second, the cusp was observed in ZFC curves; and thirdly, the peak temperature was dependent upon the frequency and applied field. Based on these data, we propose that the spin glass state probably arises from a weak ferromagnetic or ferrimagnetic transition around the T_g, or from the competition between ferromagnetic and antiferromagnetic transition.

5. Piezoresponce Force Microscopy Study of $BiFeO_3$ Thin Film

$BiFeO_3$ (BFO) material is known to be both ferroelectric and antiferromagnetic. So far more attention is given in studying ferroelectric domains of BFO thin films and their dynamics behavior because of their enhanced remnant polarization (*P*r) and magnetization compared to its single crystal [5], allowing better potential use in practical application.

Piezoresponse Force Microscopy (PFM) based on the Atomic Force Microscope (AFM) has become a powerful method for imaging nanoscale ferroelectric domains and characterizing local electromechanical properties [41-44]. In PFM, the tip detects the piezoelectric vibration signal of the ferroelectric sample caused by an ac voltage applied between the tip and the counter electrode of the sample, due to the inverse piezoelectric effect. The amplitude and the phase of the piezoresponse signal is related to the magnitude of the piezoelectric coefficient and the direction of local polarization, respectively. PFM has been employed to investigate domain structures in BTO thin films. Here our investigation through PFM in nanoscale domain structures, their polarization retention behaviors and local properties of BFO thin films will be presented.

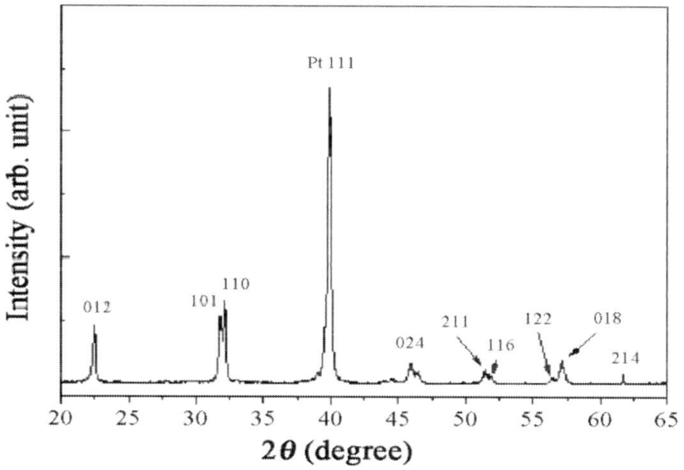

Figure 11. XRD pattern of $BiFeO_3$ thin film.

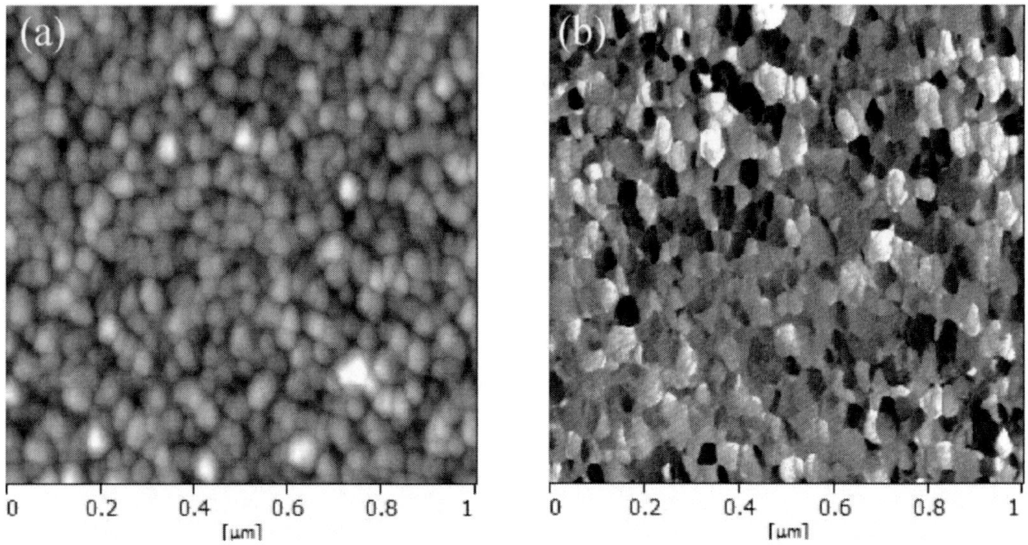

Figure 12. Topography image (a) and corresponding piezoresponse image (b) of $BiFeO_3$ thin film.

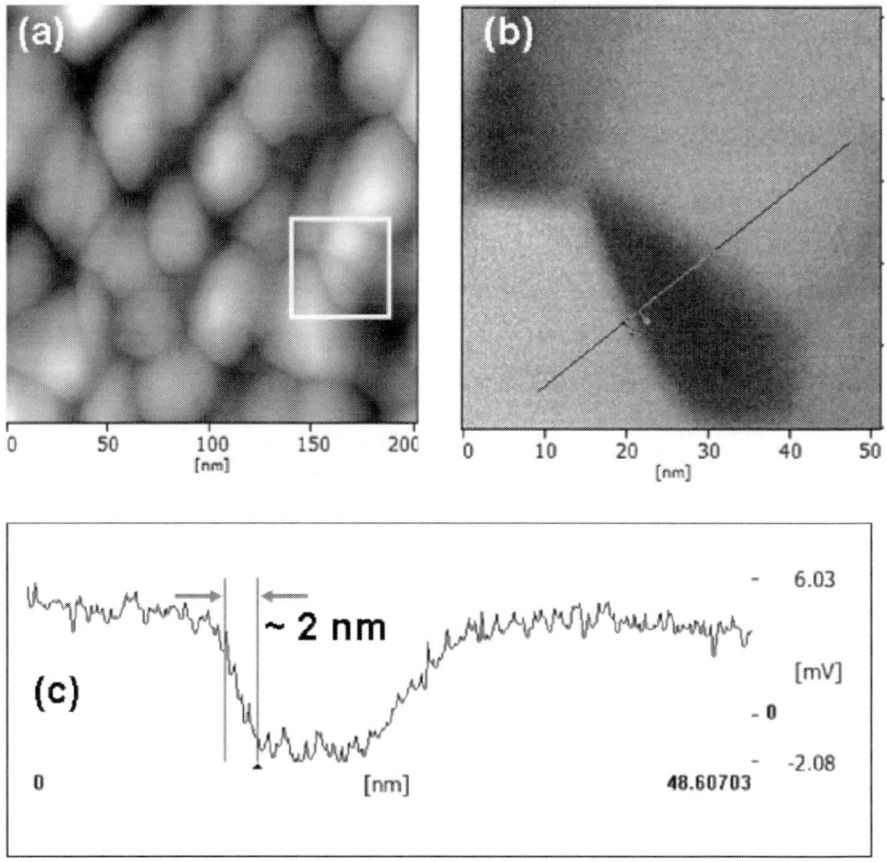

Figure 13. Topography image (a) and piezoresponse image (b) of selected area in (a) of the $BiFeO_3$ thin film. (c) shows the line of piezoresponse signal in (b) – marked by the line in (b).

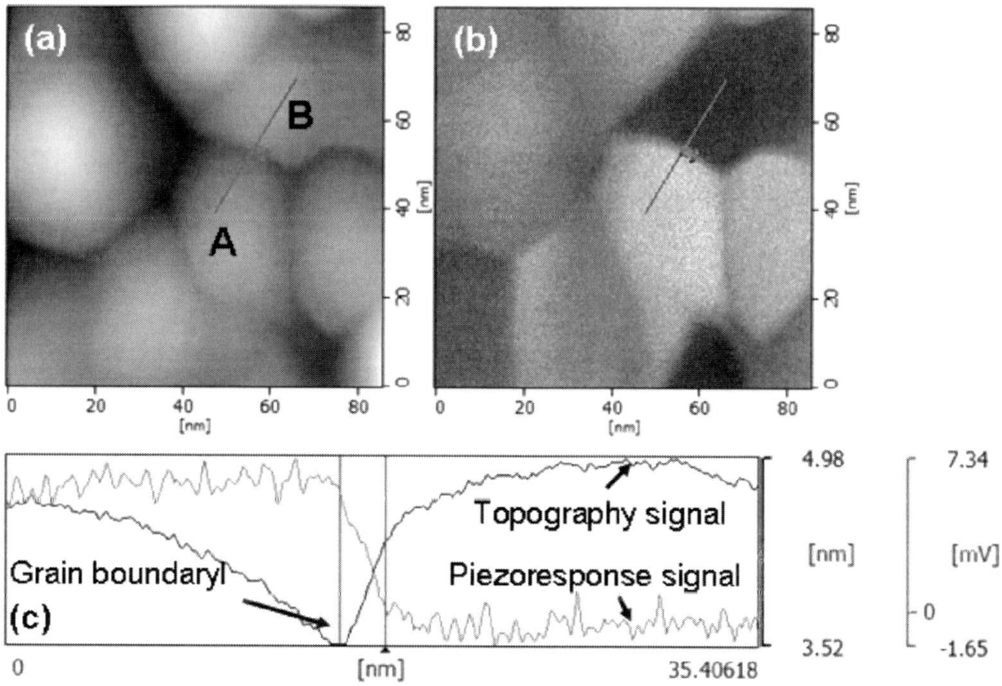

Figure 14. Topography image (a) and piezoresponse image (b) of $BiFeO_3$ thin films. (c) shows the line piezoresponse signal as marked in (a) and (b).

BFO thin film was prepared via spin-coating on the Indium Tin Oxide (ITO) bottom electrode with glass substrate. The starting reagents were Bi-nitrate and Fe-nitrate. They were dissolved in acetic acid and 2-methoxyethanol to yield a solution. Pyrolysis was conducted on hot plate at 120 °C for 10 minutes and subsequently annealed at 550 °C for 30 minutes. The final film's thickness after annealing was approximately 230 nm. The film crystalline structures were investigated by Rigaku D/MAX-2550V X-ray diffractometer with Cu target (Cu*K*α, 0.154 nm). Ferroelectric domains were imaged by a custom-built PFM, which was based on a commercial AFM (SPA 400, Seiko Inc., Japan). A commercial tip cantilever system (NSC35/Ti–Pt, MicroMash) with spring constant of 14 nm^{-1} was used. Domain visualization was performed under an externally applied ac voltage with the amplitude V_{ac} = 1.2 V and frequency f = 8.5 kHz. All piezoresponse images were taken at room temperature.

Figure 11 shows the XRD pattern of the BTO thin film on ITO/glass substrate, which shows the rhombohedra polycrystalline structure. Figure 12(a) and (b) shows the topography image and the corresponding piezoresponse image of the BTO film in 1×1 μm^2. The average grain size was about 25 nm. The topography in (b) demonstrates complicated bright, dark and gray piezoresponse patterns. Some grains with individual bright or dark contrast indicate single domain state with upward or downward polarization orientation. High-resolution piezoresponse images reveal that the 8-nm-size grain also exhibit strong piezoresponse contrast (Figure 13(a), (b)). The line scanning signal domain wall was around 2 nm in width (Figure 13(c)).

Further more, by comparing the topography signal profile of the corresponding piezoresponse signal at the grain boundary of the BFO thin film, some very interesting information about the relation between domain wall and grain boundary was revealed. Figure

14 (a)-(c) show the topography image, the corresponding piezoresponse image and the scanning signal profile, respectively. The grain A shows the single domain state, as well as the grain B, as indicated by their piezoresponse contrasts in Figure 14(a). The piezoresponse signal through grain A and B in Figure 14(c) exhibits an exact steep decrease trend as exactly at the grain boundary, which is obviously as a result of non-ferroelectric phase at the grain boundary. While most interestingly, the piezoresponse variation region was not consistent with the grain boundary extension region in Figure 14(c), the former is about two times than that of the latter. This phenomenon gives a direct, spatial imaging for understanding the role of the grain boundary in affecting the domain evolution, which was generally considered as an energy barrier for the domain growth. Our results in the BFO thin film indicates clearly that nano-grain boundary acts as an energy barrier for the domain growth, and the local piezoresponse signal here was hence subjected to a more abrupt change compared to the corresponding topology variation.

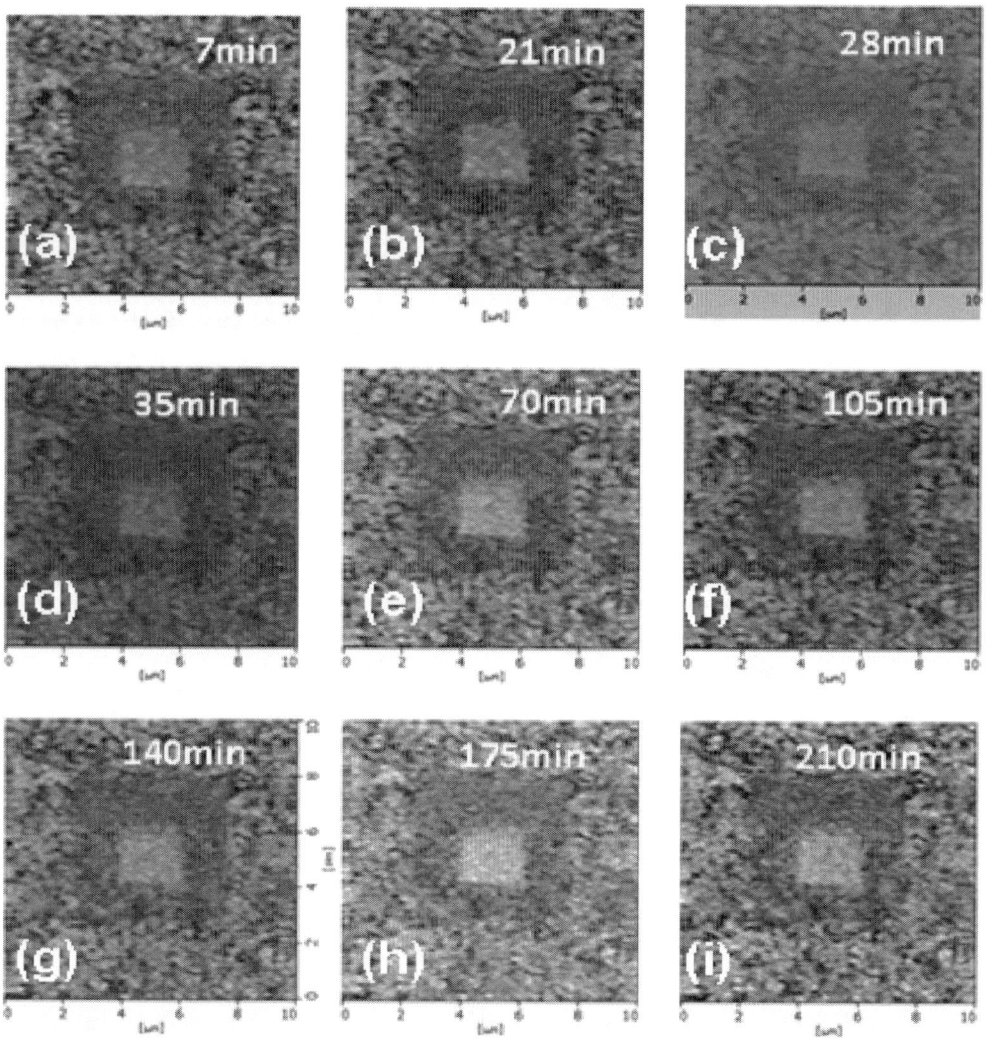

Figure 15. Piezoresponse images of $BiFeO_3$ thin films at different time.

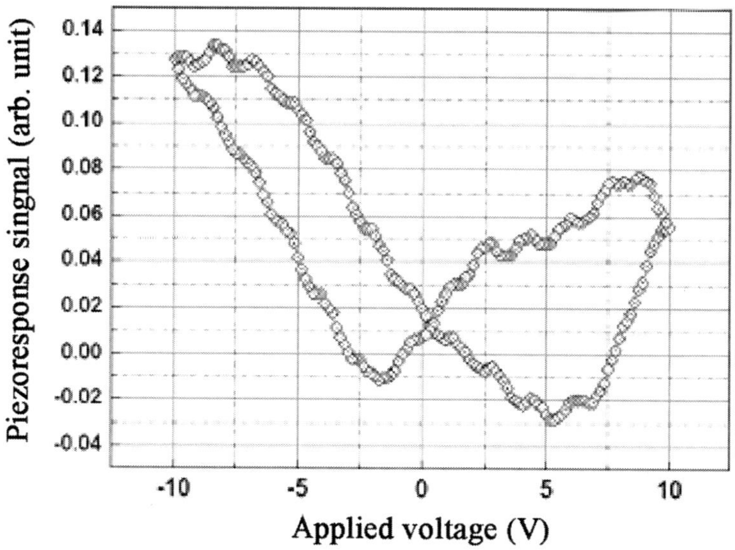

Figure 16. Local piezoresponse hysteresis loop of $BiFeO_3$ thin films.

To check the domain switching behavior of the BFO thin film, we applied DC ±10 V at the cantilever to the ITO bottom electrode with three consecutive scans. The scanned areas were 10×10 μm^2 (0 V), 6×6 μm^2 (+10 V), and 2×2 μm^2 (-10 V), in sequence. The bright region in Figure 15(a) represents the area with an upward component of polarization vector induced by a -10 V, while the area with a downward component of polarization vector induced by a +10 V appears to be dark. Domain regions are well defined, i.e., each region produces a uniform piezoresponse, indicating a homogeneous polarization state. The magnitudes of the piezoresponse signals of the bright domains and the dark domains were found to be similar (not shown here), suggesting a single polarization switching. Therefore, the PFM image shows a clear evidence of ferroelectric polarization switching of the BFO thin film at room temperature. To understand the domain stability, the piezoresponse image of the same area as Figure 15(a) was taken at different times, as shown in Figure 15(b)-(i). It can be clearly seen that the PFM tip fields-induced domain demonstrates a better stability behavior after 210 min, i.e. the polarization state exhibits a better polarization retention behavior.

Ferroelectric domain switching behavior could also be confirmed by the local piezresponse hysteresis loops, as shown in Figure 16. A local butterfly-type piezoresponse hysteresis loop in Figure 16 provides a direct evidence of the local domain switching of the BFO thin film underneath the PFM tip, which is consistent with the previous results as shown in Figure 15.

CONCLUSION

(1) La, Pb: BFO thin films were fabricated on $Pt/Ti/SiO_2/Si$ substrate. The La, Pb: BFO thin film showed greatly improved ferroelectric properties with P_r and E_c values to be 75 $\mu C/cm^2$ and 291 kV/cm. The magnetic measurement below the room temperature revealed ferromagnetic properties of bulk La, Pb: BFO and the magnetic moment

was enhanced compared to La: BFO and Pb: BFO. Hence, the La, Pb: BFO thin film is an effective way to the excellent ferroelectric and ferromagnetic properties.

(2) Nd: $BiFeO_3/Bi_2FeMnO_6$ bilayered films were fabricated on $Pt/Ti/SiO_2/Si$ substrate. The film showed good ferroelectric properties with P_r and E_c values of 54 μC/cm^2 and 237 kV/cm, respectively. The magnetic measurement revealed ferromagnetic properties of the bilayered film. The ferromagnetic transition temperature of T_c was estimated to be 440 K. The hysteresis in *P-E* loop and *M-H* loop revealed that the co-existence of ferroelectric and ferromagnetic properties at room temperature. There are some anomalies observed in *P-E* loop and fatigue measurement, which are still to be investigated thoroughly.

(3) LBFM ceramic and film which was deposited on (100) $SrTiO_3$ substrate. The LBFM film shows antiferromagnetic with a T_N above room temperature. Because of the substrate influence, whether the phenomena in the LBFM film can be classified, as a classical spin-glass still remains open. For the LBFM ceramic, all the observations indicate that the spin glass state is present at low temperature. While compared to LBFM film, the ceramic shows a T_N below room temperature. The different properties between film and ceramic are due to the delicate effects like structure, strain, valence, spin, charge and so on.

(4) Ferroelectric domains of $BiFeO_3$ thin film were investigated by PFM. High-resolution piezoresponse images show finer domain wall were 2 nm in width. Locally applied DC voltage scans and local piezoresponse hysteresis loop reveal its local ferroelctricity and domain switching behavior. Good polarization retention behavior was demonstrated in the $BiFeO_3$ thin film.

ACKNOWLEDGMENTS

The authors gratefully acknowledge Dr. Shigeki Nimori, Dr. Hideaki Kitazawa and Dr. Wenhong Wang of NIMS for the valuable discussions about magnetic properties. Part of this work was supported by grants from JSPS (Japan Society for Promotion of Science) and ARC (Australian Research Council) under the Japan-Australia Research Cooperative Program, and JSPS fellowship program (P09608, P10796). H. Zeng acknowledges financial support by the Foundation for the National Basic Research Program of China (No. 2009CB623305), the Innovative Research Groups of the National Natural Science Foundation of China (No. 50821004) and The Project-sponsored by SRF for ROCS, SEM, China.

REFERENCES

[1] Ramesh, B, Spaldin, N.A.: *Nat. Mater.* 2007, 6, 21.

[2] Fiebig, M.: *J. Phys. D* 2005, 38, R123.

[3] Dho, J., Qi, X., Kim, H., MacManus-Driscoll, J.L., Blamire, M.G.: *Adv. Mater.* 2006, 18, 1445.

[4] Cheng, Z.X, Wang, X.L., Kannan, C.V., Ozawa, K, Kimura, H., Nishida, T., Zhang, S.J., Shrout, T.R.: *Appl. Phys. Lett.* 2006, 88, 132909.

[5] Wang, J., Neaton, J.B., Zheng, H., Nagarajan, V., Ogale, S.B., Liu, B., Viehland, D., Vaithyanathan, V., Schlom, D.G., Wuttig, M., Ramesh, R.: *Science* 2003, 299, 1719.

[6] Zhao, T., Scholl, A., Zavaliche, F., Lee, K., Barry, M., Doran, A., Cruz, M.P., Chu, Y.H., Ederer, C., Spaldin, N.A., Das, R.R., Kim, D.M., Baek, S.H., Eom, B., Ramesh, R.: *Nat. Mater.* 2006, 5, 823.

[7] Zavaliche, F., Das, R.R., Kim, D.M., Eom, C.B., Yang, S.Y., Shafer, P., Ramesh, R.: *Appl. Phys. Lett.*, 2005, 87, 182912.

[8] Bai, F., Wang, J.L., Wuttig, M., Li, J.F., Wang, N.G., Pyatakov, A.P., Zvezdin, A.K., Cross, L.E., Viehland. D.: *Appl. Phys. Lett.* 2005, 86, 032511.

[9] Eerenstein, W., Morrison, F.D., Dho, J., Blamire, M.G., Scott, J.F., Mathur, N.D.: *Science*, 2005, 307, 1203a.

[10] Qi, X.D., Dho, J., Tomov, R., Blamire, M., Macmanus-Driscoll, J.L.: *Appl. Phys. Lett.* 2005, 86, 062903.

[11] Cheng, Z.X., Wang, X.L., Dou, S.X., Ozawa, K., Kimura, H.: *Phys. Rev. B* 2008, 77, 092101.

[12] Naganuma, H., Miura, J., Okamura, S.: *App. Phys. Lett.* 2008, 93, 052901.

[13] Singha, S.K., Ishiwara, H., Maruyama, K.: *Appl. Phys. Lett.* 2006, 88, 262908.

[14] Kim, J.K., Kim, S.S., Kim, W.J., Bhalla, A.S., Guo, R.: *Appl. Phys. Lett.* 2006, 88, 132901.

[15] Shannigrahi, S.R., Huang, A., Chandrasekhar, N., Tripathy, D., Adeyeye, A.O.: *Appl. Phys. Lett.* 2007, 90, 022901.

[16] Wang, Y., Nan, C.W.: *Appl. Phys. Lett.* 2006, 89, 052903.

[17] Lee, Y.H., Wu, J.M., Lai, C.H.: *Appl. Phys. Lett.* 2006, 88, 042903.

[18] Cheng, Z.X., Li, A.H., Wang, X.L., Dou, S.X., Ozawa, K., Kimura, H., Zhang, S.J., Shrout T.R.: *J. Appl. Phys.* 2008, 103, 07E507.

[19] Seshadri, R., Hill, N.A.: *Chem. Mater.* 2001, 13, 2892.

[20] Kim, J.K., Kim, S.S., Kim, W.J., Bhalla, A.S.: *J. Appl. Phys.* 2007, 101, 014108.

[21] Cheng, Z.X., Wang, X.L., Dou, S.X., Kimura, H., Ozawa, K.: *J. Appl. Phys.* 2008, 104, 116109.

[22] Singh, M.K., Prellier, W., Singh, M.P., Katiyar, R.S., Scott, J.F.: *Phys. Rev. B* 2008, 77, 144403.

[23] Yun, K.Y., Ricinschi, D., Kanashima, T, Okuyama, M.: *Appl. Phys. Lett.* 2006, 89, 192902.

[24] Lin, Y.H., Jiang, Q.H., Wang, Y., Nan, C.W., Chen, L., Yu, J.. *Appl. Phys. Lett.* 2007, 90, 172507.

[25] Cheng, Z.X., Wang, X.L., Kimura, H., Ozawa, K., Dou, S.X.: *Appl. Phys. Lett.* 2008, 92, 92902.

[26] Lebeugle, D., Colson, D., Forget, A., Viret, M.. *Appl. Phys. Lett.* 2007, 91, 022907.

[27] Zhao, H.Y., Kimura, H., Cheng, Z.X., Wang, X.L., Nishida, T.: *Appl. Phys. Lett.* 2009, 95, 232904.

[28] Zhao, H.Y., Kimura, H., Cheng, Z.X., Wang, X.L., Ozawa, K., Nishida, T.: *J. Appl. Phys.* 2010, 108, 093903.

[29] Azuma, M., Takata, K., Saito, T., Ishiwata, S, Shimakawa, Y., Takano, M: *J. Am. Chem. Soc.* 2005, 127, 8889.

[30] Langenberg, E., Garcia-Cuenca, M.V., Ferrater, C, Polo, M.C., Fina, I., Fabrega, L., Sanchez, F., Fontcuberta, J.: *J. Mag. Mag. Mater.* 2009, 321, 1748.

[31] Ramesh, R., Spaldin, N.A.: *Nat. Mater.* 2007, 6, 21.

[32] Bi, L., Taussig, A.R., Kim, H.S., Wang, L., Dionne, G.F., Bono, D., Persson, K., Ceder, G., Ross, C.A.: *Phys. Rev. B* 2008, 78, 104106.

[33] Jang, H.W., Ortiz, D., Baek, S., Foliman, C.M., Das, R.R., Shafer, P., Chen, Y.B., Nelson, C.T., Pan, X.Q., Ramesh, R., Eom, C.: *Adv. Mater.* 2009, 21, 817.

[34] Wu. J.G., Kang, G.Q., Liu, H.J., Wang, J.: *Appl. Phys. Lett.* 2009, 94, 172906.

[35] Scott, J.F.: Ferroelectric memories, Springer, Berlin 2000, Chap. 6.

[36] Bobeck, A.H., Fischer, R.F., Perneski, A.J., Remeika, J.P., Van Uitert, L.G.: *IEEE Trans. Magn.* 1969, 5, 544.

[37] Bolduc, M., Taussig, A.R., Rajamani, A., Dionne, G.F., Ross, C.A.: *I EEE Trans. Magn.* 2006, 42, 3093.

[38] Zhao, H.Y., Kimura, H., Cheng, Z.X., Wang, X.L., Ozawa, K., Nishida, T.: *Phys. Status Solidi RRL* 2010, 4, 314.

[39] Dionne, G.F.: *J. Appl. Phys.* 2007, 101, 09C509.

[40] Catalan, G., Scott, J.F.: *Adv. Mater.* 2009, 21, 2463.

[41] Zeng, H.R., Shimamura, K., Villora, E.A.G., Takekawa, S., Kitamura, K.: *J. Appl. Phys.* 2007, 101, 074109.

[42] Liu, X.Y., Kitamura, K., Terabe, K., Zeng, H.R., Yin, Q.R.: *Appl. Phys. Lett.* 2007, 91, 232913.

[43] Balke, N., Choudhury, S., Jesse, S., Huijben, M., Chu, Y.H., Baddorf, A.P., Chen, L.Q., Ramesh, R., Kalinin, S.V.: *Nat. Nanotechnol.* 2009, 4, 868.

[44] Kalinin, S.V., Budai. J.D., Jesse, S., Morozovska, A.N., Bokov, A.A., Ye, Z.G.: *Phys. Rev. B* 2010, 81, 064107.

In: Ferroelectrics: New Research
Editors: Xui Li Huang and Su Lyn Ma
ISBN: 978-1-61942-282-7

Chapter 6

THE NANOSCALE FERROELECTRIC BEHAVIOR OF GESBTE FILMS STUDIED BY PIEZORESPONSE FORCE MICROSCOPY

J. J. Gervacio Arciniega, F. J. Espinoza Beltrán, E. Prokhorov and L. G. Trapaga Martínez
CINVESTAV, Unidad Queretaro, Qro., Mexico

ABSTRACT

Ferroelectric materials offer a wide range of useful properties to develop applications for the electronic engineer. The tendency to develop smaller ferroelectric devices implies to have a deep understanding of ferroelectric materials properties at the nanoscale level. Piezoresponse force microscopy (PFM) in a scanning probe microscopy mode (SPM), shows to be an ideal non-destructive technique to study the nanoscale behavior of ferroelectric materials due to its high resolution in the range of nanometers and to its ability to manipulate ferroelectric domains. An improvement of this technique developed to investigate ferroelectric materials with small piezoelectric coefficient is the use of contact resonance PFM, which allows an amplification of the piezoresponse signal in proportion to the Q-value (10-100) of the cantilever-sample contact resonance. Moreover, switching spectroscopy PFM allows hysteresis loop measurements in a local domain achieving quantitative information about the switching processes at nanoscale resolution, including coercive bias and work of switching. The first part of this chapter will be a brief description about the principles, theoretical background, recent developments and applications of piezoresponse force microscopy. Furthermore, a new development of PFM will be described for the investigation of chalcogenide thin films, GeSbTe alloys, which have application in data storage memory. These alloys are non-typical ferroelectric materials with low electrical resistivity, which make difficult to investigate their ferroelectric properties by other conventional techniques. Different characterization techniques were employed to study GeTe, $Ge_4Sb_1Te_5$ and $Ge_2Sb_2Te_5$ thin films: contact resonance PFM and switching spectroscopy PFM for the investigation of ferroelectric domain structure and impedance spectroscopy technique to obtain Curie temperatures. Because of their ferroelectric properties, these chalcogenide materials can find applications in the super-resolution optical disks where the resolution is far smaller than the diffraction limit, for achieving ultrahigh storage density.

INTRODUCTION

Throughout modern history many and various digital storage systems have been researched, developed, manufactured, and eventually surpassed in an effort to address ever-increasing demands for density, operating speed, low latency, endurance, and economy [1].

One group of the materials widely used for data storage application is chalcogenide semiconductors. The properties of chalcogenide semiconductors were first explored as a potential memory technology by S. Ovshinsky [2] who proposed using the reversible phase-transformation from the amorphous (high resistance/ low reflectivity state) to crystalline (low resistance/ high reflectivity state). This transformation is always accompanied by abrupt changes in optical and electric properties in chalcogenide materials. Later, a next generation of high speed phase change materials based on the germanium-antimony-tellurium (GeSbTe or GST) alloys was discovered [3-5], with compositions mainly along the GeTe–Sb_2Te_3 pseudobinary line in the ternary phase diagram, such as: $Ge_2Sb_2Te_5$, $Ge_1Sb_2Te_4$ and $Ge_1Sb_4Te_7$.

For several years the GeSbTe alloy along pseudobinary GeTe–Sb_2Te_3 line in phase diagram of the GeSbTe system has been used as the main material for optical data storage devices [6] due to the high reflectivity contrast between the amorphous and crystalline phases. Many investigations have been carried out on this type of phase-change optical memory with the purpose of improving their properties and to increase its storage capacity, stability, speed and versatility.

Four generations of optical data storage devices based on the phase-change GeSbTe material have been developed: the first is the compact disc (CD), the second is the digital versatile disc (DVD), the third are the high definition digital versatile disc (HD-DVD) and the blue ray disc (BD), and the fourth, proposed during the last decades, is the super resolution near field structure (super-RENS) disc with a resolution smaller than the diffraction limit. This type of optical disks allows further increase the recording density and data transfer rate compared with the conventional optical data storage. That is why the super-RENS disc is under intensive investigations [7-9]; however, the working mechanism of super-RENS is still unknown. According to the literature, the super-resolution effect in one of the GST alloys, with composition $Ge_2Sb_2Te_5$, has been explained in terms of thermoreflectance, local thermal expansion, thermally assisted saturable absorption [10], and ferroelectric phase transition [11-13].

It is known that upon annealing, amorphous layers of $Ge_2Sb_2Te_5$ first undergo the amorphous-to-crystalline (NaCl-type) transition in the range of 130-160°C (depending on the preparation technique [14]) and at higher temperatures (around 200-250°C) the NaCl-type structure transforms into a more stable hexagonal phase. According to several references [12, 15, 16] the first crystalline structure is a distorted rock salt-like structure. Ge and Sb atoms deviate from the ideal rock salt positions not in a random way but in a strongly correlated manner with respect to the neighboring Te atoms. The off-centered location of Ge atoms generates a net dipole moment and suggests that a NaCl-type phase of the $Ge_2Sb_2Te_5$ is a ferroelectric material [16]. In the temperature range of the NaCl-type to the hexagonal phase transition, the dielectric constant and refractive index change greatly due to the ferroelectric transition in the vicinity of the Curie temperature, this could be responsible for the super-resolution effect. But in the literature most publications only focus on the first phase transition and none of them have taken into account the second phase transition, observed as a tiny discontinuity of the heat flow in Differential Scanning Calorimetry (DSC) measurements [12, 15].

As mentioned above, materials for optical data storage applications have been used with compositions mainly along the GeTe–Sb_2Te_3 pseudobinary line that undergo reversible transformations between amorphous and crystalline states. The structure of these materials is not precisely known yet but according to [17] the amorphous material consist of three-dimensional building blocks: a GeTe chain and a stibnite Sb_2Te_3. In the rock salt-like structure one sublattice is occupied by tellurium atoms, the other one by a random distribution of germanium atoms, antimony atoms, and vacancy sites [18] and the hexagonal phase consist of alternating GeTe and Sb_2Te_3-blocks [19]. GeTe lies at the end of GeTe-Sb_2Te_3 pseudobinary line and investigation of its properties is important to understand the properties of GST materials.

Similar to GST alloys, GeTe exists in amorphous and crystalline states. Upon annealing, GeTe amorphous layers, at temperatures around 200^0C, crystallize in a rhombohedral ferroelectric phase and at higher temperatures transform into a cubic rock salt structure [20]. The transition temperature occurs in the interval of 327-427^0C depending on various factors, such as exact composition, carrier density, etc. For example, changes in Te composition from 50 to 50.7 at. % shift the transition temperature from 427^0C to 356^0C [21]. GeTe shows displacive type ferroelectric-paraelectric transition from rhomhedral ferroelectric phase to a cubic rock salt type structure with paraelectric properties. There is a lot of information in the literature about the ferroelectric behavior of this material, but only a few articles give an evidence of its ferroelectrics domains [22, 23]. Nevertheless, the work reported on GeTe corresponds to bulk materials. From our knoveledge, there is no work reported in the literature on ferroelectric domains in GeTe films.

Another material, the $Ge_4Sb_1Te_5$ alloy, also undergoes phase-change transformation in a NaCl-type crystalline phase. This material is interesting for data storage applications, because it has the largest optical [24] and electrical [25] contrast between the crystalline and amorphous states, compared with other GeSbTe alloys. The structure of this alloy is unclear, in contrast to other GST compounds, since it does not belong to the GeTe – Sb_2Te_3 homologous series, although lies on (or very close to) the GeTe-Sb_2Te_3 pseudobinary line in the GeSbTe phase diagram. In ref. [26] it is suggested that this alloy is a solid solution between GeTe and a compound with a composition close to $Ge_4Sb_1Te_5$. The ferroelectric properties of this bulk material have been reported in ref. [27] based on measuring polarization as a function of applied electric fields.

As a rule ferroelectric properties have been observed in non conducting crystals or dielectrics that exhibit spontaneous electrical polarization. The essential feature of a ferroelectric is that spontaneous polarization can be reversed by the application of an electric field. In addition to the change in spontaneous polarization, temperature and electric field affects on the dielectric constant of the material. But such measurements are difficult to perform on low resistivity (about 10^{-2}-10^{-3} Ohm cm) GST crystalline films. Additional DSC and X-ray diffraction measurements do not allow observing ferroelectric-paraelectric transitions due to the presence of a second NaCl-type to hexagonal transition in $Ge_2Sb_2Te_5$,

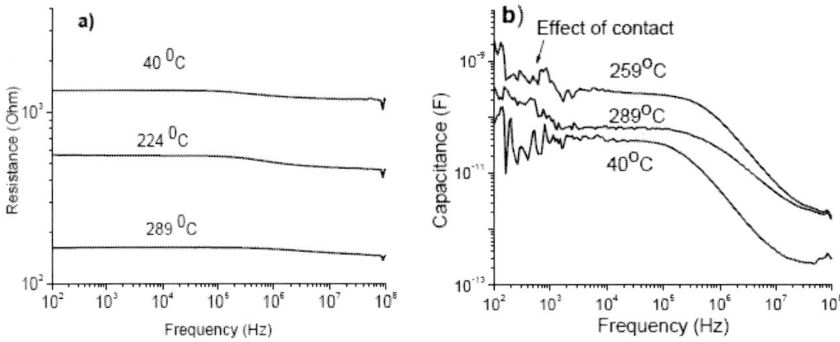

Figure 1. Resistance (a) and capacitance (b) as a function of frequency, measured on $Ge_2Sb_2Te_5$ film at the temperatures indicated on the graphs.

$Ge_1Sb_2Te_4$ and $Ge_1Sb_4Te_7$ films. Due to such difficulties, this chapter presents ferroelectrical properties of thin chalcogenide films with compositions: GeTe, $Ge_4Sb_1Te_5$, and $Ge_2Sb_2Te_5$ which have been investigated with impedance spectroscopy, X-ray diffraction techniques, and scanning probe microscopy.

Impedance Measurements

The impedance spectroscopy is a well known method that can be applied to study electrical properties, crystallization kinetics, crystallization temperatures, etc, in GST films [28]. This technique has been used to investigate the temperature dependence of the resistivity and capacitance of crystalline GeTe, $Ge_4Sb_1Te_5$ and $Ge_2Sb_2Te_5$. The amorphous GeTe, $Ge_4Sb_1Te_5$, and $Ge_2Sb_2Te_5$ thin films (with a thickness between 200-300 nm) have been prepared on glass substrates by DC magnetron sputtering. The films compositions determined by energy dispersive X-ray spectroscopy were $Ge_{58}Te_{42}$, $Ge_{38.7}Sb_{10.7}Te_{50.6}$ and $Ge_{20.7}Sb_{21.2}Te_{58.1}$. Impedance measurements using a 4294A Agilent Precision Impedance Analyzer were carried out as a function of temperature on samples previously annealed to obtain crystalline phases. Due to the low resistivity of the material in the crystalline state the capacitance and resistance measurements were carried out on samples with a planar configuration with two gold contacts at a frequency of about 50 kHz.

This frequency is convenient because it is in the saturation region of the resistance-frequency and capacitance-frequency curves and it avoids contact effects in the low frequency range (Figure 1). From impedance measurements in a planar configuration it is difficult to calculate the dielectric constant, but in ref. [29] it was shown that for this geometry, the capacitance is proportional to the dielectric constant.

Principe of Piezoresponse Force Microscopy

The scanning probe microscopy (SPM) techniques are powerful techniques developed to measure materials properties with a nanometric scale resolution [30, 31]. For ferroelectric materials, the piezoresponse force microscopy (PFM) mode is an ideal tool to study the local behavior of ferroelectric domains [32, 33]. The typical PFM technique provides simultaneous information of topography, polarization of ferroelectric domains and the boundaries of these domains through the deflection of cantilever, phase and amplitude of piezoresponse signal, as shown in Figure 2. PFM amplitude images typically show the border of the ferroelectric domain, while the phase images depict the ferroelectric domain area [34, 35].

Ferroelectric materials generate an electric response as a result of a mechanical compression; this is called the direct piezoelectric effect. In contrast, if an electric field is applied to the ferroelectric material, this material will contract or elongate, this is the called the indirect piezoelectric effect. PFM measurements are carried out by means of the indirect piezoelectric effect.

The sample is under an electric field applied through the conductive PFM probe and an electrode in the bottom of the sample (Figure 3a).

When an electric field is applied in the same direction as the polarization vector of the domain, the ferroelectric domain is expanded (Figure 3b). But, if the electric field is applied in the opposite direction of polarization, the ferroelectric domain is contracted (Figure 3c). In this configuration the tip of the probe works like a top electrode and the electric field is not uniform under the tip.

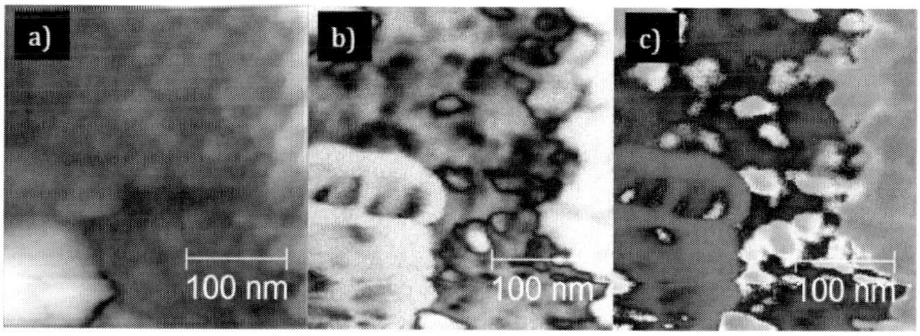

Figure 2. Images of surface topography (a), PFM amplitude (b) and PFM phase (c) of a $K_{0.5}Na_{0.5}NbO_3$ lead free ferroelectric ceramic.

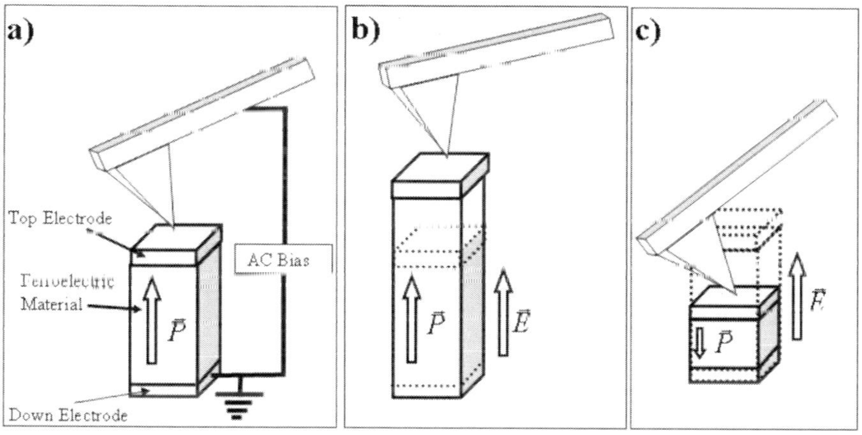

Figure 3. Schematic description of the applied electric field effect by the tip on the ferroelectric domain for E and P parallels, with the same and the opposite directions.

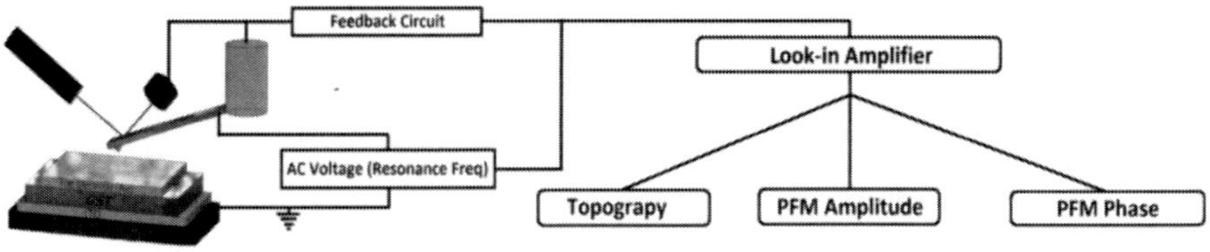

Figure 4. Schematic setup of conventional PFM measurements.

To avoid this effect, an extended top electrode is used. This top electrode, which is a thin gold film, permits to obtain uniform electric field under the tip and an improved signal.

Also, this metallic film is very useful for soft samples, softer than gold, to avoid damage on the sample due to the tip movement [36]. A schematic setup of conventional PFM measurements is shown on Figure 4.

The deflection of the cantilever by the laser beam deviation senses the changes in the surface topography of the sample. A feedback circuit maintains the cantilever deflection defining a fixed tip-sample contact force. For PFM measurements, a sinusoidal voltage signal is used to excite the ferroelectric sample through the tip. Usually, the frequency of the AC bias excitation signal is selected in the range from 10 to 20 kHz, always far from the cantilever resonance frequencies. Afterwards, a look-in amplifier processes the PFM signal and the results are stored in a computer.

Equation 1 gives a relation between the piezoresponse signal and piezoelectric coefficient d_{33} using the conventional PFM technique [37]:

$$\text{piezoresponse} = d_{33} V_{AC} \cos(\theta), \tag{1}$$

where V_{AC} is the applied AC bias, and θ is the phase between excitation signal and piezoresponse signal. The PFM signal, amplitude and phase, measured by a lock-in amplifier using the excitation signal as the reference, allow obtaining $d_{33}V_{AC}$. In addition, the piezoelectric coefficient d_{33} relates to the mechanical strain produced by the applied electric field and measured in picometers per volt (pm/V) units.

The typical values of the d_{33} coefficient are in the range between 500 pm/V for PZT materials and 0.1 pm/V for polar semiconductors [38, 39]. For materials with small d_{33}, it is impossible to obtain a well-defined piezoresponse signal by convectional PFM techniques. For this kind of materials a dynamic method called the contact resonance piezoresponse force microscopy (CR-PFM) mode [40] is employed. The experimental configuration of CR-PFM is similar to conventional PFM, but in CR-PFM the excitation signal applied through the tip has the resonance frequency of the cantilever when the tip is in contact with the sample surface. Using this frequency, the piezoresponse signal in CR-PFM technique follows the equation [40]:

$$\text{piezoresponse} = d_{33} V_{AC}\, Q \cos(\theta), \tag{2}$$

where Q is the Q-value or factor of quality of the cantilever resonance frequency. The Q-values for rectangular silicon probes are in the range from 10 to 100, being Q the amplification factor of the piezoresponse signal according to Eq. 2. CR-PFM is very sensitive to the piezoresponse signal, but also to the surface roughness of the sample that produces a contaminant signal due to the topographic crosstalk. In order to reduce the topographic crosstalk, a direct solution could be used to reduce the surface roughness of the sample. Working with hard metallic or ceramics materials, it is possible to reduce their roughness by polishing but, in general, this solution cannot be applied to soft materials. CR-PFM is a powerful technique to study ferroelectric materials with small d_{33} and smooth surfaces. Moreover, CR-PFM measurements produce qualitative images of the ferroelectric domains, but cannot give quantities information about the ferroelectric domains in the material. To obtain quantitative data on ferroelectric materials, the switching spectroscopy PFM (SS-PFM) technique has been developed [41]. Details and application of this technique for investigation of ferroelectric properties of chalcogenide materials will be describe at the end of this chapter.

GeTe

GeTe undergoes a simple phase-change and is a classical ferroelectric material which shows displacive type ferroelectric-paraelectric transition from a rhombohedral ferroelectric phase to a rock salt type structure with paraelectric properties [42-44]. The rhombohedral phase of GeTe can be viewed as a rock salt distorted along the [111]-direction [45]. Additionally, the central atom is displaced along the [111]-direction from the center of the rhombohedron. Furthermore, it has been found that GeTe exhibits a large concentration of Ge-vacancies [18]. The NaCl type crystalline structure of GeTe is an unstable phase. This crystalline structure exists for temperatures equal to or above 270°C. Two sublattices form this rock salt structure; Ge atoms compose one of the lattices and tellurium atoms the other [46].

Figure 5 shows X-ray patterns of GeTe crystalline films (annealed at 200 °C previous to XRD measurements) at the temperature indicated on the graph. At the temperature below

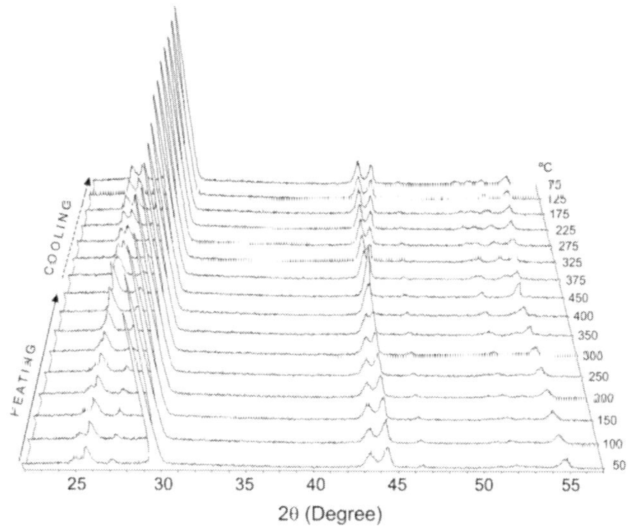

Figure 5. X-ray diffraction patterns of a GeTe film as a function of temperature, in the range from 30°C to 400°C, for the heating and the cooling process..

350^0C the material shows the presence of a rhombohedral phase. At a temperature of 350^0C two phases have been observed: rhombohedral and NaCl-type, and at higher temperatures the material transforms into a NaCl-type phase. This phase exists only at temperatures above 400^0C and, in the process of cooling, again transforms into the stable rhombohedral phase. These results confirm the reversibility of the transformation observed in the XRD as a function of temperature measurements and the displacive type ferroelectric-paraelectric transition from rhombohedral ferroelectric phase to a rock salt type structure with paraelectric properties reported in the literature [42-44]. Additionally, these measurements allow to determine the temperature interval of the ferroelectric-paraelectric transition or T_c, the Curie temperature (between 350-400^0C) at which the material changes from a low temperature polarized state to a high temperature unpolarized state [47].

Additionally, the value of the Curie temperature can be obtained from the temperature dependence of the capacitance. It is well known that in the vicinity of the Curie temperature the dielectric constant ε rises sharply and above this temperature ε follows the Curie-Weiss equation [27, 47]:

$$\varepsilon = \frac{A}{T - T_0}$$

where A is a positive constant, T_o is called the Curie–Weiss temperature, which is lower than the transition temperature.

Figure 6a shows the capacitance as a function of temperature for crystalline GeTe films (previously annealed at 200^0C) measured at a frequency of 50 kHz. From impedance measurements in a planar configuration it is difficult to calculate the dielectric constant, but in ref. [29] was shown that for this geometry, the capacitance is proportional to the dielectric constant. The temperature dependence of capacitance shows an abrupt change at the temperature T_c equal to 362^0C, in the same temperature range as obtained from XRD measurements. In addition, the reciprocal capacitance as a function of temperature (Figure 6b) shows a typical Curie-Weiss behavior for temperatures above T_c. This behavior is characteristic of ferroelectric materials and T_c is related to the ferroelectric–paraelectric transition [47].

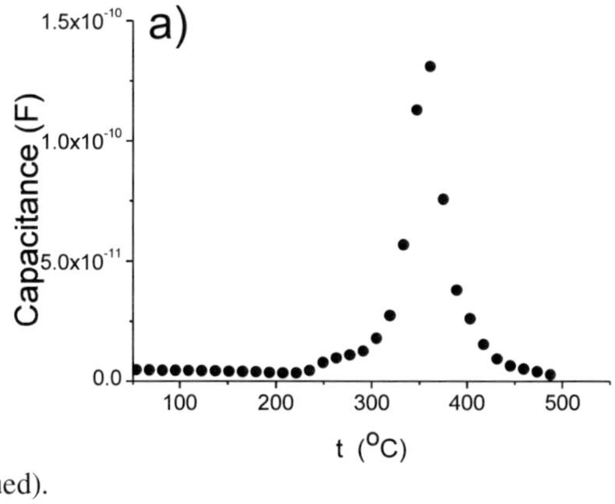

Figure 6. (Continued).

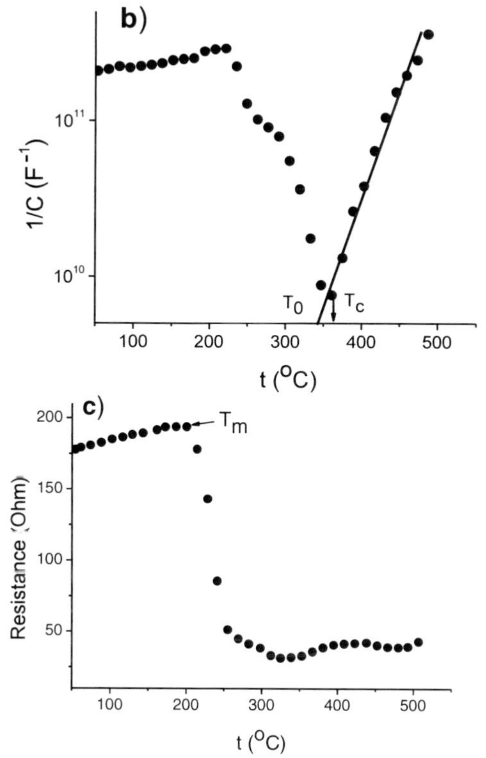

Figure 6. Dependencies of (a) capacitance, (b) reciprocal capacitances and (c) resistance as a function of temperature for GeTe film.

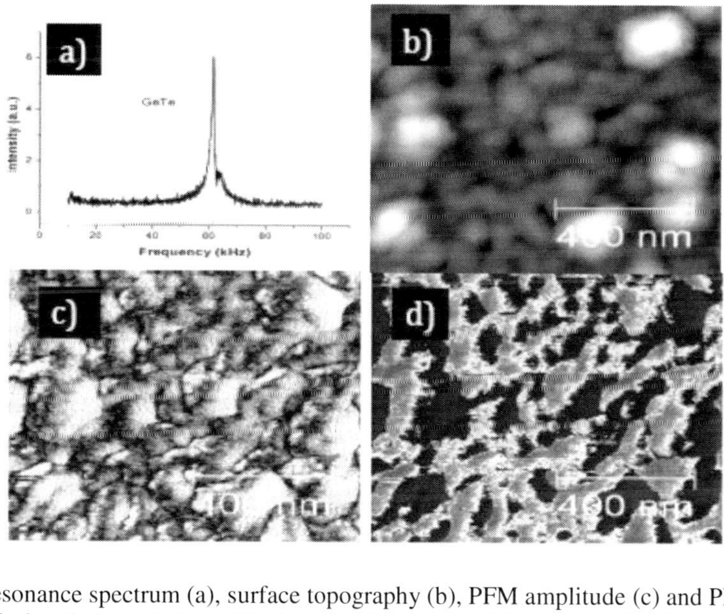

Figure 7. The resonance spectrum (a), surface topography (b), PFM amplitude (c) and PFM phase (d) of a GeTe film with rhombohedral phase. Black and white areas in (d) correspond to ferroelectric domains in the opposite direction.

The resistance measurements in GeTe crystalline films as a function of temperature show a maximum (T_m) at the temperature of 200^0C (Figure 6c), which is typical of a metal-insulator transition. For temperatures below T_m, the resistance increases with the increase in temperature (metallic behavior), whereas for temperatures above T_m, the resistance decreases with the increase in temperature (semiconductor behavior). T_m is much lower than the temperature value for the ferroelectric-paraelectric transition (362^0C). This effect has been observed in many ferroelectric-ferromagnetic materials and it depends on the properties of grain boundaries, which play an important role in the carrier transport (see for example refs. [20, 48]).

Ferroelectric materials generally consist of ferroelectric domains in which the polarization is in the same direction. Ferroelectric domains images of GeTe thin films have been obtained by CR-PFM which due to its high sensitivity allows measuring the low piezoelectric signals (low d_{33} values). Piezoelectric measurements were performed in the contact mode with a commercially available SPM system (Veeco Dimension 3100 Nanoscope IV), equipped with a Signal Access Module (SAM), a high frequency lock-in amplifier (Stanford Research SR844), an additional lock-in amplifier for contact resonance spectra measurements (Anfatec LIA), a function generator (HP 3314A) and a home-made electronic setup for an automatic control and data acquisition software created using LabVIEW. Additionally, The PFM data was processed using the Matlab program. The contact resonance frequency was 61.1 kHz for GeTe films. The resonance spectrum is shown in the Figure 7a. The conditions include 2V amplitude of AC bias with the resonance frequency being applied to the sample between the Au top contact electrode and a lower grounded contact electrode.

Figure 7b, 7c and 7d shows the topography, PFM amplitude, and PFM phase, respectively, for the GeTe thin film. The topography image shows grains with average size of 117 nm. The PFM amplitude typically shows the border of the ferroelectric domain, while the phase images depict the ferroelectric domains area with directions in-plane and out-of-plane (Figure 7c and 7d). The dark areas correspond to inward directions and the white areas correspond to outward directions. The size of the domains is about 176 nm and consists from only a few grains.

$Ge_4Sb_1Te_5$

As mentioned in the introduction, $Ge_4Sb_1Te_5$ materials are a solid solution close to the GeTe-Sb_2Te_3 pseudobinary line. It has only one NaCl type crystalline phase [49] and its properties, as it will be shown below, vary similar to the properties of GeTe. Additionally, in ref. [27] has been reported that bulk alloys of $Ge_4Sb_1Te_5$ prepared using a melt-quench technique shows ferroelectric properties and ferroelectric–paraelectric transitions. But DSC measurements have shown that ferroelectric properties can be observed not only in the bulk material but also in thin films [50].

Figure 8 shows DSC measurements on a crystalline $Ge_4Sb_1Te_5$ bulk alloy and on the amorphous material obtained by removing the as-deposited films from the glass substrates (Figure 1b). In the bulk alloys, a nonsymmetrical endothermic peak at a temperature of 329^0C has been observed. In contrast, DSC measurements on amorphous material showed an exothermic peak at 176^0C which is associated with crystallization of the sample and a broad endothermic peak, with a maximum at approximately 327^0C, and that corresponds with the

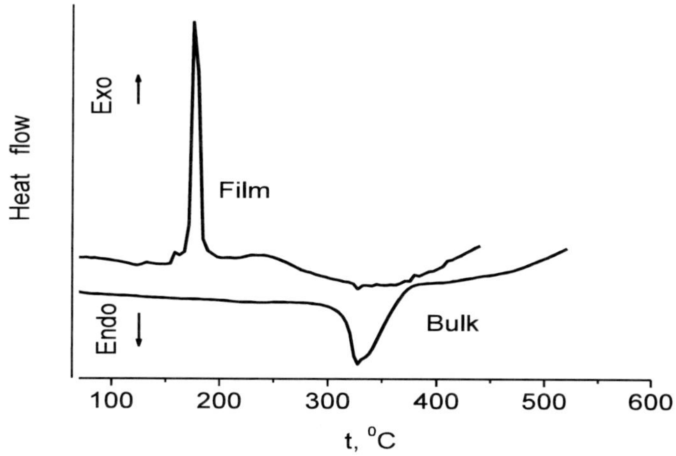

Figure 8. DSC measurements on a bulk crystalline alloy and on the amorphous material, obtained by removing the as-deposited films from the glass substrates.

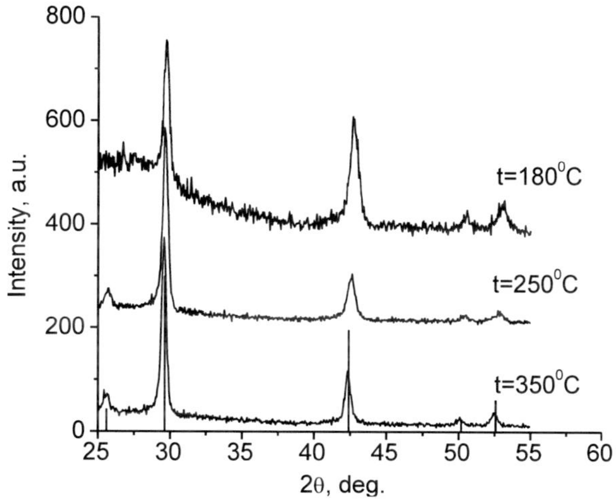

Figure 9. XRD patterns of $Ge_4Sb_1Te_5$ film measured at the temperature indicated on the graph.

same temperature as in the bulk alloy. The observed endothermic peak can be associated with the ferroelectric-paraelectric transition.

XRD measurements at different temperatures on crystalline film (annealed previously to 200^0C) show that as the temperature is increased the position of all $Ge_4Sb_1Te_5$ peaks shift to lower 2θ values (Figure 9), reaching saturation at a temperature of approximately 327^0C at the positions close to the NaCl type of GeTe structure (ICSD cart #602124).

The position of the NaCl type of GeTe is marked on the graph with the horizontal line. The same behavior as for GeTe, the temperature dependence of the capacitance (Figure 10a) for crystalline films shows a sharp maximum at the same temperature T_c at which the endothermic reaction occurs in the DSC measurements and which relates to the ferroelectric–paraelectric transition. The reciprocal capacitance follows a typical Curie-Weiss behavior (Figure 10b). Also similar to the behavior for GeTe, the resistance measurements in crystalline $Ge_4Sb_1Te_5$ films as a function of temperature show a maximum (T_m) at a temperature of about 200^0C which is lower than the Curie temperature and is typical for a metal-insulator transition in ferroelectric materials.

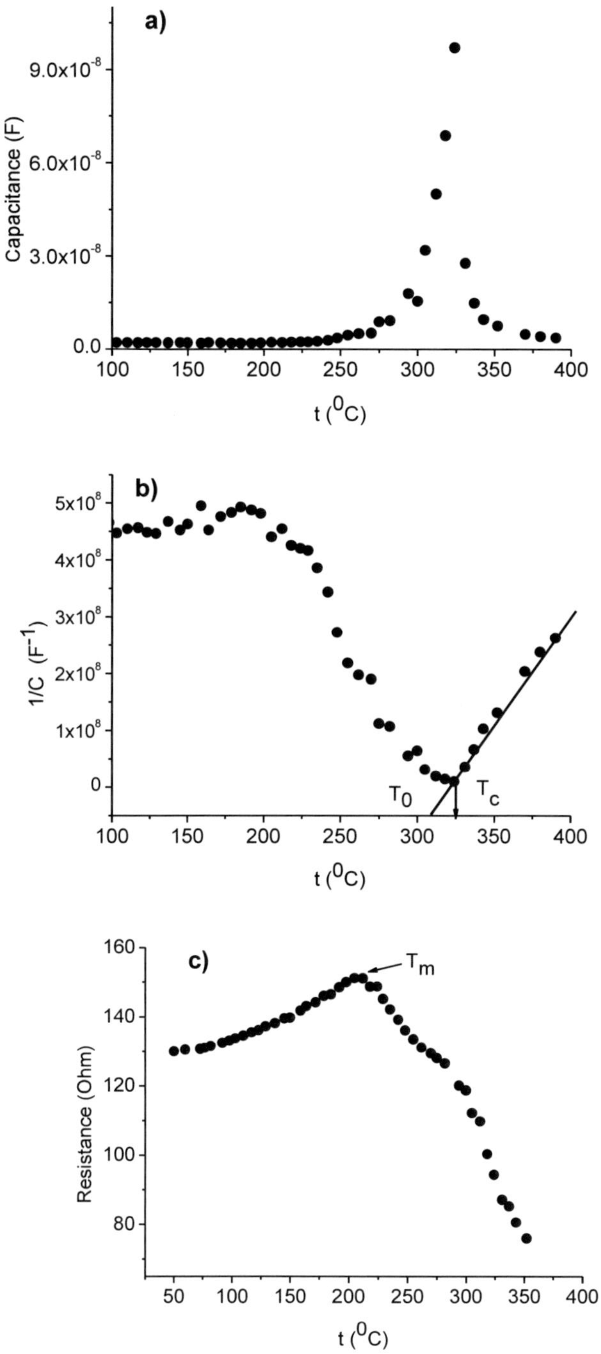

Figure 10. Temperatures dependencies of capacitance (a), reciprocal capacitances (b) and electrical resistance (c) for $Ge_4Sb_1Te_5$ films.

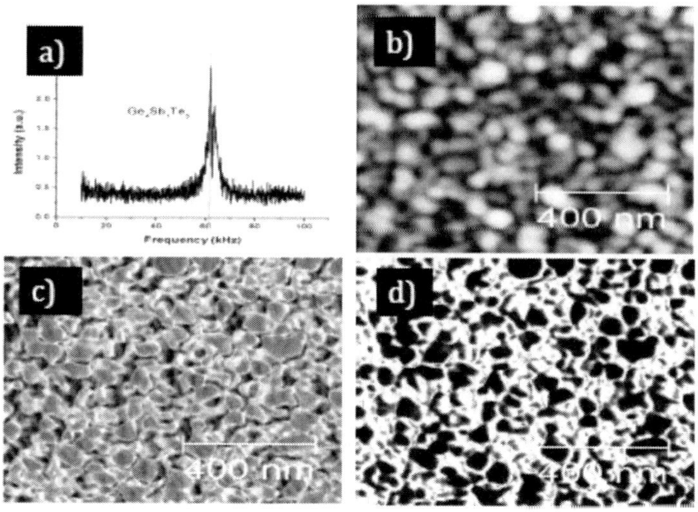

Figure 11. The resonance spectrum (a), surface topography (b) PFM amplitude (c) and PFM phase (d) of $Ge_4Sb_1Te_5$. Black and white areas correspond to ferroelectric domains in the opposite direction.

In the same way as for GeTe films, CR-PFM measurements (with a resonance frequency of 62.1 kHz) on crystalline $Ge_4Sb_1Te_5$ films have shown ferroelectric domains (Figure 11). But comparing to GeTe films, the grains values are smaller (approximately 75 nm) and the domains dimension correspondent to the dimension of grains (approximately 86 nm).

$Ge_2Sb_2Te_5$

The crystalline structure of $Ge_2Sb_2Te_5$ materials in the NaCl type phase consists of two sublattices [51]: one of these is formed by tellurium atoms and the other is occupied by germanium and antimony atoms and 20% of vacancies. This NaCl structure is slightly distorted, the germanium positions have a small local displacement with respect to the tellurium sublattice, for this reason, the $Ge_2Sb_2Te_5$ can demonstrate ferroelectric properties [52]. But the existence of a second hexagonal phase does not allow observing ferroelectric-paraelectric transitions using standard DSC or X-ray methods as in the case of GeTe and $Ge_4Sb_1Te_5$, which permit the direct observation of ferroelectric-paraelectric transitions in these materials [53].

Figure 12 shows XRD measurements on $Ge_2Sb_2Te_5$ thin film as a function of temperature. It is seen that below 150°C, the sample only shows wide bands, which are characteristic of the amorphous materials. In the temperature range between 150°C and 200°C, only a NaCl type crystalline phase was observed.

A mixture of the hexagonal and the NaCl type crystalline phases are observed in the temperature range from 250°C to 260°C, and above 270°C only the hexagonal phase is detected. During cooling the hexagonal phase remains without changes. So XRD measurements do not allow detecting ferroelectric properties of material.

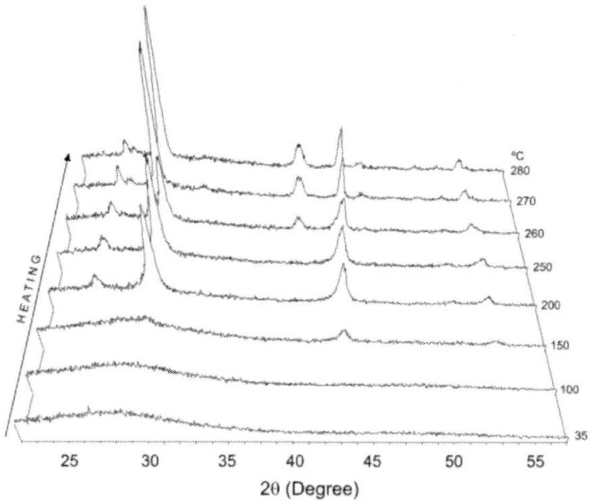

Figure 12. XRD patterns of $Ge_2Sb_2Te_5$ film measured at the temperatures indicated on the graph.

But the temperature dependency of capacitance of NaCl type $Ge_2Sb_2Te_5$ film (annealed previously at 200^0C), as well as for the case of GeTe and $Ge_4Sb_1Te_5$, shows an abrupt change with a maximum at $T_c = 258^oC$ which corresponds to the end of the fcc-hexagonal transition (Figure 13a).

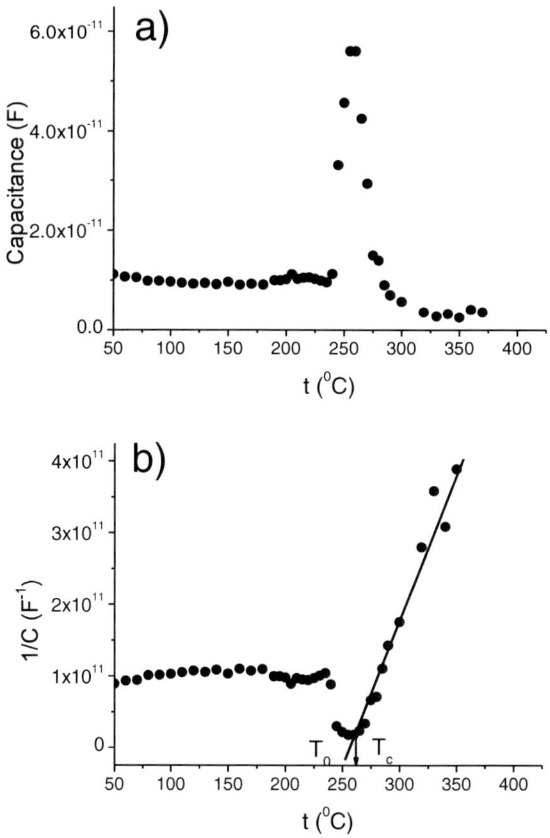

Figure 13. (Continued).

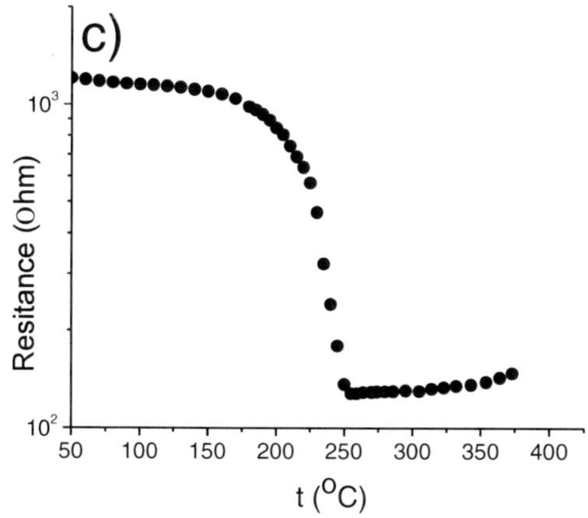

Figure 13. Dependencies of capacitance (a), reciprocal capacitances (b) and resistance (c) as a function of temperature for $Ge_2Sb_2Te_5$ films.

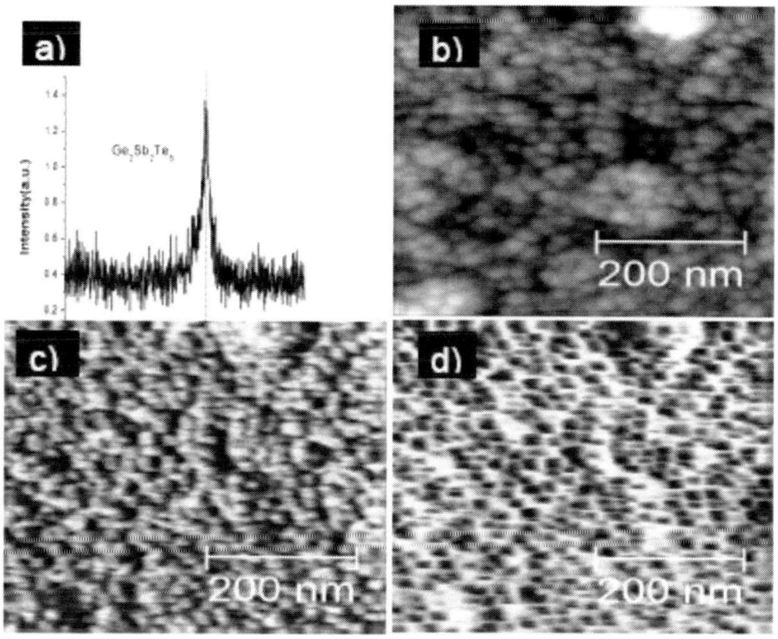

Figure 14. The resonance spectrum (a), surface topography (b) PFM amplitude (c) and PFM phase (d) of $Ge_2Sb_2Te_5$ films with NaCl-type phase. Black and white areas correspond to ferroelectric domains in opposite directions.

The reciprocal capacitance (Figure 13b) for temperatures above T_c follows a typical Curie-Weiss behavior, the same as for the GeTe and $Ge_4Sb_1Te_5$ films. By comparing the capacitance measurements obtained in these three materials it is possible to conclude that NaCl type structures of $Ge_2Sb_2Te_5$ exhibits ferroelectric properties.

It is necessary to note that, in contrast to GeTe and $Ge_4Sb_1Te_5$, in $Ge_2Sb_2Te_5$ films the resistance as a function of temperature demonstrates a typical semiconductor behavior, with a resistivity drop at a temperature around 200°C, which corresponds to the beginning of the hexagonal phase formation (Figure 13c).

But the polycrystalline nature of the films and the grain boundaries (similar to the case of GeTe and in $Ge_4Sb_1Te_5$) affected the transport properties of the material. It was shown [54] for $Ge_2Sb_2Te_5$ films that the conductivity and the mobility is limited by scattering of charge carries at the grain boundaries and this can be described by a conduction model based on thermionic emission over the potential barriers at the grains boundaries.

The same as for the case of GeTe and in $Ge_4Sb_1Te_5$ films, CR-PFM measurements in NaCl type crystalline structures (at the resonance frequency 66.5 kHz) have shown ferroelectric domains in $Ge_2Sb_2Te_5$ materials (Figure 14). But compared with another investigated materials, the grain sizes are the smallest (approximately 28 nm), the same as the domains with dimensions corresponding to the dimensions of the grains.

The results obtained confirm the calculations presented in ref. [55] discussing that the off-center location of Ge atoms in $Ge_2Sb_2Te_5$ NaCl type structures is responsible for the appearance of dipole moments causing that the $Ge_2Sb_2Te_5$ NaCl type phase be a ferroelectric material.

SWITCHING PFM SPECTROSCOPY (SS-PFM)

A ferroelectric material is defined as a crystal that shows spontaneous electric polarization which can be reversed by an electric field [56]. The SS-PFM permits to study the ferroelectric behavior of ferroelectric materials through the amplitude and the phase signals. The experimental configuration used for SS-PFM measurements is the same as for the case of traditional PFM techniques (Figure 3). But in the SS-PFM method the sum of the AC signal (with tip-sample contact resonance frequency) and the DC bias applied on the tip as is shown in figure 15. The DC bias, generated as squared pulses with variable amplitude following a triangular waveform envelope, produces an electric field across the sample in order to switch the ferroelectric domains [41]. The shape for the bias was selected to avoid an electrostatic contribution, and to investigate only the electromechanical properties of the ferroelectric material [33]. The piezoresponce signal is acquired at null bias between two consecutive DC bias pulses. The maximum amplitude of this DC bias depends on the sample characteristics. It is necessary to adequately select the frequency of the triangular envelope and the frequency of the DC squared pulses in order to prevent saturation of the hysteresis loops.

In order to calibrate the SS-PFM measurements a ferroelectric PZT thin film used as a reference sample, measured in a calibrated PFM Asylum system (Cypher Scanning Probe Microscope) and in our system. For this, it is necessary to plot the amplitude signal versus the DC bias excitation and the phase against the DC bias excitation.

The piezoelectric response of the amplitude and phase on the DC bias for a PZT thin film is shown in Figure 16. The dependency of the PFM amplitude versus the DC bias shows a typical butterfly hysteresis loop.

The calculation of the d_{33} value according to the method suggested in ref. [57], a linear fitting was applied to the data set, only in the range marked on Figure 16a, where a linear relationship between the PFM amplitude and the electric field is observed. The value of d_{33} corresponds to the slope of this linear relationship and is equal to 40.1 pm/V. The same measurement on the same PZT sample was carried out on our SS-PFM system, and this allowed calibrating the system.

Figure 16b shows the DC bias dependency of the PFM phase for the PZT film. This figure clearly shows the changing in domains direction. This changing corresponds to a coercive polarization voltage V_c equal to 8.2V, which is the same for both figures (16a and 16b).

The SS-PFM measurements for the GeTe sample were carried out in the same way as for the PZT film. The DC bias dependencies of the PFM amplitude and phase are shown in Figure 17. The PFM amplitude versus DC bias (Figure 17a) shows a hysteresis loop. The calculated value of d_{33} for GeTe is 4.4pm/V. The PFM phase versus DC bias (Figure 17b) shows a clear switching of the ferroelectric GeTe domains approximately at 180º. The obtained coercive voltage V_c of this material is approximately 0.23V.

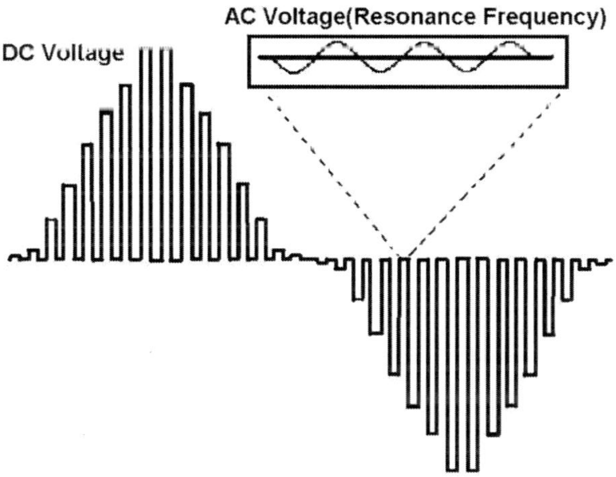

Figure 15. Schematic illustration of the setup used for SS-PFM.

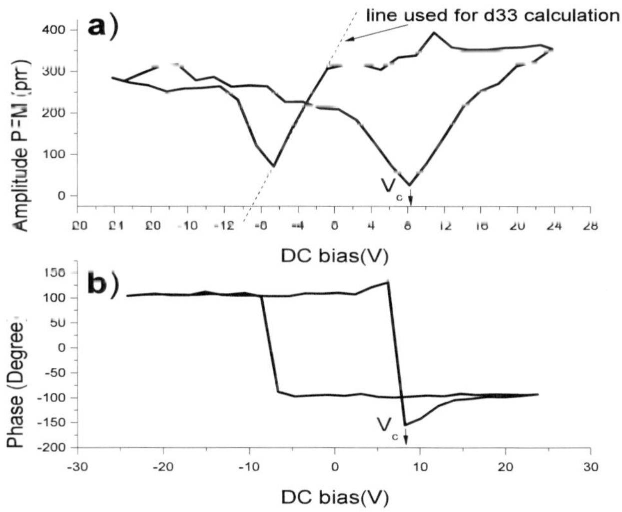

Figure 16. Dependencies of the PFM amplitude (a) and the phase (b) on DC bias for PZT thin film.

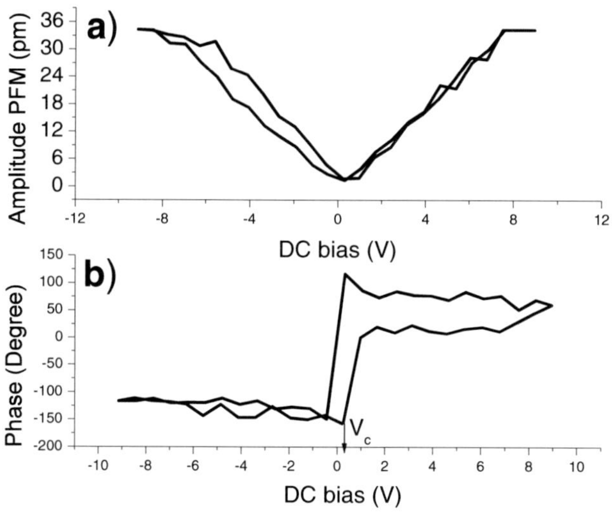

Figure 17. Dependencies of PFM amplitude (a) and phase (b) on DC bias for GeTe thin film.

Table 1. Parameters used for SS-PFM measurements on PZT thin films

DC bias	24 Vpp
Triangular Frequency DC bias	0.1Hz
Frequency of DC bias pulses	5Hz
AC bias	2V
AC frequency (tip-sample contact resonance)	359kHz

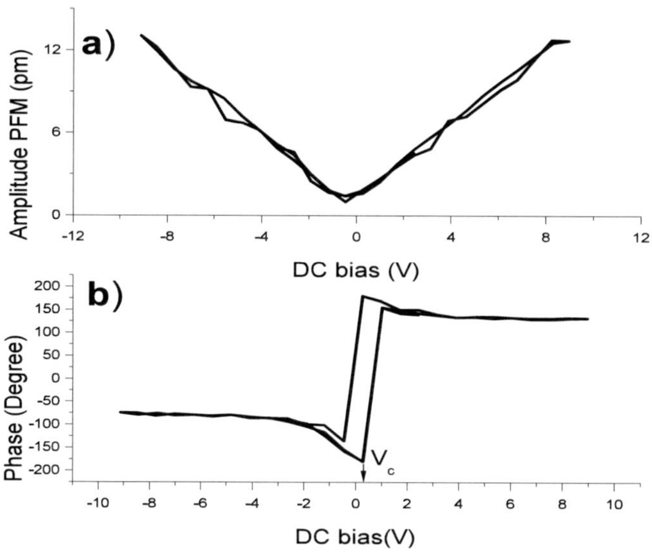

Figure 18. DC bias dependencies of PFM amplitude (a) and phase (b) for $Ge_2Sb_2Te_5$ thin film.

Figure 18 shows the SS-PFM measurements carried out on the $Ge_2Sb_2Te_5$ thin film sample. As in the case of GeTe, a thin hysteresis loop has been observed. The value of d_{33}

calculated for $Ge_2Sb_2Te_5$ thin film is 1.35 pm/V. The same as for the GeTe sample, the PFM phase of $Ge_2Sb_2Te_5$ thin film shows the switching of the ferroelectric domains approximately at 180º. The coercive voltage for $Ge_2Sb_2Te_5$ is about 0.3V.

The parameters used for the SS-PFM measurements on GeTe and $Ge_2Sb_2Te_5$ thin films are given in Table 1, but the AC bias used was 200 mV and the DC bias 12Vpp.

The d_{33} values for the $Ge_2Sb_2Te_5$ and GeTe films are sufficiently smaller than for the PZT material, but are similar to the d_{33} values for semiconductor materials (d_{33}=2pm/V for AlN, d_{33}=3pm/V for GaN/AlN [21], and for polar semiconductors d_{33} is in the range 0.1-2pm/V [58]).

It is necessary to note that the SS-PFM measurements for semiconductor materials are difficult due to their low resistivity, making difficult to generate an electric field under the tip and to induce contractions or expansions on the ferroelectric material by the inverse piezoelectric effect. But in spite of these issues, the achieved SS-PFM measurements on the chalcogenide materials gave additional confirmation of their ferroelectric nature: the spontaneous electric polarization can be change on reverse direction by applying an electric field.

CONCLUSION

Impedance, CR-PFM and SS-PFM measurements have shown that crystalline GeTe and NaCl-type crystalline $Ge_4Sb_1Te_5$ and $Ge_2Sb_2Te_5$ thin films show ferroelectric properties and ferroelectric-paraelectric transitions. For the first time, ferroelectric domains have been observed in chalcogenide semiconductors using CR-PFM measurements. These alloys are non-typical ferroelectric materials with low electrical resistivity, which makes difficult to investigate their ferroelectrical properties by other conventional techniques.

The existence of a second hexagonal phase in $Ge_2Sb_2Te_5$ does not allow observing ferroelectric-paraelectric transitions that occur at the end of the NaCl type-hexagonal transition using DSC or X-ray methods. Only comparisons of capacitance measurements in GeTe, $Ge_4Sb_1Te_5$ and $Ge_2Sb_2Te_5$ materials as a function of temperature (the former demonstrating a classical ferroelectric properties and the ferroelectric nature of the later has been recently reported in the literature) allow concluding about ferroelectric nature of NaCl type $Ge_2Sb_2Te_5$ films. It was found that these materials have ferroelectric domains with dimension approximately equal to the dimension of the grains. GeTe has the highest dimension of grains and ferroelectric domains. The lowest dimension of grains and domains has been observed in $Ge_2Sb_2Te_5$ film. SS-PFM measurements on chalcogenide materials gave additional confirmation of their ferroelectric nature. For the first time, hysteresis loop and switching effects of domains under electrical field have been observed in chalcogenide materials.

Transport properties of chalcogenide materials strongly depend on the grain boundaries.

Due to their ferroelectric properties, these chalcogenide materials may find application in the super-resolution optical disks where the resolution is far smaller than the diffraction limit, for achieving ultrahigh storage density.

References

[1] Klein, D. History of Digital Storage, *Micron Technology, Inc*. 2008.

[2] Ovshinsky, S.R. *Phys. Rev. Lett.*, 1968, vol. 21, 1450-1462.

[3] Ohta, T.; Uchida, M.; Yoshioka, K.; Inoue, K.; Akiyama, T., Furukawa, S.; Kotera, K.; Nakamura, S. *Proc., SPIE* 1989, vol. 1078, 61-64.

[4] Yamada, N.; Ohno, E.; Nishiuchi, K.; Akahira, N.; Takao M., *J. Appt. Phys.* 1991, *vol.*, 69, 2849-2856.

[5] Ohta, T.; Nishiuchi, K.; Kitaoka, Y.; Ishibashi, H.; Yamada, N.; Kozaki, T. *Jpn. J. Appl. Phys.*, 2000, *vol.* 39, 770-774.

[6] Kolobov, A. V.; Tominaga, J.; Fons, P.; Uruga, T. *Appl. Phys. Lett.*, 2003 Vol. 82, 382-384.

[7] Tominaga, J.; Nakano, T.; Atoda, N. *Appl. Phys. Lett.*, 1998, 73, 2078-2080.

[8] Kasap, S.; Capper, P.; Springer handbook of electronic and photonic materials, *Springer*, 2006.

[9] Shima, T.; Nakano T.; Kim J.; Tominaga J. *Jpn. J. Appl. Phys.*, 2005, *vol.* 44, 3631-3633

[10] Lee, H. S.; Cheong, B.; Lee, T. S.; Jeong, J.; Lee, S.; Kim, W. M.; Kim, D. *Jpn. J. Appl. Phys.* 2007, *vol.* 46, L277-L279.

[11] Tominaga, J. In: Surface Plasmon Nanophotonics. Brongersma, M.L; Kik, P.G. *(eds.)*; *Springer*, 2007, 235.

[12] Tominaga, J.; Shima, T.; Kuwahara, M.; Fukaya, T.; Kolobov, A.; Nakano, T. *Nanotechnology*, 2004, Vol. 15, 411-415.

[13] Tominaga, J. *European\Phase Change and Ovonics Symposium,* 2004.

[14] Morales-Sanchez, E.; Prokhorov, E.; Mendoza-Galvan, A.; Gonzalez-Hernandez, J.

[15] *Vacuum*, 2003, *vol.* 68, 361-364.

[16] Lee, H. S.; Lee, T. S.; Lee, Y.; Kim, J.; Lee, S.; Huh, J.; Kim, D.; Cheong, B. *Appl. Phys. Lett.*, 2008, *vol.* 93, 221108.

[17] Kolobov, A. V.; Fons, P.; Tominaga, J.; Frenkel, A. I.; Ankudinov, A. L.; Yannopoulos, S. N.; Andrikopoulos, K. S.; Uruga, T. *Jpn. J. Appl. Phys.* 2005, *vol.*, 44, 3345-3349.

[18] Im, J.; Eom, J.H.; Park, C.; Park, K.; Suh, D.S.; Kim, K.; Kang, Y.S.; Kim, C.; Lee, T.Y.; Khang, Y.; Yoon, Y.G.; Ihm, J., *Phys. Rev.*, 2008, *vol.* 78, 205205.

[19] Raoux, S.; Wełnic, W.; Lelmini, D. *Chem. Rev.* 2010, *vol.* 110, 240–267.

[20] Karpinsky, O.; Shelimova, L.; Kretova, M.; Fleurial, J. *J. Alloys. Compd.* 1998, *vol.*, 268, 112-117.

[21] Ciucivara, A.; Sahu, B. R.; Kleinman, L. *Phys. Rev. B*, 2006, *vol.* 73, 214105.

[22] Okura; Naito, H.; Matsushita, T. *Jpn. J. Appl. Phys.*, 1992, *vol.* 31, 466-470.

[23] Snykers, M.; Delavignette, P.; Amelinck, S. *Mat. Res. Bull.*, 1972, *vol.* 7, 831-840.

[24] Polking, M. J.; Zheng, H.; Ramesh, R.; Alivisatos, A. *J. Am. Chem. Soc.*, 133 2011, *vol.* 133, 2044–2047.

[25] Kato N.; Takeda Y.; Fukano, T.; Motohiro, T.; Kawai, S.; Kuno, H. *Jpn. J. Appl. Phys.*, 1999, *vol.* 38, 1707-1708.

[26] Morales-Sanchez, E.; Prokhorov, E.; Gonzalez-Hernandez, J.; Mendoza-Galvan, A. *Thin Solid Films* 2005, *vol.* 471, 243-247.

[27] Coombs, J. H.; Jongenelis, A. P. J. M.; Van Es-Spiekman, W.; Jacobs, B. A. J. *J. Appl. Phys.,* 1995, *vol.* 78, 4918-4956.
[28] Bahgat, A. A.; Mahmoud, E. A.; Abd Rabo, A. S.; Mahdy, I. A. *Physica, B*, 2006 *vol.* 382, 271-278.
[29] Prokhorov, E.; González-Hernández, J.; Ge:Sb:Te Stoichiometric and Eutectic Films for Phase-Change Memory Technology In: Amorphous Materials: Research, Technology and Application, Editors: Telle, J. R.; Pearlstine, N. A. *Nova Science Publishers, Inc.,* 2009, 147-184.
[30] Morales-Sánchez, E.; Prokhorov, E.; Mendoza-Galván, A.; González-Hernández, J., *J. Appl. Phys.*, 2002, *vol.* 91, 697-702.
[31] Barth, C.; Adam S. Foster; Claude R. Henry; Alexander L. Shluger, *Adv. Mater.* 2011, *vol.* 23, 477–501.
[32] Morita, S.; *Jornal of ferroelectron microscopy*, 2011, *vol.* 60, Supplement 1.
[33] Ivry, Y.; Ping Chu, D.; Durkan, C. *Appl. Phys. Lett.*, 2009, *vol.* 94, 162903.
[34] Balke, N.; Bdikin, I.; Kalinin, S. V.; Kholkin, A. L.; *J. Am. Ceram. Soc.*, 2009 *vol.* 8, 1629–1647.
[35] Kalinin, S.V.; Rodriguez, B.J.; Jesse, S., Mirman, B.; Karapetian, E.; Eliseev, E. A.; Morozovska, A.N., *Annu. Rev. Mat. Sci.,* 2007, *vol.* 37, 189-238.
[36] Shin, J.; Rodriguez, B.; Baddorf, A.; Thundat, T.; Karapetian, E.; Kachanov, M; Gruverman, A.; Kalinin, S. *J. Vac. Sci. Technol., B*, 2005, *vol.* 23, 2102-2108.
[37] Meng-wei, L.; Wei-jie, D.; Jian-hua, T.; Jing, W.; Yana, W.; Tian-honga, C.; Li-ding W. *3rd International Symposium on Instrumentation Science and Technology*, Xi'an. China, Aug.18-22, 2004.
[38] Jesse, S.; Mirman, B.; Kalinin, S. V. *Appl. Phys. Lett.,* 2006, *vol.* 89, 022906.
[39] Lan L.; Xie, S.; Jiangyu, L. T.; *J. Mater. Sci. Technol.,* 2010, *vol.* 26, 439-444.
[40] Rodriguez, J.; Gruverman, A.; Kingon, A.; Nemanich, R. *Appl. Phys. Lett.,* 2002, *vol.* 246, 252–258.
[41] Stephen, J.; Boris, M.; Kalinin, S. *Appl. Phys. Lett.*, 2006, *vol.* 89, 022906.
[42] Stephen, J.; Baddorf, A.; Kalinin, S. *Appl. Phys. Lett.,* 2006, *vol. 88*, 062908.
[43] Andrikopoulos, K.; Yannopoulos, S.; Voyiatzis, G.; Kolobov, A.; Ribes, M.; Tominaga, J. J. Phys. Condens. Matt., 2006, *vol.* 18, 965-979.
[44] Huang, B.; Robertson, J. *Phys. Rev. B* , 2010, *vol.* 81, 081204
[45] Caravati, S.; Bernasconi, M.; Parrinello, M. *J. Phys. Condens., Matt.* 2010, *vol.* 22, 315801.
[46] Chattopadhyayt, T.; Boucherlet, J.; Von Schneringt, H. *J. Phys. C: Solid State Phys.,* 1987, *vol.* 20, 1431-1440.
[47] Rabe, K.; Joannopoulos, J. *Phys. Rev., B,* 1987, *vol.* 36, 6631-6639.
[48] Kittel C. Introduction to Solid State Physics, *John WileyandSons, Inc.,* 1998.
[49] Lee, S.; Yoon, S.; Kim, H.; *J. Europ. Ceramic Soc.,* 2004, *vol.* 24, 2465-2470.
[50] Gonzalez-Hernandez, J.; Chao, B.; Strand, D.; Ovshinsky, S.; Pawlik, D.; Gasiorowski, P.; *Appl. Phys. Com.,* 1992, *vol.* 11, 557-569.
[51] Ruiz, R.; Prokhorov, E.; Espinoza, F.; Trapaga, L.; Gonzalez-Hernandez, J. *J. of Non-Crystalline Sol.,* 2010, *vol.* 356, 3026–3031.
[52] Yamada, N.; Matsunaga, T.; *J. Appl. Phys.,* 2000, *vol.* 88, 7020-7028.
[53] Kolobov, A.; Fons, P.; Frenkel, A. L.; Ankudinov, A. L.; Tominaga, J.; Uruga, T.; *Nat. Mat.,* 2004, *vol.* 3, 703-708.

[54] Gervacio Arciniega, J. J.; Prokhorov, E.; Espinoza Beltran, F. J.; Gonzalez-Hernandez, F. J. *Appl. Phys. Lett.*, 2010, *vol.* 97, 063504.
[55] Prokhorov, E.; Trapaga, G.; González-Hernández, J. *J. Appl. Phys.*, 2008, *vol.* 104, 103712.
[56] Kolobov, A. V.; Fons, P.; Tominaga, J.; Frenkel, A. I.; Ankudinov, A. L.; Yannopoulos, S. N.; Andrikopoulos, K. S.; Uruga, T. *Jpn. J. Appl. Phys.*, 2005, *vol.* 44, 3345-3349.
[57] Mitsui, T. *Springer Handbook of Condensed Matter and Materials Data.*, 2005, Part 4, 903-938.
[58] Zeng, H. R.; Yu, H. F.; Tang, X. G.; Chu, R. Q.; Li, G. R.; Yin, Q. R. *Mat. Sci. and Eng., B* 2005, *vol.* 120, 104-108.
[59] Proksch, R.; Kalinin, S. *Asylum Research*, Application note 10.

In: Ferroelectrics: New Research
Editors: Xui Li Huang and Su Lyn Ma
ISBN: 978-1-61942-282-7

Chapter 7

COMPOSITION DESIGN AND ELECTRICAL PROPERTIES' INVESTIGATION OF SILVER-CONTAINING PEROVSKITE LEAD-FREE PIEZOELECTRIC CERAMICS

Lang Wu
School of Materials Science and Engineering, Southwest University of Science and Technology, Sichuan Mianyang, China

ABSTRACT

Recent theoretical calculations indicate that the situation of Ag in the A site of ABO_3 perovskite is very similar to that of Pb. Good piezoelectrics may be promisingly realized in the Ag-containing perovskite system. In recent years, we have developed some new Ag-containing $(Bi_{0.5}Na_{0.5})TiO_3$ (BNT)- and $(K_{0.5}Na_{0.5})NbO_3$ (KNN)-based perovskite lead-free piezoelectric ceramics, and the piezoelectric and ferroelectric properties of the ceramics were investigated. The design of BNT-based ceramics focuses on the effects of Ag ion substitution and the effects of $Ag_{0.9}Li_{0.1}NbO_3$ (ALN10) addition on BNT–$BaTiO_3$ (BNT–BT) system. The design of KNN-based ceramics focuses on the effects of Ag ion substitution on KNN–$LiTaO_3$ system and the effects of $AgTaO_3$ or $AgSbO_3$ addition on KNN–$LiNbO_3$ system. All the ceramics were prepared by the conventional ceramic technique. Results show that both the substitution of Ag^+ for Na^+ and the addition of ALN10 can improve the piezoelectric and ferroelectric properties of the BNT–BT ceramics. The temperature dependence of electrical properties shows that the depolarization temperature (T_d) of the BNT-based ceramics shifts towards lower temperatures with increasing amount of Ag or ALN10. For the KNN-based ceramics, the orthorhombic–tetragonal polymorphic phase transition (PPT) was modified by the substitution of Ag^+ for A-site ions and the addition of $AgTaO_3$ or $AgSbO_3$. The KNN-based ceramics with a desired PPT near room temperature exhibit the enhanced piezoelectric and ferroelectric properties. In addition, the Curie temperature (T_c) of KNN-based ceramics increases with increasing amount of Ag. Moreover, we also synthesized alkaline elements (Li, Na, K)-modified $AgNbO_3$ (AN)-based ceramics by the conventional ceramic technique under air atmosphere. The results show that AN-based solid solution with pure perovskite structure can be available by optimizing the compositions and using a proper processing, such as low-temperature sintering. A high T_c of 340 °C and a large remnant polarization (P_r) of 14.5 ìC/cm^2 were obtained for the CuO-doped $(Ag_{0.75}Li_{0.1}Na_{0.1}K_{0.05})NbO_3$ ceramics. These results show that Ag-containing perovskite is a promising candidate for lead-free piezoelectric materials.

1. INTRODUCTION

Perovskite $Pb(Zr,Ti)O_3$ (PZT)-based piezoelectric ceramics have dominated the commercial market of piezoelectric devices over the past 50 years, owing to their excellent electrical properties. Recently, the demand of the sustainable development of the world and the environmental regulations has induced a new surge in developing various lead-free piezoelectric ceramics to replace the PZT-based ones [1, 2].

It is known that the hybridization between Pb 6*p* and O 2*p* orbits is curial for realizing the high piezoelectric properties in Pb-based materials. Recent theoretical calculations indicate that there is strong hybridization between Ag 4*d* and O 2*p* orbits in $AgNbO_3$ (AN), leading to a short covalent Ag–O bonds (~2.44 Å) in the A-site of perovskite [3, 4]. Such bonding causes a large off-centering displacement (~ 0.5 Å) of Ag, which is expected to drive a large local polarization [4]. Therefore, high piezoelectric performance may be promisingly realized in the Ag-based system.

We have concentrated on the research on the composition design and electrical properties of Ag-containing perovskite lead-free piezoelectric ceramics, especially on the $(Bi_{0.5}Na_{0.5})TiO_3$ (BNT)- and $(K_{0.5}Na_{0.5})NbO_3$ (KNN)-based ceramics. The design of BNT-based ceramics focuses on the effects of Ag ion substitution and the effects of $Ag_{0.9}Li_{0.1}NbO_3$ addition on BNT-based ceramics. The design of KNN-based ceramics focuses on the effects of Ag ion substitution and the effects of $AgTaO_3$ or $AgSbO_3$ addition on KNN-based ceramics. The piezoelectric and ferroelectric properties of the ceramics were mainly studied. Furthermore, we also tried to prepare alkaline elements (Li, Na, K)-modified $AgNbO_3$ (AN)-based ceramics by the conventional ceramic technique under air atmosphere. In this chapter, our recent research results on the composition design and electrical properties' study of these Ag-containing perovskite lead-free piezoelectric ceramics were presented.

2. THE DESIGN OF BNT-BASED CERAMICS

2.1. Effects of Ag ion Substitution on BNT-based Ceramics

2.1.1. Background

BNT has a perovskite structure with a rhombohedral symmetry at room temperature, which is considered to be an excellent candidate for lead-free piezoelectric ceramics because of its large remnant polarization (P_r = 38 μC/cm^2) and high Curie temperature (T_c = 320 °C). However, because of their high coercive field (E_c = 73 kV/cm), pure BNT ceramics are very difficult to polarize and their piezoelectric properties (d_{33} = 72.9 pC/N) are not desirable [5]. It has been shown that BNT can form a morphotropic phase boundary (MPB) with other perovskites, such as $BaTiO_3$ (BT) [6], $Bi_{0.5}K_{0.5}TiO_3$ (BKT) [7], and $KNbO_3$ (KN) [8], and these MPB compositions exhibit much better piezoelectric properties than pure BNT ceramics. It is noted that the BNT–$BaTiO_3$ system seems more interesting. The MPB composition of 0.94BNT–0.06$BaTiO_3$ has relatively good piezoelectric properties. The coupling factor and piezoelectric constant of the composition are $k_{33} = 0.55$, $k_{31} = 0.19$, and $d_{33} = 125$ pC/N, respectively [6].

Based on the design of the multiple complex in the A-site of ABO_3 compounds, new $[Bi_{0.5}(Na_{1-x}Ag_x)_{0.5}]_{0.94}Ba_{0.06}TiO_3$ (abbreviated as BNABT-x, x = 0–0.08) lead-free piezoelectric ceramics were proposed [9], and the piezoelectric and ferroelectric properties were investigated [10].

2.1.2. Electrical Properties of Ag-modified BNT-based Ceramics

A conventional ceramics fabrication technique was used to prepare the BNABT-x ceramics. The detailed preparation process of the ceramics was reported in ref [10]. The densities (ρ) of the obtained samples are about 5.72~5.82 g/cm^3, which are more than 95.6 % of the theoretical values. X-ray diffraction (XRD) analysis indicates that all the samples possess a pure perovskite phase. In addition, no obvious change of the crystal structure is observed for the BNABT-x ceramics with increasing Ag content. Similar to the 0.94BNT–0.06$BaTiO_3$ ceramics, the BNABT-x ceramics with x = 0–0.08 situate near the rhombohedral–tetragonal MPB.

Table 1 presents the electrical properties of some typical compositions of the BNABT-x ceramics. The d_{33} of the ceramics increases with increasing x and reaches maximum value of 172 pC/N at $x = 0.06$. Further increasing x causes a decrease in the d_{33}. The k_p and k_t of the ceramics decreases slightly with increasing x. It can also be observed in Table 1 that both ε_r and tan δ of the BNABT-x ceramics increase with increasing x. In addition, the P_r of the BNABT-x ceramics increases slightly with increasing x and reaches maximum value of 42.5 μC/cm^2 at $x = 0.05$. The E_c of the ceramics remains almost unchanged (~ 32 kV/cm) with increasing x. These results indicate that the BNABT-x ceramics with x = 0.05–0.06 possess the optimum properties (d_{33} ~ 172 pC/N, k_p ~ 31.2 %, k_t ~ 51.5 %, P_r ~ 42.5 ìC/cm^2, and E_c ~ 32.3 kV/cm, respectively.

Table 1. Electrical properties of BNABT-*x* ceramics

Compositions	$x = 0$	$x = 0.04$	$x = 0.05$	$x = 0.06$	$x = 0.07$	$x = 0.08$
ρ (g/cm^3)	5.72	5.73	5.74	5.80	5.81	5.82
d_{33} (pC/N)	154	162	168	172	165	160
k_p (%)	32.3	32.6	31.0	31.2	30.0	28.0
k_t (%)	52.6	51.7	51.6	51.5	50.0	49.0
ε_r (1 kHz)	763	857	919	972	1050	1120
tan δ (1 kHz)	0.024	0.027	0.029	0.031	0.037	0.040
P_r (μC/cm^2)	39.6	40.7	42.5	40.1	40.2	39.6
E_c (kV/cm)	32.6	32.0	32.3	32.2	32.1	32.3

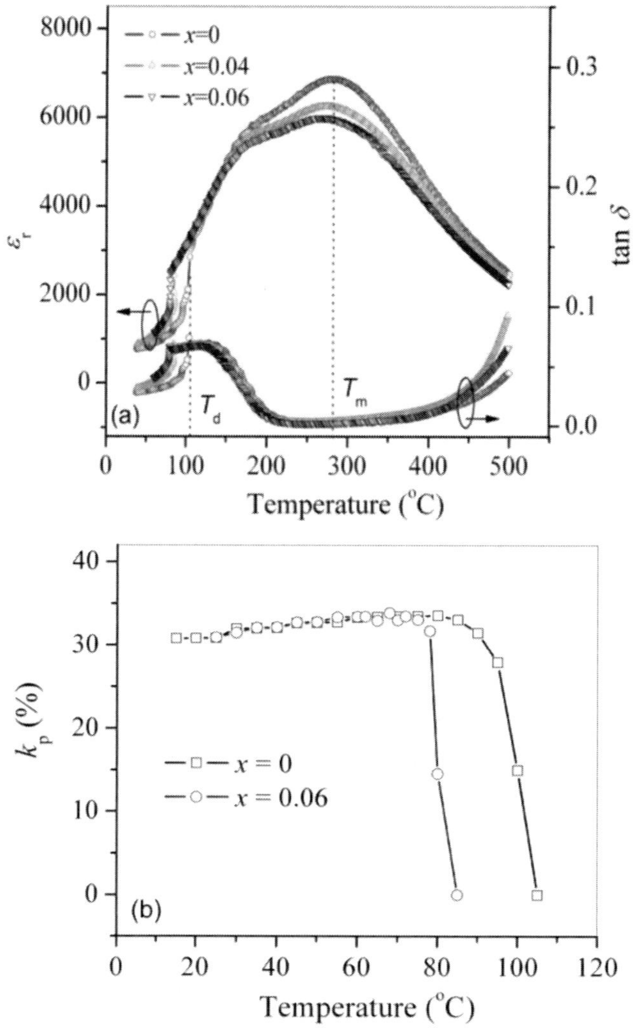

Figure 1. Temperature dependence of (a) ε_r and tan δ of poled BNABT-x ceramics at 10 kHz and (b) k_p of the ceramics with $x = 0$ and 0.06.

As mentioned above, the addition of Ag has no obvious change on the crystal structure of the BNT–BT ceramics, confirmed by the XRD analysis. It is thought that the effect of the structural distortion caused by Ag on the ferroelectric and piezoelectric properties might not be so significant, owing to the low concentration level of Ag ($\leq$ 0.0376 mol) in this system. Compared with pure BNT–BT ceramics (d_{33} = 154 pC/N), Ag-modified BNT–BT ceramics exhibit the higher piezoelectric properties (d_{33} = 172 pC/N), which should be mainly attributed to the improvement of densification. Another possible reason for the increase of d_{33} with increasing x may partly be attributed to the higher dielectric constant (ε_r ~ 972) and larger P_r (~ 42.5 μC/cm^2) as compared with the BNT–BT ceramics (ε_r ~ 763, P_r ~ 39.6 μC/cm^2), because the d_{33} is proportional to both the polarization and dielectric constant.

2.1.3. Temperature Dependence of Electrical Properties

It is very important to study the depolarization temperature (T_d) of BNT-based ceramics from the viewpoint of device applications. The T_d could be determined from the temperature dependence of piezoelectric and dielectric properties. Figure 1(a) shows the temperature dependence of dielectric constant ε_r and dielectric loss tan δ of poled BNABT-x ceramics at 10 kHz. For the pure BNT–BT ceramics (i.e., $x = 0$), a sharp increase of the ε_r takes place at the temperature T_d (about 105 °C) where the tan δ reaches a high value, as shown in Figure 1(a). In addition, the dielectric loss shows a broad peak near the T_d. The T_d of the BNABT-x ceramics decreases slightly with increasing x. For the ceramics with $x = 0.06$, the T_d decreases to about 85 °C. Figure 1(b) shows the temperature dependence of k_p of the BNABT-x ceramics with $x = 0$ and 0.06. It can be observed in Figure 1(b) that the k_p of the ceramics with $x = 0$ remains almost unchanged with increasing temperature up to 95 °C. Further increasing temperature causes a sharp decline in the k_p. It was found that when the temperature reached 105 °C, the resonance and antiresonance peaks disappeared suddenly and the ceramics were depolarized. The T_d determined from this method corresponds to the temperature of tan δ's first peak. Similar behavior of k_p versus temperature was also observed in other BNABT-x ceramics.

In order to investigate the feature of T_d further, Figure 2(a) shows the polarization versus electric field (*P–E*) hysteresis loops of the BNABT-x ceramics with $x = 0.06$ at different temperatures. It can be noted in Figure 2(a) that the sample exhibits deformed *P–E* hysteresis loops at temperatures near T_d (~80 °C). The loop of the sample appears to form double-like *P–E* hysteresis loop and the ferroelectricity becomes very weak ($P_r = 4.73$ μC/cm^2) when the temperature reaches 100 °C. Similar double-like *P–E* hysteresis loops were also observed in other BNABT-x ceramics. T_d is traditionally thought to indicate a phase transition from ferroelectric to anti-ferroelectric [6]. At higher temperatures, the BNABT-x ceramics still exhibit ferroelectric-like characteristics (P_r = 2–5 μC/cm^2), not typically antiferroelectric-like (double *P–E* loops). Recently, Tai *et al* [11] confirmed that the ferroelectric-antiferroelectric transition does not occur in the BNT–BKT–$Bi_{0.5}Li_{0.5}TiO_3$–BT ceramics near T_d. Similar deformed hysteresis loops have also been observed in other BNT-based ceramics, such as BNT–BT–$Bi_{0.5}Li_{0.5}TiO_3$ [12] and $(Na_{0.5}Bi_{0.5})_{0.70}Ba_{0.30}TiO_3$ [13] system, which was suggested to result from the coexistence of polar and microscopically non-polar phases [11-13]. Figure 2(b) shows the d_{33} of the BNABT-x ceramics with $x = 0$, 0.04, and 0.06 as a function of annealing temperature. It was found that for all of the samples, the d_{33} had a steep fall near the T_d. The samples still had d_{33} values of 5–15 pC/N after they were annealed above T_d for 30 min, which is also an indication for the existence of polar phases at temperatures above T_d. So the results suggest that the polar and non-polar phases may coexist in the BNABT-x ceramics at temperatures above T_d.

2.2. Effects of $Ag_{0.9}Li_{0.1}NbO_3$ on BNT-based Ceramics

2.2.1. Background

For BNT-based solid solutions, the T_d usually shifts downward from 187 °C for pure BNT to approximately 100 °C for the MPB composition [6-10]. According to our previous work [10], the tetragonal side of the MPB composition in the BNT-BT system possesses both relatively high T_d (~ 190 °C) and high d_{33} values. However, the piezoelectric properties of the ceramics are still unfavorable for application (d_{33} ~ 130 pC/N).

It has been reported that a partial substitution of Li for Ag can induce a ferroelectric rhombohedra distortion [14], which can be attributed to the Ag displacement by the large off-centering nature of Li in the perovskite structure [15]. In particular, a large P_r of 23 μC/cm^2 was observed for the $Ag_{0.9}Li_{0.1}NbO_3$ (ALN10) ceramics at room temperature [14]. Therefore, it can be expected that by incorporating Ag and Li into A site of perovskite structure, it would show high ferroelectric and piezoelectric properties owing to the large distortion of the structure.

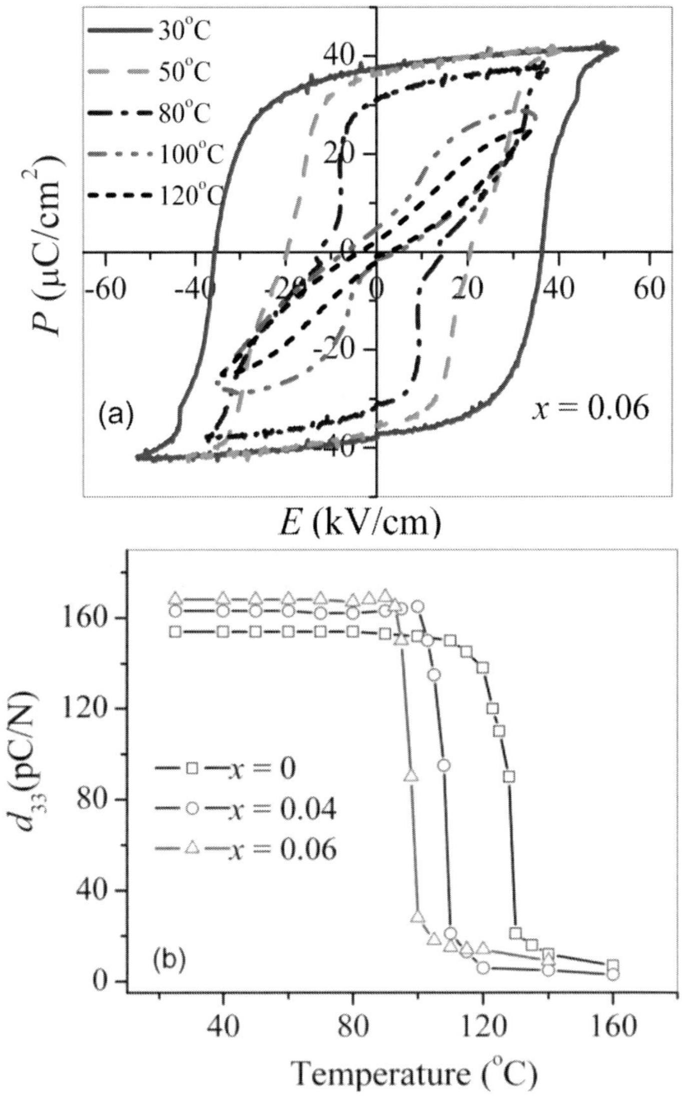

Figure 2 (a) *P–E* hysteresis loops of BNABT-*x* ceramics with $x = 0.06$ at different temperatures and (b) d_{33} of the ceramics with x=0, 0.04, and 0.06 as function of annealing temperature.

Based on the above consideration, a new BNT-based solid solution between $(Bi_{0.5}Na_{0.5})_{0.88}Ba_{0.12}TiO_3$ (BNBT12) and ALN10, possessing tetragonal and rhombohedral phases, respectively, was proposed [16]. It is expected to form a MPB composition in the BNBT12−ALN10 system. It is reasonable to anticipate that the BNBT12−ALN10 solid solution would possess high piezoelectric properties due to the formation of MPB and the large distortion of Ag and Li in the A site.

2.2.2. Electrical Properties of $Ag_{0.9}Li_{0.1}NbO_3$-Modified BNT-Based Ceramics

The (1−x)BNBT12−xALN10 ceramics were prepared by the conventional oxide-mixed method. The detailed preparation process of the ceramics was reported in ref [16]. All ceramic samples have high densities about 5.71−5.80 g/cm^3, which are more than 95 % of the theoretical values. XRD analysis indicates that all of the samples have a pure perovskite structure. ALN10 appears to have diffused into the BNBT12 lattice forming a solid solution. For $x < 0.01$, the crystal structures of the ceramics are tetragonal symmetry. The tetragonality of the ceramics decreases with increasing x. When $x \geq 0.02$, the lattice anisotropy is very small (~1.0006) and no peak splitting can be observed. A MPB between tetragonal and pseudocubic phases of (1−x)BNBT12−xALN10 ceramics may exist at x = 0.01−0.02.

Figure 3(a) shows the *P–E* hysteresis loops of the (1−x)BNBT12−xALN10 ceramics with x = 0, 0.02, 0.03, and 0.04. P_r and E_c of the samples as a function of x are shown in Figure 3(b). It was found that normal *P–E* hysteresis loops were observed for the ceramics with $x \leq$ 0.025. However, the *P–E* loop becomes slightly deformed at x = 0.03. The loop appears to form a double-like *P–E* hysteresis loop at x=0.04 [Figure 3(a)]. It can be observed in Figure 3(b) that the E_c decreases almost linearly with increasing x. The P_r increases slightly from 19.2 to 23.0 μC/cm^2 as x increases from 0 to 0.03. With further raise of x, the P_r decreases significantly and the ferroelectricity becomes very weak as the *P–E* hysteresis loop becomes "pinched". For the ceramics with x = 0.04, the P_r and E_c of the sample are only 6.33 μC/cm^2 and 1.18 kV/mm, respectively. It is thought that ferroelectric phase and nonpolar phase may coexist as a matastable phase in the samples with x = 0.03 and 0.04 [17]. The ferroelectric phase dominates in the sample with x = 0.03 while the nonpolar one becomes dominant at x = 0.04. On the other hand, these results also confirm that the crystal structures of the samples with x = 0.02−0.04 are not true cubic symmetry but a pseudocubic structure because the pervoskite material with cubic symmetry has no spontaneous polarization.

Table 2 presents the piezoelectric and dielectric properties of (1−x)BNBT12−xALN10 ceramics. It can be noted in Table 2 that d_{33} increases with increasing x and reaches the maximum value of 155 pC/N at x = 0.025. At x = 0.04, the d_{33} drops to about 9 pC/N. In addition, k_p and k_t increase slightly with increasing x. The maximum values of k_p (14.8 %) and k_t (45.1 %) are observed at x = 0.025 and 0.01, respectively. These results indicate that the ceramics with x = 0.025 possess the optimum piezoelectric properties: d_{33} = 155 pC/N, k_p = 14.8 %, k_t = 40.2 %.

It is thought that the lower E_c values can make the ceramics polarize sufficiently in poling process, which is very helpful to the increase in piezoelectric properties. Additionally, for the compositions near tetragonal-pseudocubic MPB in this system, the improvement of symmetry can reduce the internal stress of the domain reorientation and make the domains transform more easily in the poling process, which can also increase the piezoelectric properties [8]. As mentioned above, the strong hybridization between Ag and O atoms and the large off-centering nature of Li were expected to improve the ferroelectric and piezoelectric properties of BNBT12 ceramics. However, the non-polar phases become dominant and the T_d is shifted towards below room temperature when $x > 0.03$, resulting in the sharp decline of the piezoelectric properties. Therefore, the amount of Ag and Li additive has been limited. The effect of the structural distortion caused by Ag and Li on the ferroelectric and piezoelectric properties might not be so significant, owing to their low concentration level in this system.

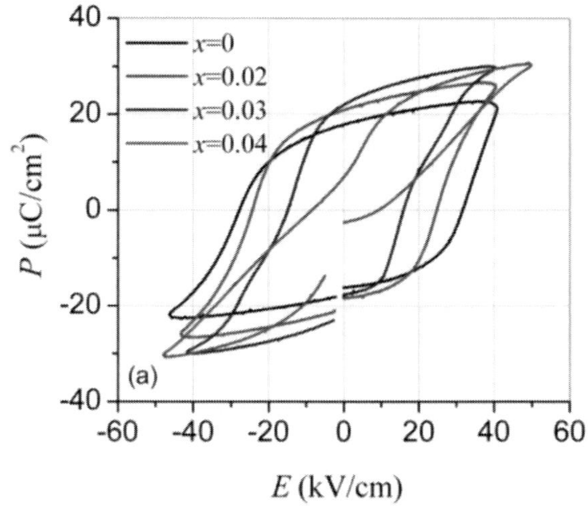

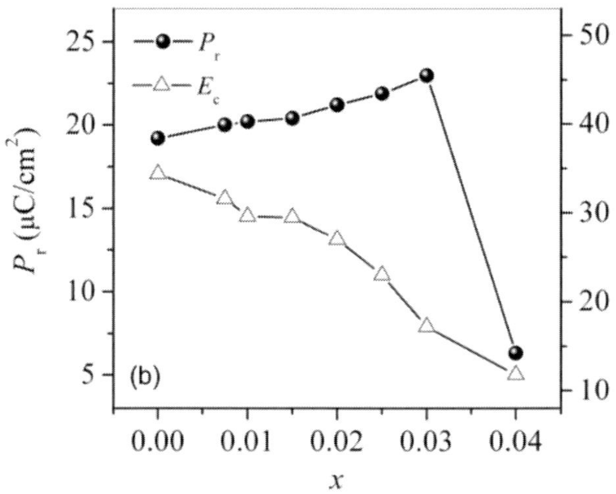

Figure 3. (a) *P–E* hysteresis loops of the (1−*x*)BNBT12−*x*ALN10 ceramics with $x = 0$, 0.02, 0.03, and 0.04, and (b) P_r and E_c of the ceramics as function of *x*.

Table 2. Piezoelectric and dielectric properties of (1−*x*)BNBT12−*x*ALN10 ceramics

Samples	x=0	x=0.01	x=0.015	x = 0.02	x=0.025	x=0.03	x=0.04
d_{33} (pC/N)	126	140	143	150	155	152	9
k_p (%)	13.7	14.1	14.3	14.6	14.8	13.2	–
k_t (%)	41.2	45.1	44.6	44.1	40.2	36.9	–
ε_r (1 kHz)	775	872	888	1010	1150	1350	1530
tan δ (1 kHz)	0.029	0.031	0.036	0.036	0.041	0.050	0.061

2.2.3. *Temperature Dependence of Dielectric Properties*

Figures 4(a)–(d) show the temperature dependence of ε_r and tan δ of poled (1−*x*)BNBT12−*x*ALN10 ceramics with *x* = 0, 0.01, 0.02, and 0.04, respectively, at different frequencies ranging from 100 Hz to 100 kHz. For BNBT12 ceramics (i.e., *x* = 0), a sharp increase in the ε_r and tan δ near 195 °C is observed, which corresponds to the T_d. At temperatures higher than T_d, the ε_r increases gradually with increasing temperature and reaches the maximum value at a temperature T_m (~ 272 °C). In addition, after an abrupt jump in the ε_r versus temperature curves, a relaxor-like behavior can be observed near T_d [Figure 4(a)]. This indicates that a transformation between a normal and a relaxor ferroelectric state occurs near T_d, which is very similar to the results obtained for $(Bi_{0.5}Na_{0.5})_{0.70}Ba_{0.30}TiO_3$ ceramics [13]. It can also be noted in Figure 4 that the maximum values of ε_r decline with increasing *x* and the ε_r versus temperature curves become more and more flat. Furthermore, both T_d and T_m are shifted towards low temperatures with increasing *x*. The ceramics with *x* = 0.01 show similar dielectric response as the BNBT12 ceramics, as shown in Figure 4(b). The T_d and T_m of the sample (*x* = 0.01) are 150 °C and 265 °C, respectively.

It is interesting that a different dielectric behavior is observed for the ceramics with $x \geq 0.02$ [Figures 4(c)–(d)]. Firstly, it shows a typical relaxation behavior for these samples from room temperature to about 200 °C. It suggests that the pseudo-cubic symmetry of the samples with $x \geq 0.02$ is a relaxor-ferroeletric phase. Similar pseudocubic relaxor-ferroelectric phase has also been observed in (1−*x*)BNT−*x*KN ($x \geq 0.08$) [18] and (1−*x*)BNT−*x*$Ba(Al_{1/2}Sb_{1/2})O_3$ system ($x \geq 0.045$) [19]. Secondly, the anomaly of tan δ exhibits obvious frequency dispersion and it does not show a sharp peak, owing to the relaxation behavior of the samples with pseudocubic structure. So it is very difficult to determine the T_d accurately from the temperature dependence of dielectric properties. The T_d of the samples with $x>0.01$ was determined from the temperature dependences of piezoelectric properties. The T_d is about 130 °C and 92 °C for the ceramics with *x*=0.02 and 0.03, respectively. As the ferroelectric and piezoelectric properties of the sample with *x*=0.04 is very weak at room temperature, it is difficult to determine its T_d, which might be already below room temperature. Additionally, it should be noted that the poled (1−*x*)BNBT12−*x*ALN10 samples also possess d_{33} values of 5–15 pC/N after they were annealed above T_d. This result also indicates that the polar and non-polar regions may coexist at temperatures between T_d and T_m [11-13].

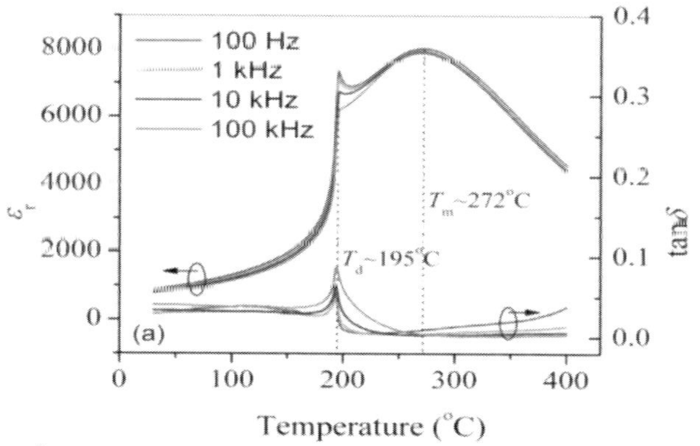

Figure 4. (Continued).

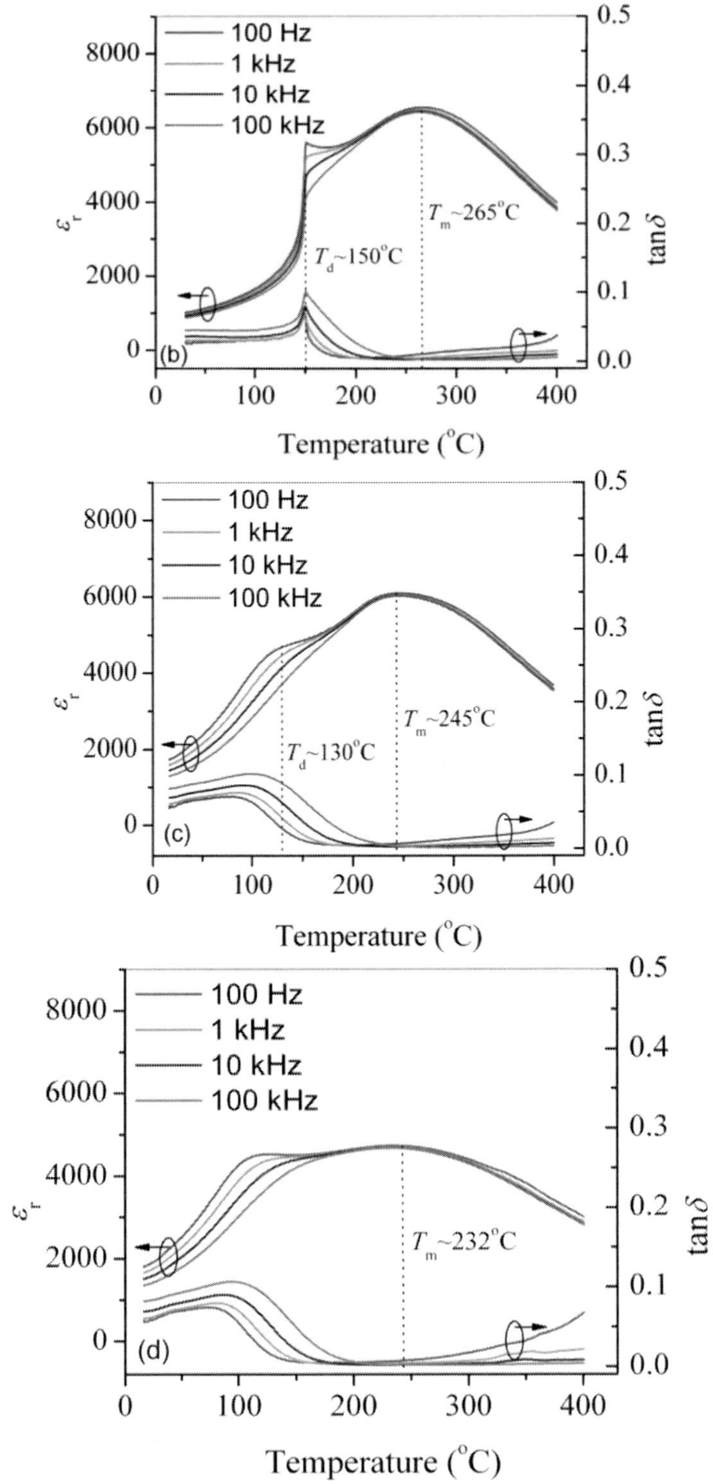

Figure 4. Temperature dependence of ε_r and tan δ of poled $(1-x)$BNBT12$-x$ALN10 ceramics at different frequencies ranging from 100 Hz to 100 kHz: (a) $x = 0$; (b) $x = 0.01$; (c) $x = 0.02$; and (d) $x = 0.04$.

3. THE DESIGN OF KNN-BASED CERAMICS

3.1. Background

Over the past few years, considerable attention has been paid to the KNN family following the report of textured (Li, Ta, Sb) modified (K, Na)NbO_3 ceramics with comparable piezoelectric properties (d_{33}~416 pC/N)to a hard PZT [2]. With different additions, such as $LiNbO_3$ [20], $LiTaO_3$ [21], and $LiSbO_3$ [22], the KNN-based ceramics were reported to exhibit enhanced piezoelectric properties owing to presence of the orthorhombic–tetragonal polymorphic phase transition (PPT) near room temperature [23].

Our previous study showed that Ag^+ could effectively improve the piezoelectric properties of BNT-based ceramics [9]. However, there are few reports on studying the effects of Ag^+ on KNN-based ceramics. The authors' group has concentrated the research on the piezoelectric and ferroelectric properties of (Li, Ag, Ta, Sb) modified KNN ceramics [24-26]. In this chapter, the main results are reviewed as follows.

3.2. Effects of Ag Ion Substitution on KNN-based Ceramics

Lead-free piezoelectric ceramics $[(K_{0.50}Na_{0.50})_{0.95-x}Li_{0.05}Ag_x](Nb_{0.95}Ta_{0.05})O_3$ (KNLANT–x) were prepared by the conventional mixed oxide method. The detailed preparation process of the ceramics was reported in ref [24]. XRD analysis indicates that all of the ceramics exhibit pure perovskite phase, suggesting that the Ag^+ has diffused into the KNLNT lattice to form a solid solution in the composition range of $0.01 \leq x \leq 0.06$. In addition, it was found that the positions of the diffraction peaks shifted to higher angles slightly.

The temperature dependence of ε_r of the KNLANT–x ceramics measured shows that the $T_{O\text{-}T}$ is about 99 and 68 °C for the ceramics with $x = 0$ and 0.03, respectively. The $T_{O\text{-}T}$ is below room temperature when $x = 0.05$. The T_c of the ceramics with $x = 0$ is about 427 °C. The T_c shifts slightly to higher temperatures with increasing x. The ceramics with $x = 0.03$ possess a high T_c of 438 °C. Table 3 presents the electrical properties of the KNLANT–x ceramics as a function of x. It is found that the d_{33}, k_p, and ε_r increase with increasing x, and reach maximum values of 252 pC/N, 45.4 %, and 1030 at $x = 0.03$, respectively. Further increasing x causes a decrease in the piezoelectric and dielectric properties. Similarly, the P_r reaches the maximum value of 30.1 μC/cm² at $x = 0.03$. These results indicate that the

Table 3. Electrical properties and Curie temperature of KNLANT–x ceramics

Compositions	$x = 0.00$	$x = 0.01$	$x = 0.02$	$x = 0.03$	$x = 0.04$	$x = 0.05$
d_{33} (pC/N)	195	216	223	252	200	190
k_p (%)	40.0	43.5	44.5	45.4	41.2	38.0
ε_r (1 kHz)	900	870	930	1030	1000	975
P_r (μC/cm²)	21.0	21.3	23.5	30.1	29.0	28.0
E_c (kV/cm)	13.2	12.4	12.7	13.8	12.1	11.0

ceramics with $x = 0.03$ possess enhanced piezoelectric and dielectric properties at room temperature, which may be attributed to the PPT near room temperature.

3.3. Effects of $AgTaO_3$ on KNN-Based Ceramics

Lead-free piezoelectric ceramics $(1-x)[(K_{0.4725}Na_{0.4725})Li_{0.055}]NbO_3-xAgTaO_3$ [(1−x)KNLN–xAT, $x = 0–0.05$] were prepared by the conventional mixed oxide method. The detailed preparation process of the ceramics was reported in ref [25]. XRD analysis indicates that the ceramics with $x \leq 0.03$ possess pure perovskite structure. A small amount of secondary phase of $K_3Li_2Nb_5O_{15}$ was observed when $x \geq 0.04$. Furthermore, the orthorhombic and tetragonal phases coexist in the (1−x)KNLN–xAT ceramics with $0.015 \leq x \leq 0.025$.

Table 4 presents the electrical properties and T_c of the (1−x)KNLN–xAT ceramics as a function of x. It is found that both d_{33} and k_p increase with increasing x, and reach maximum values of 237 pC/N and 39 % at $x = 0.02$, respectively. Further increasing x causes a decrease in the d_{33} and k_p. In addition, it can be noted that the P_r increases slightly with increasing x and reaches maximum value of 29.4 μC/cm^2 at $x = 0.02$. The P_r decreases gradually with further increasing x.

It is known that for pure KNN ceramics, the orthorhombic−tetragonal phase transition temperature ($T_{O\text{-}T}$) and T_c are 200 and 420 °C, respectively. It was found that the $T_{O\text{-}T}$ of the ceramics with $x = 0.015–0.025$ shifted downward to near room temperature, suggesting that the coexistence of the orthorhombic and tetragonal phases is formed in these ceramics around room temperature. This is consistent with the results of XRD analysis. It is thought that the improvement in piezoelectric and ferroelectric properties of the ceramics with $x = 0.02$ can be mainly attributed to the more possible polarization states resulting from the coexistence of the two ferroelectric orthorhombic and tetragonal phases at room temperature.

It is interesting to note in Table 4 that the T_c of the ceramics increases to 470 °C as x increases to 0.02, and then it drops with a further increase of x. It was reported that the incorporation of Li or Ag in the A-site of KNN could improve the T_c of the ceramics, while the incorporation of Ta^{5+} or Sb^{5+} in the B-site could decrease the T_c [22, 27]. It is thought that Li^+ and Ag^+ play a dominant role in the T_c for the (1−x)KNLN–xAT ceramics with a high level of AT ($x \leq 0.02$). When $x > 0.02$, the Ta^{5+} in the ceramics plays a dominant role in the T_c, and thus decreases the T_c gradually with increasing x. These results indicate that by adding an appropriate amount of Ag and Ta, good electrical properties (d_{33}~237 pC/N, k_p~39 %, and P_r~29.4 μC/cm^2) and high T_c (~ 470 °C) can be realized in the (1−x)KNLN–xAT ceramics.

Table 4. Electrical properties and Curie temperature of (1−*x*)KNLN−*x*AT ceramics

Compositions	$x = 0.01$	$x = 0.02$	$x = 0.03$	$x = 0.04$	$x = 0.05$
d_{33} (pC/N)	165	237	180	155	116
k_p (%)	34	39	33	35	31
P_r (μC/cm^2)	25.3	29.4	23.6	15.5	14.8
E_c (kV/cm)	12.5	13.6	10.0	12.8	10.3
T_c (°C)	462	470	440	435	410

Table 5. Electrical properties and Curie temperature of (1−x)KNLN–xAS ceramics

Compositions	$x = 0.01$	$x = 0.02$	$x = 0.03$	$x = 0.04$	$x = 0.05$
d_{33} (pC/N)	145	230	185	195	165
k_p (%)	32	39	33	37	32
P_r (μC/cm^2)	32.5	40.3	22.5	32.5	23.5
E_c (kV/cm)	13.5	12.3	9.40	13.5	12.0
T_c (°C)	445	430	395	370	350

3.4. Effects of $AgSbO_3$ on KNN-based Ceramics

Lead-free piezoelectric ceramics $(1-x)[(K_{0.4725}Na_{0.4725})Li_{0.055}]NbO_3$–$xAgSbO_3$ [$(1-x)$KNLN–xAS, x = 0.01–0.06] were prepared by the conventional mixed oxide method. The detailed preparation process of the ceramics was reported in ref [26]. XRD analysis indicates that the ceramics with $x \leq 0.04$ have pure perovskite structure. When $x < 0.01$, the ceramics possess the orthorhombic symmetry. The tetragonal phase appears and increases continuously with the increasing AS content ($x \geq 0.02$). The coexistence of orthorhombic and tetragonal phases was identified in the composition range of $0.02 \leq x \leq 0.04$.

Table 5 presents the electrical properties and T_c of the $(1-x)$KNLN–xAS ceramics as a function of x. It is found that both d_{33} and k_p increase with increasing x, and reach maximum values of 230 pC/N and 39 % at x = 0.02, respectively. Further increasing x causes a decrease in the d_{33} and k_p. Similar to the d_{33} and k_p, a maximum P_r of 40.3 μC/cm^2 was also obtained for the ceramics with x = 0.02. This value is much higher than that of the 0.94KNN–$0.06LiNbO_3$ ceramics (P_r = 20.0 μC/cm^2) reported by Du et al [28]. It is thought that the improvement in the piezoelectric and ferroelectric properties of the ceramics can also be mainly attributed to the coexistence of orthorhombic and tetragonal phases near room temperature. Additionally, the Pauling electronegativity of Sb (2.05) is much larger than Nb (1.5). Therefore, the partial substitution of Sb^{5+} for Nb^{5+} in the $(1-x)$KNLN–xAS ceramics can make the structure more covalent, resulting in the higher ferroelectricity and piezoelectricity.

The temperature dependence of ε_r of $(1-x)$KNLN–xAS ceramics measured shows that the $T_{O\text{-}T}$ of the ceramics with x = 0.02–0.04 shifts downward to near room temperature. The T_c of the ceramics decreases gradually from to 446 to 330 °C as x increases from 0.01 to 0.06. For the ceramics with x = 0.02, the sample maintains a relative high T_c of about 430 °C, while it possesses good electrical properties (d_{33} ~ 230 pC/N, k_p ~ 39 %, and P_r ~ 40.3 μC/cm^2).

4. Synthesis and Properties of (Li, Na, K)-modified $AgNbO_3$ Ceramics

4.1. Background

It is known that pure $AgNbO_3$ exhibits very weak ferroelectricity with remnant polarization (P_r) of 0.041 μC/cm^2 at zero field, which is believed to be derived from the small displacement of Nb^{5+} in the B site [29]. Recently, an extremely large local polarization (~ 52 μC/cm^2) and a large electromechanical response were observed in the $AgNbO_3$ ceramics under an electric field of 220 kV/cm, which is attributed to the large local displacement of Ag^+ in the perovskite structure [30].

In addition to the high electric field, chemical modification is another approach to induce strong ferroelectric state in AN. Sakabe et al [14] reported that a partial substitution of Li for Ag can induce a ferroelectric rhombohedra distortion in AN. In particular, a large P_r of 23 μC/cm^2 was obtained for the composition of $Ag_{0.9}Li_{0.1}NbO_3$ (ALN10) ceramics. It also exhibit good piezoelectric properties (d_{33} ~ 52 pC/N, k_{33} ~ 44 %) along with a high T_c of about 295 °C. The piezoelectric properties were improved (d_{33} ~ 66 pC/N, k_{33} ~ 48 %) when incorporating a few percent of K and Na into Ag site of ALN10 [31]. It is considered that the substitution of Na and K ions for Ag ions in (Ag, Li)NbO_3 is useful for wide application in piezoelectric devices. However, it should be noted that the synthesis of AN-based ceramics is very difficult due to a thermodynamic instability of the Ag^+ at elevated temperatures. Furthermore, the start of the decomposition of AN was determined to be at about 1120 °C in air [32]. Therefore, in order to avoid the presence of silver precipitates, an oxygen atmosphere is essential during the calcination and sintering. But this makes the preparation conditions extremely complicated and expensive.

4.2. Preparation of (Li, Na, K)-modified AN Ceramics

Our recent research work has showed that CuO as a sintering aid could reduce the sintering temperature and improve the densification of AN-based ceramics. We tried to prepare $(Ag_{0.75}Li_{0.1}Na_{0.1}K_{0.05})NbO_3$ (ALNKN) ceramics doped with 1 mol % CuO by solid state reaction method under air atmosphere. The detail preparation process is as follows.

Ag_2O (99 %) K_2CO_3 (98 %), Na_2CO_3 (99.8 %), Li_2CO_3 (97 %), Nb_2O_5 (99.5 %), and CuO (99 %) were used as starting raw materials. The stoichiometric powders were mixed in alcohol by ball milling for 12 h. After calcination at 880 °C for 3 h in air, the calcined powders were again ball milled, granulated, and pressed into discs by dry pressing with

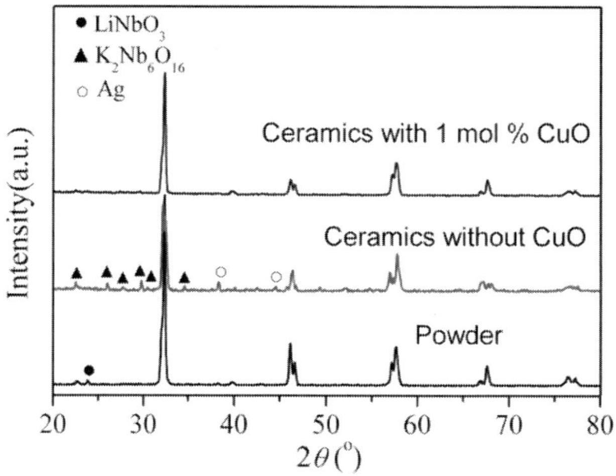

Figure 5. XRD patterns of the ALNKN powders and ceramics.

diameters of 12 mm and thicknesses of 1.1–1.2 mm. The disk samples of ALNKN without and with CuO doping were sintered in air at 1060 and 980 °C for 2 h, respectively. After the deposition of silver electrodes, the specimens were poled in silicone oil at 120 °C under 4 kV/mm for 20 min.

4.3. Structure and Electrical Properties of (Li, Na, K)-modified AN Ceramics

Figure 5 shows the XRD patterns of the ALNKN powders and ceramics. It can be observed from Figure 5 that the ALNKN calcined powders possess an orthorhombic phase. In addition, a small amount of $LiNbO_3$ phase is observed. For the ALNKN ceramics without CuO doping, in addition to the perovskite phase, $K_2Nb_6O_{16}$-type phase and metal silver phase are also observed in the compounds. It seems that the formation of the $K_2Nb_6O_{16}$ phase is may be due to the easy precipitation of metal silver at high temperature (~ 1060 °C). After the doping of 1 mol % CuO, the ALNKN ceramics can be well sintered at a relatively low temperature (~ 980 °C). The bulk density of the ceramics is 6.05 g/cm^3, almost 96.1% of the theoretical density. Pure perovskite structure with the orthorhombic phase is observed for the CuO-doped ALNKN ceramics. Apparently, it is difficult to form a single-phase perovskite structure for the ALNKN solid solution in a wide composition range because of the large difference of ionic radius among Ag^+ (1.28 Å), K^+ (1.64 Å), and Li^+ (0.76 Å) [33]. However, the above results suggest that ALNKN solid solution with pure perovskite structure can be available for the optimized compositions by a proper process, such as low-temperature sintering.

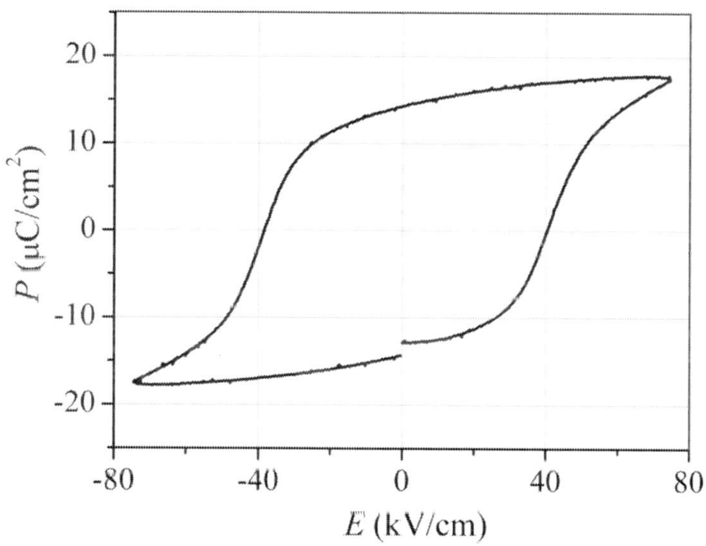

Figure 6. *P–E* hysteresis loop of CuO-doped ALNKN ceramics.

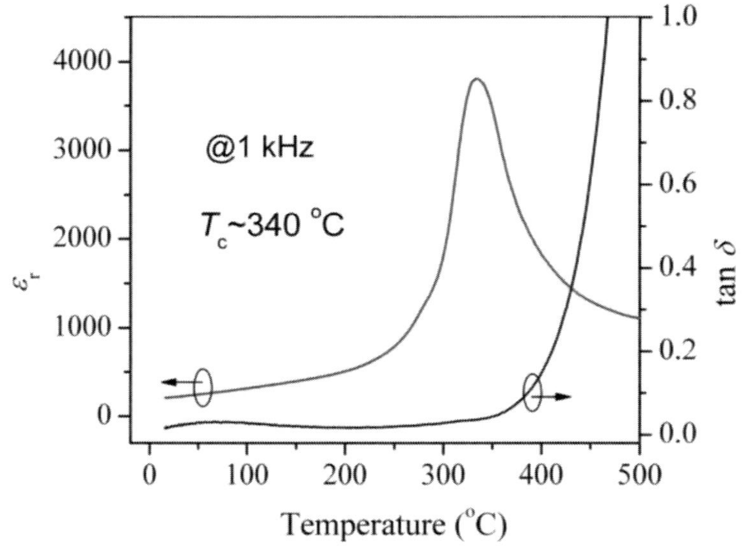

Figure 7. Temperature dependence of ε_r and tan δ of poled CuO-doped ALNKN ceramics at 1 kHz.

Figure 6 shows *P–E* hysteresis loop of CuO-doped ALNKN ceramics at room temperature. It can be noted that the sample shows a normal hysteresis loop and possesses a large P_r value of 14.5 μC/cm^2. This indicates that a strong ferroelectricity is realized in ALNKN through a chemical modification of AN structure with alkaline ions substitution. Figure 7 shows the temperature dependence of ε_r and tan δ of poled CuO-doped ALNKN ceramics at 1 kHz. At room temperature, the ε_r and tan δ of the ceramics are 210 and 0.015, respectively. The ceramics possess a high T_c of 340 °C and exhibit good temperature stability of the dielectric properties in a wide temperature range. This is very important from the viewpoint of application. These results indicate that alkaline elements modified AN-based ceramics are promising lead-free piezoelectric materials. Further investigations on the composition, structure, processing, and electrical properties in AN-based ceramics are necessary in the future works.

CONCLUSION

Some new Ag-containing $(Bi_{0.5}Na_{0.5})TiO_3$ (BNT)- and $(K_{0.5}Na_{0.5})NbO_3$ (KNN)-based perovskite lead-free piezoelectric ceramics were developed, and the piezoelectric and ferroelectric properties of the ceramics were investigated. The design of BNT-based ceramics focuses on the effects of Ag^+ ion substitution and $Ag_{0.9}Li_{0.1}NbO_3$ addition on the BNT–$BaTiO_3$ system. The design of KNN-based ceramics focuses on the effects of Ag ion substitution on the KNN–$LiTaO_3$ system and the effects of $AgTaO_3$ or $AgSbO_3$ addition on the KNN–$LiNbO_3$ system. All the ceramics were prepared by the conventional ceramic technique.

Results show that the $[Bi_{0.5}(Na_{1-x}Ag_x)_{0.5}]_{0.94}Ba_{0.06}TiO_3$ system exhibit good piezoelectric and ferroelectric properties, and the optimum properties are d_{33} = 172 pC/N, k_p = 31.2 %, k_t = 51.5 %, P_r = 42.5 ìC/cm^2, and E_c = 32.3 kV/cm, respectively. The T_d shifts towards lower temperatures with increasing amount of Ag. For the $Ag_{0.9}Li_{0.1}NbO_3$-modified BNT–$BaTiO_3$ system, the MPB between tetragonal and pseudocubic phases was formed and the E_c was reduced simultaneously, resulting in the higher piezoelectric properties of the ceramics. For the KNN-based ceramics, the PPT was modified by the substitution of Ag^+ for A-site ions and the addition of $AgTaO_3$ or $AgSbO_3$. The KNN-based ceramics with a desired PPT near room temperature exhibit the enhanced piezoelectric and ferroelectric properties. In addition, the (Li, Ag, Ta, Sb)-modified KNN ceramics also possess high T_c. In particular, the $(1-x)[(Na_{0.4725}K_{0.4725})Li_{0.055}]NbO_3$–$xAgTaO_3$ ceramics with x = 0.02 exhibit the optimum electrical properties (d_{33}~237 pC/N, k_p~39 %, P_r~29.4 μC/cm^2) along with a high T_c of 470 °C. Moreover, alkaline elements (Li, Na, K)-modified $AgNbO_3$ (AN)-based ceramics were prepared by the conventional ceramic technique under air atmosphere. The results show that AN-based solid solution with pure perovskite structure could be available by optimizing the compositions and using a proper processing, such as low-temperature sintering. A high T_c of 340 °C and a large P_r of 14.5 ìC/cm^2 were obtained for the CuO-doped $(Ag_{0.75}Li_{0.1}Na_{0.1}K_{0.05})NbO_3$ ceramics. These results show that Ag-containing perovskite is a promising candidate for lead-free piezoelectric materials.

REFERENCES

[1] Cross, E. *Nature* 2004, 432, 24-25.

[2] Saito, Y., Takao, H., Tani, T., Nonoyama, T., Takatori, K., Homma, T., Nagaya, T., & Nakamura, M. *Nature* 2004, 432, 84-87.

[3] Shigemi, A., & Wada,T. *Mol. Simul.* 2008, 34, 1105-1114.

[4] Grinberg, I., & Rappe, A. M. *Appl. Phys. Lett.* 2004, 85, 1760-1762.

[5] Hiruma, Y., Nagata, H., & Takenaka, T. *J. Appl. Phys.* 2009, 105, 084112.

[6] Takenaka, T., Maruyama, K., & Sakata, K. *Jpn. J. Appl. Phys.* 1991, 30, 2236-2239.

[7] Sasaki, A., Chiba, T., Mamiya, Y., & Otsuki, E. *Jpn. J. Appl. Phys.* 1999, 38, 5564-5567.

[8] Fan, G., Lu, W., Wang, X., Liang, F., & Xiao, J. *J. Phys. D: Appl. Phys.* 2008, 41, 035403.

[9] Wu, L., Xiao, D. Q., Lin, D. M., Zhu, J. G., & Yu, P. *Jpn. J. Appl. Phys.* 2005, 44, 8515-8518.

[10] Wu, L., Xiao, D. Q., Zhou, F., Teng, Y. C., & Li, Y. X. *J. Alloys Compd.* 2011, 509, 466-470.

[11] Tai, C. W., Choy, S. H., & Chan, H. L.W. *J. Am. Ceram. Soc.* 2008, 91, 3335-3341.

[12] Lin, D. M., Kwok, K.W., & Chan, H.L.W. *Solid State Ionics* 2008, 178, 1930-1937.

[13] Suchanicz, J., Kusz, J., Böhm, H., Duda, H., Mercurio, J. P. *J. Eur. Ceram. Soc.* 2003, 22, 1559-1564.

[14] Sakabe, Y., Takeda, T., Ogiso, Y., & Wada, N. *Jpn. J. Appl. Phys.* 2001, 40, 5675-5678.

[15] Bilc, D. I., Singh, D. J. *Phys. Rev. Lett.* 2006, 96, 147602.

[16] Wu, L., Liu, N. M., Zhou, F., Wu, W. J., Teng, Y. C., & Li Y. X. *J. Alloys Compd.* 2010, 507, 479-483.

[17] Jo, W., Granzow, T., Aulbach, E., Rödel, J., & Damjanovic, D. *J. Appl. Phys.* 2009, 105, 094102.

[18] Hiruma, Y., Nagata, H., & Takenaka, T. *J. Appl. Phys.* 2009, 104, 124106.

[19] Hiruma, Y., Nagata, H., & Takenaka, T. *Appl. Phys. Lett.*, 2009, 95, 052903.

[20] Guo, Y., Kakimoto, K., & Ohsato, H. *Appl. Phys. Lett.*2004, 85: 4121-4123.

[21] Guo, Y., Kakimoto, K., & Ohsato, H. *Mater. Lett.* 2005, 59, 241-244.

[22] Zhang, S. J., Xia, R., Shrout, T. R., Zang, G. Z., & Wang, J. F. *J. Appl. Phys.* 2006, 100, 104108.

[23] Shrout, T. R., & Zhang, S. J. *J. Electroceram.* 2007, 19, 111-124.

[24] Wang, Y. Y., Wu, J. G., Xiao, D. Q., Wu, W. J., Zhang, B., Wu, L., & Zhu, J. G. *J. Am. Ceram. Soc.* 2008, 91, 2772-2775.

[25] Wang, Y. Y., Wu, J. G., Xiao, D. Q., Zhu, J. M., Jin, Y., Zhu, J.G., Yu, P., Wu, L., & Li, X. *J. Appl. Phys.* 2007, 102, 054101.

[26] Wang, Y. Y., Wu, J. G., Xiao, D. Q., Zhu, J.G., Yu, P., Wu, L., & Li, X. *J. Alloys Compd.* 2008, 462, 310-314.

[27] Hollenstein, E., Davis, M., Damjanovic, D., & Setter, N. *Appl. Phys. Lett.* 2005, 87, 182905.

[28] Du, H. L., Tang, F. S., Liu, D. J., Zhu, D. M., Zhou, W. C., & Qu S. B. *Mater. Sci. Eng. B* 2007, 136, 165-169.

[29] Kania, A., Roleder, K., & Lukaszewski, M. *Ferroelectrics* 1984, 52, 265-269.

[30] Fu, D., Endo, M., Taniguchi, H., Taniyama, T., & Itoh, M. *Appl. Phys. Lett.* 2007, 90, 252907.

[31] Takeda, T., Takahashi, Y., Wada, N., & Sakabe, Y. *Jpn. J. Appl. Phys.* 2003, 42, 6023-6026.

[32] Valant, M., Axelsson, A. K., & Alford, N. *J. Eur. Ceram. Soc.* 2007, 27, 2549-2560.

[33] Shannon, R. D. *Acta. Cryst.* 1976, A32, 751-767.

In: Ferroelectrics: New Research
Editors: Xui Li Huang and Su Lyn Ma
ISBN: 978-1-61942-282-7

Chapter 8

RECENT DEVELOPMENTS IN THE AREA OF SODIUM POTASSIUM NIOBATE (NKN) THIN FILMS BY CHEMICAL SOLUTION DEPOSITION METHODS

Anirban Chowdhury
Department of Materials Engineering, Indian Institute of Science, Bangalore, India

ABSTRACT

Over the past few years, environmental concerns have stimulated interest in developing lead-free ferroelectric and piezoelectric ceramic compositions as replacements for lead zirconate titanate (PZT). Sodium potassium niobate (NKN) based allied solid solutions have been investigated as one of the most promising alternative candidates for PZT systems. Given the technological potential of NKN-based compositions, recently, a number of synthesis routes have been developed to synthesize thin films and powders of NKN. Due to its distinct advantages, e.g., lower temperature processing, homogeneous mixing at the molecular level, large area deposition etc, sol-gel and chemical solution deposition methods have found wide applicability in modern advanced technological applications for thin films. The success of these efforts primarily depends on the stability and homogeneity of the NKN precursor solutions, stability of the film substrate with respect to the crystallisation temperatures (interfacial reactions, if any), extent of reactivity of the alkaline earth metal ions (Na^+, K^+) when exposed to high-temperature crystallisation and in ambient conditions. This article reviews the recent developments in NKN thin films via chemical solution-based routes and also sheds light on the fundamental issues involved in the materials chemistry of the NKN system when approached via chemical solution deposition (CSD) routes.

1. INTRODUCTION

For a long time, lead zirconate titanate ($Pb(Zr_{1-x}Ti_x)O_3$ or PZT) compositions has dominated the piezo- and ferroelectric sectors both in academia and industry [1]. This perovskite ferroelectric series of compositions is noted for its useful piezoelectric coefficients, particularly for compositions close to the morphotropic phase boundary around x=0.48, where properties such as piezoelectric coefficients, dielectric permittivity, and coupling factors are maximized [1, 2]. The increasing success of PZT also brings forward the issue of environmental safety due to the evaporation of toxic PbO during calcination and sintering process. As a consequence, the European Union in 2003 included lead-based compositions in its legislature to be included as a hazardous substances [3]. Since then, there has been a huge drive to develop environmentally benign lead-free piezo- and ferro- compositions ceramics with properties comparable to those of traditional lead-based materials [4]. One of the promising alternatives is a solid solution series based on sodium potassium niobate. The $Na_xK_{1-x}NbO_3$ system has a perovskite structure for compositions, x=0.973. Compositions around, x=0.5, $Na_{0.5}K_{0.5}NbO_3$ (NKN) show the most favourable ferro and piezoelectric parameters [5-7].

Compositions of NKN modified by lithium and tantalum ions constitute one of the most promising known lead free piezoelectric systems [8, 9]. In 2004, Saito and his group at Toyota Central Research Laboratory reported on NKN materials modified by Li, Ta, and Sb with piezoelectric constants comparable to those of PZT at room temperature.

Given the technological potential of NKN-based lead-free piezoceramics, it is important to develop suitable ceramic fabrication strategies and hence there have been significant efforts from the researchers to synthesise NKN compositions both in bulk and thin films. This article would mainly concentrate on the NKN thin films synthesised via chemical solution deposition routes and the fundamental issues involved in the synthesis of such systems.

2. Structure and Properties of Pure $Na_xK_{1-x}NbO_3$ System

Ferroelectric sodium potassium niobate is the continuous solid solution of $NaNbO_3$ and $KNbO_3$, and is perovskite structured for $x < 0.97$. Both $NaNbO_3$ and $KNbO_3$ are orthorhombic at room temperature. Pure sodium niobate, $NaNbO_3$, has an antiferroelectric (AFE) structure at room temperature and exhibits a complicated sequence of temperature and pressure driven structural phase transitions which are not clearly understood [10]. $KNbO_3$ exhibits the same sequence of phases as barium titanate, but all phase transitions appear at higher temperatures than in $BaTiO_3$[7].

Bulk ceramic $Na_xK_{1-x}NbO_3$ exhibits a complex phase diagram with several structural phase transitions as shown in Figure 1 [1, 11, 12]. At room temperature, phase transitions lie at 17.5%, 32.5%, and 47.5% $NaNbO_3$ content, the latter commonly taken as 50/50 equivalent to PZT. In the vicinity of one of its phase boundaries, when x = 0.50, $Na_{0.5}K_{0.5}NbO_3$ (NKN) shows the highest electromechanical coupling coefficient and moderate dielectric susceptibility [5, 7].

Also notable, is the almost composition-independent phase transitions between ferroelectric phases at ~200°C and between ferroelectric and paraelectric phases at ~400°C, Figure 1. Recently, theoretical first principles calculations of this phase diagram have been carried out to explore the possibility of a relaxor composition from this system [13].

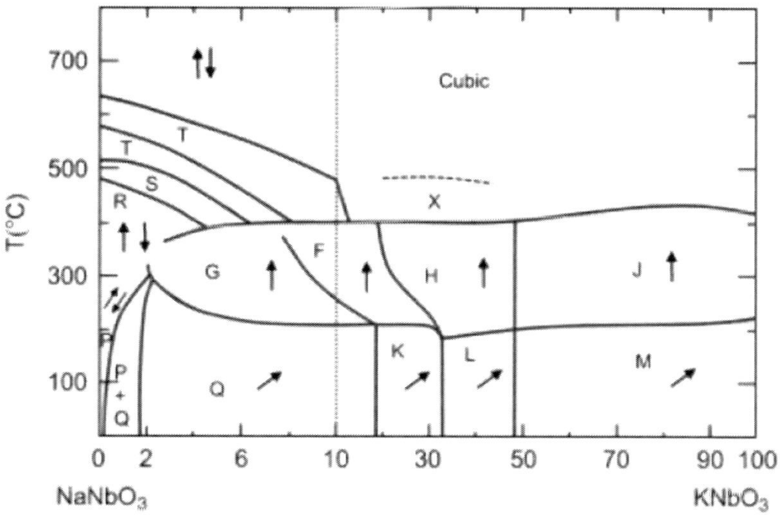

Figure 1. Phase diagram of $Na_xK_{1-x}NbO_3$[1, 4]. Noteworthy phase regions are: monoclinic (regions labelled with Q, K, L), orthorhombic ferroelectric (M, G), tetragonal ferroelectric (F, H, J) and orthorhombic antiferroelectric (P) [ACerS-NIST Phase Equilibria Diagram Database].

The physical properties of sodium potassium niobate have been studied in detail [6, 7, 14, 15]. The remanent polarisation reaches a maximum for the 50/50 composition and decreases towards both end members ($NaNbO_3$ and $KNbO_3$), whereas the coercive field has its minimum for the 50/50 composition [5]. In $Na_{0.5}K_{0.5}NbO_3$ ceramics, the dielectric permittivity peaks at around 400 °C exhibiting values in the order of 4000 [7], whereas the values are moderate at room temperature. Typical relaxor-like dielectric properties, such as broad frequency dispersion in the complex relative permittivity, have been detected in the lead-free $Na_{0.5}K_{0.5}NbO_3$ – $SrTiO_3$ ceramic system [16]. Due to the absence of lead in the compound, it has attracted interest in biotechnology. In 1999, these ceramics were patented as a biocompatible implant material by Pacesetter [17]. The application of NKN has been suggested as part of a pressure sensor in pacemakers.

2.1. $Na_xK_{1-x}NbO_3$ Thin Films via Physical Vapour Deposition Methods

Interest in thin films of the perovskite phase NKN has increased in the past few years. Margolin [18] have used rf cathode sputtering of a NKN powder target in order to obtain NKN films on stainless steel substrates. The crystalline nature of the films was confirmed using X-ray diffraction and transmission electron microscopy, but no data for the film composition were presented in the study.

In the last ten years, a lot of the research on synthesising NKN through various physical vapour deposition methods has been carried out by Swedish research groups of A. Grishin and S. Gevorgain. NKN thin films have been grown by Wang et al. [19] on $LaAlO_3(001)$ substrates using rf magnetron sputtering from a Na and K enriched target. However, a major drawback was the NKN films irreproducibility as the use of a nonstoichiometric target always makes the sputtering process very 'fragile'. This issue was resolved in further work by their group [20]. Cho et al. [21] reported high-performance NKN films on polycrystalline

$Pt_{80}Ir_{20}$ and SiO_2/Si(111) substrates by KrF pulsed-laser ablation of a stoichiometric ceramic target. The films were found as single-phase and highly c-axis orientated with the strong effect of self-assembling both on slightly textured and even amorphous substrates. Further work [22], revealed that for growth at a substrate temperature of 650 °C and a total pressure of 53 Pa, NKN films form with a relative permittivity between 500 and 550 as measured in the 0.4–100 kHz frequency range. A strong effect of the background oxygen pressure was also noticed with morphology and electrical properties of the NKN films improving for higher oxygen pressures.

NKN films have also been studied as promising candidates for low loss electrically tunable microwave devices [23]. Wang et al. [24] reported on the growth of NKN films by rf magnetron sputtering on the same substrates, for both on- and off-axis geometry, resulting in a relative permittivity as high as 590 at 10 kHz and hydrostatic piezoelectric constants of 14.5 pC/N. Cho et al. [25] prepared highly c-axis orientated NKN films on single crystal quartz substrates. Dielectric spectroscopy measurements in the frequency range of 1 kHz – 1 MHz indicate that the films have a relatively high dielectric permittivity of about 200 with a low dielectric loss. Abadei et al. [26] prepared NKN films by a laser-ablation technique on high resistivity silicon substrates with native SiO_2 buffer (~about 10 Å-thick) films. C-axis orientated, smooth NKN films have been synthesised on Si substrates using metal organic chemical vapour deposition (MOCVD) technique for electro-optic application by Cho et al.[27, 28]. Khartsev et al.[29] did a comparative study between two film synthesis techniques, namely, pulsed laser deposition and rf-magnetron sputtering, and concluded that two different kinds of compressive stresses exist for the films, which, in turn, showed different piezo and ferroelectric properties. Excellent ferroelectric properties (rectangular hysteresis loop with $P_r = 26\ \mu C/cm^2$ and fatigue-free characteristics up to 10^{10} switching cycles) for NKN thin films fabricated using radio frequency magnetron sputtering have also been reported by Lee et. al. [30]. These results were obtained for the film annealed at 700 ºC.

The doping of manganese has also been successfully tried in NKN and allied systems [31, 32]. Single phase, epitaxial, (001) oriented, Mn-doped (1 mol%) $(K_{0.44},Na_{0.52},Li_{0.04})(Nb_{0.84},Ta_{0.10},Sb_{0.06})O_3$ thin films of 400 nm thickness were synthesized on $SrRuO_3$ coated $SrTiO_3$. Films exhibited well developed saturated hysteresis loops and had a spontaneous polarization (P_s) of 10 $\mu C/cm^2$, which was reported to be 150% higher over the P_s of the undoped composition. Mn-doping resulted in three orders of magnitude decrease in leakage current above 50 kV/cm electric field, which the authors attributed to the suppression of intrinsic p-type conductivity of undoped films by Mn donors [31].

2.2. Initial Reports on the Synthesis Of $Na_xK_{1-x}NbO_3$ Thin Films by Chemical Solution Deposition Methods

There have been few reports in recent years on the sol-gel synthesis of $Na_xK_{1-x}NbO_3$ films. Söderlind et al. [33] have synthesised NKN thin films via three different sol–gel methods: the alkoxide method, a modified Pechini method and an oxalate method based on 2-methoxy ethanol as the solvent with oxalic acid and ethylene glycol as the chelating ligand. They found that only the Pechini method gave a pure NKN phase whereas the other two methods gave extra peaks in their X-ray diffraction patterns. In some cases SEM images depicted a typically porous film with few cracks. No information about the electrical properties of ferroelectric NKN was given.

Lai et al. [34, 35] used a metal (Na, K) dissolution route to make sols for NKN films. They prepared a starting solution by reacting solid Na and solid K with ethanol in a solvent of 2-methoxyethanol; acetic acid and acetylacetone added as chelating agents. The mixture was stirred and heated at 120 ºC for 4 h followed by the addition of the stoichiometric amount of niobium pentaethoxide $[Nb(OC_2H_5)_5]$. There was no evidence of ferroelectricity in their reported films. Nakashima et al. [36, 37] examined the structural and electrical properties of the NKN films with different Na and K excess compositions. The leakage current and ferroelectric properties of the perovskite KNN thin films were found to be strongly affected by the excess amounts of Na and K as well as the heating conditions of the precursor films. Tanaka et. al. [38] synthesised NKN thin films at 600 °C from the (Na,K)-excess precursor solutions at those films depicted typical lossy non-ferroelectric loops, Figure 2b.

In instances where P-E loop measurements were carried out on these sol-gel NKN films [34, 35, 37, 38], no evidence of hysteresis was observed. This differs from the situation for physical vapour deposition [21, 29, 30], which showed saturated P-E hysterisis loops. A comparative figure has been shown in Figure 2a-b. The P-E response from the sol-gel films was consistent with a simple (non-ferroelectric) RC element i.e., typical of a lossy dielectric. All these facts indicate that there are significant difficulties in applying sol-gel processing to alkali niobate films. This gave an added motivation for the researchers to investigate fundamental reasons for such events.

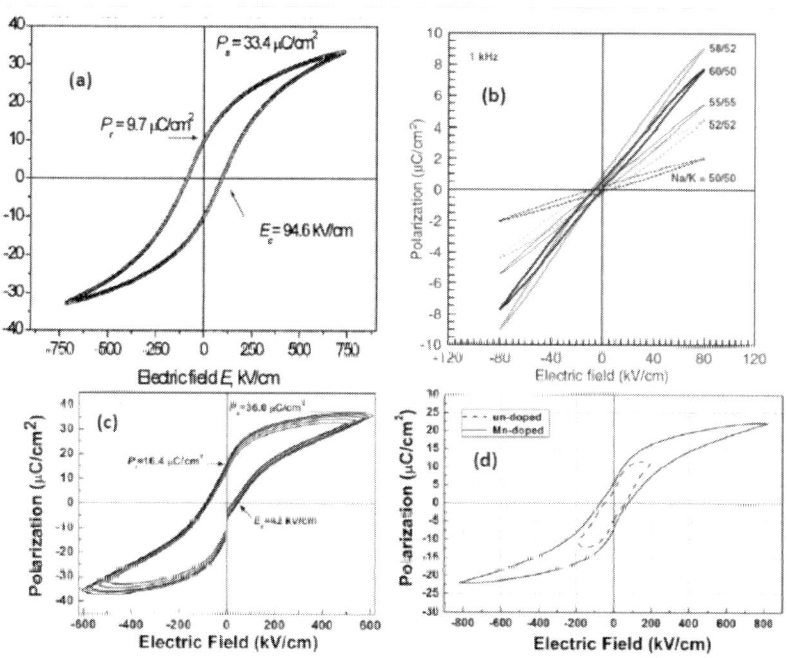

Figure 2. The P-E loops for NKN thin films when (a) synthesised by rf-magnetron sputtering [29] (b) synthesised by chemical solution deposition (CSD) using alkoxide precursors [38] (c) deposited via CSD route using polyvinylpyrrolidone [45] (d) fabricated by CSD route using Mn doping [47]. The only NKN films which depicts ferroelectric saturation is the one synthesised by physical vapour deposited route (Fig. 2a) [29].

3. Ferroelectricity in NKN Films Deposited by CSD Methods

It was reported that although the NKN 50/50 composition is orthorhombic at room temperature [15], but when it was synthesised from the conventional sol-gel route evidence of mixed phase exist as the tetragonal phase co-exist with the orthorhombic phase at intermediate temperatures below 850 °C [39]. It was also confirmed that the secondary hydrated carbonate phases present in those NKN samples are stable at high temperatures (up to 700 °C and beyond) and the reasons were attributed to the reactions with atmospheric moisture and carbon dioxide [39, 40]. These experiments, when conducted in NKN thin films, showed a very similar trend with the typical non-ferroelectric response [41]. The X-ray diffraction analysis of the thin films was not good enough to provide any noteworthy explanation on the extra phases present. The combined thermogravimetry (TG)–differential thermal analysis (DTA) - mass spectrometry (MS) experiments clearly pointed out the presence of two crystallisation peaks at 509 °C and 524 °C for the NKN system; signifying probable phase segregation/heterogeneity within the NKN gel. Even at this point, a major concern was the purity of the commercial alkoxides, due to which these phase segregation, was believed to occur in sol-gel NKN systems.

To investigate this issue further, experiments were conducted with purer precursors. Alkoxides of sodium and potassium were made in the laboratory from their metallic ingots to avoid the impurity issues; the commercially available Nb-ethoxide was very pure (>99.5%) and therefore was used as it is [42]. The system was kept 'pure' to the maximum possibility limit, i.e., no other organic additives/solvents or doping was used. The sole intention was to focus on the issue of phase heterogeneity and therefore only the effect of refluxing for three different time-scales (0, 24 and 70 hours) was investigated for this sol-gel NKN system [42, 43]. For all three NKN gels, an important feature was the presence of multiple exothermic DTA peaks in the range 480−540 °C. Also in all these gels, there was an endothermic peak in the DTA heating cycle around 825−850 °C denoting possible secondary carbonate phase melting. All these facts suggest that crystalline, single phase NKN was not formed directly from the gel decomposition product (metal oxides/carbonates), signifying phase segregation in the decomposing gels. These could be Na-rich and K-rich NKN solid solutions as opposed to the desired $Na_{0.5}K_{0.5}NbO_3$ phase. The changes in peak temperatures and peak positioning indicated a convergence in composition as reflux time increase, but, even after 70 h of prolonged refluxing, a single crystallization peak for the NKN gels could not be obtained [42]. Furthermore, thermal decomposition of binary units, e.g., $NaNbO_3$ and $KNbO_3$ gels and unary units (all the precursor alkoxides) were examined, which indicated that the endothermic peak in the DTA heating cycle around 825−850 °C could probably be due to the K-rich second phase system/s.

In a related study, effect of refluxing was correlated to the extent of oligomeric –Metal–Oxy–Metal– chain link formation and it was concluded that, in the non-refluxed case the individual binary end member (Na and K niobate gels) units persist, but with Na and/or K rich niobate regions in clusters. Another possibility could be that of a small but significant amount of the precursor chemicals which could remain individually as Na/K methoxyethoxides in the solution. Both of these factors could lead to heterogeneity and segregation in the parent sol [43].

This whole study actually points towards two main issues for obtaining ferroelectricity in NKN thin films by CSD methods:

1) The problem of volatility of the alkaline earth cations (Na^+, K^+) at high temperatures during film crystallisation stage
2) The reactivity and the propensity of mixing of these individual alkoxides towards each other; especially the issue of mixing of Nb-precurssor with the Na and K precursors.

The first issue is a typical synthesis issue encountered very often in ferroelectric systems and has been taken into consideration by optimising the process parameters, adding excess alkali ion/s precursor to compensate for the loss and/or use of dopants. Wang et al studied the phenomena of volatilization of alkali ions for the synthesis of NKN thin films by CSD methods and concluded that the volatilization of the alkali ions mainly happened at moderate temperatures (300–500 °C) [44]. Their results further showed that once perovskite NKN phase was formed, the volatilization of the alkali ions in the lattice was not substantial even at much higher temperature, at least until 850 °C. This result is completely contradictory with the recently reported results on the thermochemistry of the NKN gels [41-43] where the evidence of K-rich second phase melting was noted around 850 °C. The presence of secendary carbonates were found to be one of the fundamental reasons for not obtaining ferroelectricity in such systems.

Wang et. al. used polyvinylpyrrolidone (PVP) of different molecular weights as a chelating agent and found that PVP with the appropriate molecular weight could effectively promote the crystallization of the NKN perovskite phase at reduced temperature and substantially suppress the loss of the alkali ions before crystallization. However, they did not observe any well saturated P-E hysteresis loop and correlated it with the porous morphology caused by the excessively large molecular weight of PVP [44]. In a continuing work, the same group used PVP to make perovskite NKN thick films at 600 °C by CSD method which showed ferroelectricity, Figure 2c, (remanent polarization, P_r of 16.4 μC/cm^2) [45]. Although, this value was slightly higher compared to the reported values of P_r for NKN films by other physical vapour deposition methods, but, the evidence of ferroelectric saturation was much more pronounced in that reported work by Grishin and his co-workers [29] In a further study, researchers used both diethanolamine and ethylenediaminetetraacetic acid (EDTA) as the chelating/stabilizing agents for NKN thin films and found that the leakage current was significantly reduced (5 orders of magnitude compared to the film with no chelating agent/s) [46]. They correlated this improvement of leakage current with the control of volatilization of K and Na due to the addition of the chelating agents. The P-E hysteresis response improved from the film with no chelating agent/s but there was not enough evidence of ferroelectric saturation in the NKN film [46].

The effect of Mn-doping (2%) has also been tried on NKN films and it was found to increase the dielectric constant and decrease the dielectric loss, as well as significantly decrease the leakage current [47]; although the P_r value (7.2 μC/cm^2) was low compared to their previously published works [44, 45], Figure 2d. The prime reason for this change was attributed to the decrease in the concentration in oxygen vacancies and free carrier-holes by increasing the valence of Mn during thermal treatment.

Another approach to obtain better ferroelectric response in NKN films by CSD methods was carried out by the introduction of tantalum [48]. The partial substitution of Ta^{5+} for the Nb^{5+} ion was reported to decrease the Curie temperature, which, in turn, led to the existence of a polymorphic phase transition near room temperature and further improvement in the ferro- and piezoelectric properties. A recent work by Kondo et al. examined the effects of Mn doping on such Ta-substituted NKN films [49]. Ta-substituted NKN films showed poor ferroelectric polarizations due to the insufficient insulating resistance at room temperature. Mn-doping (up to 1%) improved the remanent polarization (P_r) values significantly (21 $\mu C/cm^2$ at 1 kHz). These films also exhibited well-shaped saturated ferroelectric polarization–electric field (P–E) hysteresis loops along with a typical field-induced butterfly loop at room temperature.

3.1. Effects of Mn Doping on Ferroelectric NKN Systems

The loss of volatile cations at high temperature during sintering is a common problem for all electroceramic compositions. PZT ceramics and thin films are usually found to have p-type conductivity, believed to result from an excess of lead vacancies [50]. Donor substituents for lead, titanium and/or zirconium reduce the conductivity by electron-hole compensation, but n-type conductivity is not observed and is probably prevented by the formation of lead site vacancies. For years, manganese doping has been popularly tried in PZT compositions [51, 52]. Electron Spin Resonance (ESR) studies have shown that manganese coexist mainly in the way of Mn^{2+} and Mn^{3+} in PZT ceramics. When the concentration of Mn ion is below 0.5 mole%, it is preferentially incorporated in the lattice Pb site in Mn^{2+}/Mn^{3+}, which gives rise to the increase of the piezoelectric co-efficients [52].

One of the major problems that the ferroelectric NKN system has to confront is a relatively high leakage current due to lattice defects [7, 53], which, eventually lead to mixed phases and non-stoichiometric compositions in the final product [41-43, 53]. Leakage current interferes with polarization switching and poling process, and a poor insulating property makes the material unsuitable for electronic applications. Leakage current properties are closely related not only to defect structure in grains but also to microstructure of ceramics and thin films. Kizaki et al. [54] studied the leakage current properties of NKN single crystals and concluded that the substitution of Mn at the Nb site and annealing under moderate oxidation condition was effective for suppressing leakage current of NKN crystals. Electron spin resonance measurements demonstrate that oxidation of Mn during annealing plays an essential role in lowering leakage current of the NKN system. The researchers found that when the as-grown crystals of NKN are oxidized gradually (first-stage oxidation), the leakage current density (J) decreases rapidly due to the valence increase of niobium from Nb^{4+} to Nb^{5+}. Immediately after completing Nb oxidation (all Nb^{4+} ions become Nb^{5+}), further oxidation (second-stage oxidation) increases J due to the formation of holes. In contrast, for Mn-doped crystals, Mn acts as a buffering species for suppressing an increase in the formation of holes (p-type conductivity). After Nb^{4+} is completely oxidized to Nb^{5+}, Mn^{2+} absorbs the holes generated in further oxidation process and the average valance of Mn increases.

In case of NKN-based thin films, K and Na ions easily volatilize yielding cation site vacancies during heat treatment [31, 47, 49]. This volatility of these alkaline ions triggers the formation of oxygen vacancies, which maintain charge neutrality. Therefore, cation and anion vacancies are formed in the thin films and the distribution of the oxygen vacancies that originate from the compositional fluctuation of K and Na is a significant problem with NKN-based thin films. The oxidation process is accompanied by the formation of electron–hole. Hole acts as a carrier of the leakage current [electron–hole (h) conduction (p-type)]. In the case of the NKN thin films, the h carrier concentration may be relatively high, due to the vaporization of K and Na ions. For the Mn-doped KNNT thin films, since Mn can have several valencies, Mn may act as a buffering species for suppressing the increase in the number of h's caused by oxidation. Kondo et al. [49] reported that the occupation site of Mn was considered as a substituent at Nb or Ta ion sites in the perovskite structure.

3.2. Fundamental Issue Related to the Materials Chemistry of NKN System

The second most important issue (Issue 2, as mentioned in section 3) for NKN system involves a deeper understanding of the materials chemistry of the system The sole reason for putting this issue in the second place is because of the fact that there has not been much emphasis on this arena. At the moment, the volatilisation of Na and K ions seems to be the main issue and the researchers have investigated the problem with the effects of Mn-doping [47, 49]. Mn-doping was inevitable since adding excess of Na and K ions in the starting precursor solution was not good enough to obtain ferroelectricity in such films [34, 38, 55] synthesised by CSD routes. But, previous researches in other systems have shown that just adding right amount/excess of these precursors was sufficient to obtain ferroelectricity in CSD thin film compositions [56-59]. In fact, there are reports suggesting that, the fatigue performance is significantly influenced by the composition, and non-stoichiometric thin films which exhibit low fatigue compared to stoichiometric thin films [60]. Moreover, it has already been reported that the issue of phase segregation in sol-gel NKN systems probably could be due to the extent of mixing between the individual precursors and their rates of hydrolysis [42, 43]. Therefore, it is important to understand the materials chemistry of the NKN system for future developments.

The alkoxides of Na and K are extremely reactive and therefore need special attention while handling. Due to their reactive nature, other ligands (acetate, carboxyllates etc) have also been tried [45, 47]. Another approach to reduce their reactivity could be to perform refluxing with the Na and K alkoxides prior to introducing the Nb reagent [42]. Refluxing ensures the completion of metal-oligomeric ligand-exchange reactions and hence extensively used for making any starting precursor sol. In a recent study, Nakashima et al. [61] reported the synthesis of NKN thin films with pure alkoxide reagents only. They used excess (10%) Na and K precursors and refluxed the solution for 18 hours. It is noteworthy to mention that, a very similar sol processing method in their earlier works [55, 62] could only show a hint of ferroelectricity in the NKN films and the leakage current density was also quite high compared to the same for any saturated ferroelectric films. In the recent work, ferroelectric saturation is not observed and in higher fields the leakage current density was reported to be as high as 10^{-4} A/cm^{-2} [61]. Nakashima et al. investigated the NKN precursor solution by NMR spectroscopy and revealed that the NKN precursor consists of complex metal alkoxides of potassium and sodium hexaalkoxy niobates with highly symmetric Nb-O octahedra [61]. They further assumed that the tailored complex NKN alkoxides dissolve uniformly at the molecular level in solution. However, considering the issue of phase heterogeneity issues in NKN systems, it has been clearly pointed out that to what extent the individual Na and K niobate gel units and/or some individual unreacted Na/K alkoxides persist in the final NKN sol, is unknown [43]. A similar issue was more thoroughly investigated by Kato et. al. [57]. They made precursors for layer-structured perovskite thin films of $SrBi_2Ta_2O_9$ (SBT) and $SrBi_2Nb_2O_9$ (SBN) by the reactions of a strontium-bismuth double methoxyethoxide and/or tantalum/niobium methoxyethoxide in methoxyethanol, followed by partial hydrolysis with a variety of water/alcohol ratio. Spectroscopic investigations to analyze the arrangement of the metals and oxygen in the precursor molecules revealed that the precursors contained Sr-O-M (where M is Ta or Nb) bonds (i.e., a strontium is connected to two MO_6 octahedra) and Sr-O-Bi bonds with a bismuth atom bonded to the oxygens of the MO_6 octahedron. The arrangement of metals and oxygens was considered to be similar to the layer-structured perovskite crystal sublattice. Similar in-depth analysis on the NbO_6 octahedra and its extent of correlated arrangements with the Na and K ions will be crucial for any future progress on this issue.

The reaction between Nb ethoxide and the alkali metal reagents could be quite slow and therefore a more fundamental chemical investigation of these reactions could be informative [42]. It could also be possible that, for the two alkali metal alkoxides, the affinity between K and Na alkoxides inhibits reaction with Nb alkoxide, which probably promotes some Na/K carbonate formation.

In the recent studies by the researchers in RWTH, Aachen [63, 64], important information have been provided for the impact of precursor chemistry and processing conditions upon the distribution of metal ion species in chemical solution deposition-derived alkaline niobate thin films. Their work clearly concluded that, the addition of excess alkali metal species and/or doping observed in literature beforehand should only be considered as a mediation of the main problem. The main issue lies in the understanding of the allocation of alkaline metal complexes in the NKN precursor solutions and the inhomogeneities that it could generate which will then finally lead to an incorrect local stoichiometry in the final film. This is the same concern voiced by Chowdhury et. al. in their works on NKN systems [42, 43]. It has already been reported that using sol-gel processes, solely the chemical processing route and more exactly the chemical modification of the niobium component determines the electrical properties of the niobium oxide thin-films [63]. This suggests that the reactivity of the niobium component will influence the electrical properties of the NKN systems significantly.

Considering the propensity to hydrolysis and condensation reactions of the niobium alkoxide system a certain level of stabilization or pre-hydrolysis seems to be necessary in order to obtain homogeneous films. One such commonly practised method for stabilization is to use the chelating agent acetylacetone [65] and have been widely used for making PZT thin films [58, 66]. The main idea here was to make the alkoxy-precursor relatively less prone to hydrolysis; ligands used for such exchange reactions are shown in Figure 3.

A detailed study has been carried out the past on various types of ligand-stabilised nioubium precursor for Nb-doped PZT thin films and it was found that, reduction, but not complete elimination, of the hydrolyzable ligands is desirable for obtaining superior ferroelectric properties [67]. Furthermore, the study also revealed that the precursor solution temperature will influence the nuclearity of the precursors, and thus the properties of the final

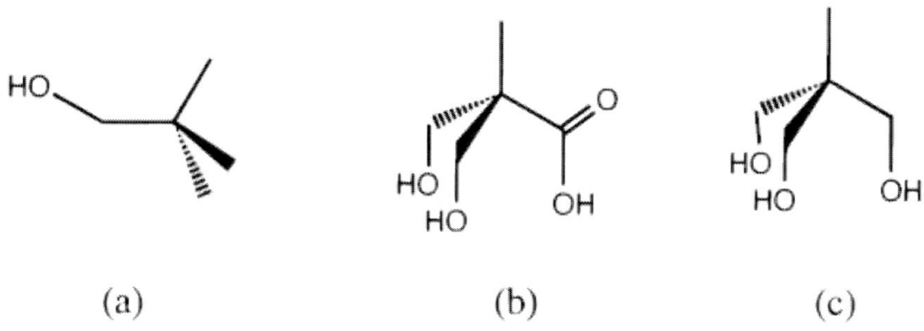

Figure 3. Schematic figures for the ligands used for exchange by Boyle et. al. [67] (a) neopentanol, (b) bis(hydroxymethyl) propionic acid, and (c) tris(hydroxymethyl) ethane.

thin films made using these compounds. This finding also suggests that the selection of solvent and refluxing temperature will play a vital role in the final precursor sol formation.

Therefore, it seems that the main focus needs to be put on these following issues:

1) The chemistry of the niobium precursor and its extent of reactivity (due to the hydrolysable groups attached to it) towards the Na and K precursor. Spectroscopic investigations will be necessary to find how these NbO_6 octahedra are arranging themselves in space with the Na and K ions.
2) Mixing of Na and K precursor: will it be beneficial to make a double alkoxide first before performing reaction with Nb-reagent?
3) Pre-hydrolysis and condensation issues: Will the pre-hydrolysis (water/alcohol stabilisation) somehow positively influence the final condensation of the Na-O-Nb and K-O-Nb chain links?
4) The role of stabilising agents: Exactly, how the stabilising agents (PVP, EDTA etc) influence the final sol formation? Is it possible to make the complex of these agents beforehand and get them reacted with Na and K precursors?
5) Role of refluxing: How long should it require to make a homogeneous NKN sol? Do the solvent (which eventually governs the refluxing temperature) also have an additional role to play in this context?

CONCLUSION

The present status on the NKN thin films via CSD methods is in a transitory phase. The way forward would only be the understanding of the underlying basics of the materials chemistry issues in NKN systems. The intermediate carbonates of Na and K (if formed) are much more stable than the Pb-carbonates (for PZT system) and therefore remedial steps needs to be ascertained for making a homogeneous NKN sol as a starting point. The issue of Mn doping has provided a temporary solution to the problem but deeper understanding of the niobium reagent will seem to be the key point for solving this issue.

Moreover, extreme precautions need to be undertaken while processing any such alkali-niobate or allied systems; especially considering the extreme reactive nature of metal ions (Na, K) involved. Utmost care has also to be taken to permit a sufficient degree of condensation in the niobium oxide matrix while suppressing a possible phase separation. This is an area where the attached hydrolysable groups with niobium and other methods of stabilisation would play a vital role. The future advanced recipe, therefore, is expected to take into account all these fundamental issues to be able to make NKN thin films stable and suitable for advanced technological applications.

REFERENCES

[1] Jaffe, B., Cook Jr., W.R., Jaffe, H, *Piezoelectric Ceramics*. 1971, New York: Academic Press.

[2] Jaffe, B., R.S. Roth, and S. Marzullo, *J. Appl. Phys.,* 1954. 25 809-10.

[3] *EU- Directive 2002/95/EC: on the restriction of the use of certain hazardous substances in electrical and electronic equipment (RoHS). Off. J. Eur. Union [L37] 46: 19–23.*

[4] Rodel, J., W. Jo, K.T.P. Seifert, E.M. Anton, T. Granzow, and D. Damjanovic, *J. Am. Ceram. Soc.,* 2009. 92(6) 1153-1177.

[5] Haertling, G.H., *J. Am. Ceram. Soc.,* 1967. 50(6) 329-330.

[6] Jaeger, R.E. and L. Egerton, *J. Am. Ceram. Soc.,* 1962. 45 209-13.

[7] Shirane, G., R. Newnham, and R. Pepinsky, *Phys. Rev.,* 1954. 96 581-8.

[8] Guo, Y., K.-i. Kakimoto, and H. Ohsato, *Appl. Phys. Lett.,* 2004. 85(18) 4121-4123.

[9] Saito, Y., H. Takao, T. Tani, T. Nonoyama, K. Takatori, T. Homma, T. Nagaya, and M. Nakamura, *Nature,* 2004. 432(7013) 84-87.

[10] Mishra, S.K., N. Choudhury, S.L. Chaplot, P.S.R. Krishna, and *R. Mittal, Phys. Rev.,* B 2007. 76(2) 024110/1-024110/8.

[11] Ahtee, M. and A.M. Glazer, *Ferroelectrics,* 1974. 7(1-4, Pt. 1) 93-5.

[12] Ahtee, M. and A.W. Hewat, *Acta. Cryst.,* A 1978. A34(2) 309-17.

[13] Burton, B.P. and T. Nishimatsu, *Appl. Phys. Lett.,* 2007. 91(9) 092907/1-092907/3.

[14] Egerton, L. and D.M. Dillon, *J. Am. Ceram. Soc.,* 1959. 42 438-42.

[15] Tennery, V.J. and K.W. Hang, *J. Appl. Phys.,* 1968. 39(10) 4749-53.

[16] Bobnar, V., J. Bernard, and M. Kosec, *Appl. Phys. Lett.,* 2004. 85(6) 994-996.

[17] Nilsson, K., J. Lidman, K. Ljungstrom, and C. Kjellman, *Biocompatible material for implants*. 1999: Application: WO (Pacesetter AB, Swed.). p. 19 pp.

[18] Margolin, A.M., Z.S. Surovyak, I.N. Zakharchenko, V.A. Aleshin, L.K. Chernysheva, M.G. Radchenko, and V.P. Dudkevich, *Soviet Physics Technical Physics*, 1988. 33(12) 1435-38.

[19] Wang, X., U. Helmersson, S. Olafsson, S. Rudner, L.-D. Wernlund, and S. Gevorgian, *Appl. Phys. Lett.,* 1998. 73(7) 927-929.

[20] Blomqvist, M., J.-H. Koh, S. Khartsev, A. Grishin, and J. Andreasson, *Appl. Phys. Lett.,* 2002. 81(2) 337-339.

[21] Cho, C.-R. and A. Grishin, *Appl. Phys. Lett*. 1999. 75(2) 268-270.

[22] Cho, C.-R. and A. Grishin, *J. Appl. Phys*. 2000. 87(9, Pt. 1) 4439-4448.

[23] Abadei, S., S. Gevorgian, C.R. Cho, A. Grishin, J. Andreasson, and T. Lindback, *Appl. Phys. Lett.,* 2001. 78(13) 1900-1902.

[24] Wang, X., S. Olafsson, L.D. Madsen, S. Rudner, I.P. Ivanov, A. Grishin, and U. Helmersson, *J. Mater. Res.,* 2002. 17(5) 1183-1191.

[25] Cho, C.-R., I. Katardjiev, M. Grishin, and A. Grishin, *Appl. Phys. Lett.,* 2002. 80(17) 3171-3173.
[26] Abadei, S., S. Gevorgian, C.R. Cho, and A. Grishin, *J. Appl. Phys.,* 2002. 91(4) 2267-2276.
[27] Cho, C.-R., *Mater. Lett.,* 2002. 57(4) 781-786.
[28] Cho, C.-R. and B.-M. Moon, *Int. Ferro.,* 2002. 45 39-48.
[29] Khartsev, S., A. Grishin, J. Andreasson, J.-H. Koh, and J.-S. Song, *Int. Ferro.,* 2003. 55 769-779.
[30] Lee, J.S., H.J. Lee, J.Y. Lee, S.H. Kang, I.W. Kim, C.W. Ahn, and G.S. Chung, *J. Kor. Phys. Soc.,* 2008. 52(4) 1109-1113.
[31] Abazari, M., E.K. Akdogan, and A. Safaria, *Appl. Phys. Lett.,* 2008. 92(21) 3.
[32] Abazari, M. and A. Safari, *J. Appl. Phys.,* 2009. 105(9) 3.
[33] Soederlind, F., P.-O. Kaell, and U. Helmersson, *J. Cryst.,* Growth, 2005. 281(2-4) 468-474.
[34] Lai, F. and J.-F. Li, *J Sol-Gel Sci. Tech.,* 2007. 42(3) 287-292.
[35] Lai, F. and J.-F. Li, *Ferroelectrics,* 2007. 358 181-187.
[36] Nakashima, Y., W. Sakamoto, H. Maiwa, T. Shimura, and T. Yogo, *Jpn. J. Appl. Phys.,*2 2007. 46(12-16) L311-L313.
[37] Nakashima, Y., W. Sakamoto, T. Shimura, and T. Yogo, *Jpn. J. Appl. Phys.,*1, 2007. 46(10B) 6971-6975.
[38] Tanaka, K., H. Hayashi, K.-i. Kakimoto, H. Ohsato, and T. Iijima, Jpn. *J. Appl. Phys.,*1 2007. 46(10B) 6964-6970.
[39] Chowdhury, A., J. Bould, Y. Zhang, C. James, and S. Milne, *J. Nanopart. Res.,* 2010. 12(1) 209-215.
[40] Chowdhury, A., S. O'Callaghan, T.A. Skidmore, C. James, and S.J. Milne, *J. Am. Ceram. Soc.,* 2009. 92(3) 758-761.
[41] Chowdhury, A., J. Bould, M.G.S. Londesborough, E. Vecerníková, and S.J. Milne, *Mater. Chem. Phys.,* 2010. 124(1) 159-162.
[42] Chowdhury, A., J. Bould, M.G.S. Londesborough, and S.J. Milne, *Chem. Mater.,* 2010. 22(13) 3862-3874.
[43] Chowdhury, A., J. Bould, M.G.S. Londesborough, and S.J. Milne, *J. Solid State Chem.,* 2011. 184(2) 317-324.
[44] Wang, L., K. Yao, P.C. Goh, and W. Ren, *J. Mater. Res.,* 2009. 24(12) 3516-3522.
[45] Wang, L., K. Yao, and W. Ren, *Appl. Phys. Lett.,* 2008. 93(9) 092903/1-092903/3.
[46] Goh, P.C., K. Yao, and Z. Chen, *Appl. Phys. Lett.,* 2010. 97(10) 102901/1-3.
[47] Wang, L.Y., W. Ren, P. Shi, X.F. Chen, X.Q. Wu, and X. Yao, *Appl. Phys. Lett.,* 2010. 97(7) 072902/1-3.
[48] Lee, S.Y., C.W. Ahn, J.S. Kim, A. Ullah, H.J. Lee, H.I. Hwang, J.S. Choi, B.H. Park, and I.W. Kim, *J. Alloys Compd.,* 2011. 509(20) L194-L198.
[49] Kondo, N., W. Sakamoto, B.Y. Lee, T. Iijima, J. Kumagai, M. Moriya, and T. Yogo, *Jpn. J. Appl. Phys.,* 2010. 49(9).
[50] Gerson, R. and H. Jaffe, *Journal of Physics and Chemistry of Solids,* 1963. 24(8) 979-984.
[51] Thomann, H., *Ferroelectrics,* 1972. 4(3) 141-146.
[52] He, L.-X. and C.-E. Li, *J. Mater. Sci.,* 2000. 35(10) 2477-2480.
[53] Tashiro, S. and K. Nagata, *Jpn. J. Appl. Phys.,*1 2004. 43(9B) 6711-6715.

[54] Kizaki, Y., Y. Noguchi, and M. Miyayama, *Appl. Phys. Lett.*, 2006. 89(14) 142910/1-3.
[55] Nakashima, Y., W. Sakamoto, T. Shimura, and T. Yogo, *Jpn. J. Appl. Phys.*,1 2007. 46(10B) 6971-6975.
[56] Budd, K.D., S.K. Dey, and D.A. Payne, *Brit. Ceram. Proceedings,* 1985. 36 107-21.
[57] Kato, K., C. Zheng, J.M. Finder, S.K. Dey, and Y. Torii, *J. Am. Ceram. Soc.,* 1998. 81(7) 1869-1875.
[58] Naksata, M., R. Brydson, and S.J. Milne, *J. Am. Ceram. Soc.,* 2003. 86(9) 1560-1566.
[59] Sriprang, N., D. Kaewchinda, J.D. Kennedy, and S.J. Milne, *J. Am. Ceram. Soc.,* 2000. 83(8) 1914-1920.
[60] Noguchi, T., T. Hase, and Y. Miyasaka, *Jpn. J. Appl. Phys.*, 1 1996. 35(9B) 4900-4904.
[61] Nakashima, Y., W. Sakamoto, and T. Yogo, *J. Eur. Ceram. Soc.,* doi:10.1016/j.jeurceramsoc.2011.04.031 2011. In Press, Corrected Proof.
[62] Nakashima, Y., W. Sakamoto, H. Maiwa, T. Shimura, and T. Yogo, *Jpn. J. Appl. Phys.*,2 2007. 46(12-16) L311-L313.
[63] Röscher, M., T. Schneller, and R. Waser, *J. Sol-Gel Sci. Tech.,* 2010. 56(3) 236-243.
[64] Röscher, M., S. Tappertzhofen, and T. Schneller, *J. Am. Ceram. Soc.,* 2011. 94(7) 2193-2199.
[65] Schwartz, R.W., T. Schneller, and R. Waser, *Comptes Rendus Chimie,* 2004. 7(5) 433-461.
[66] Chowdhury, A., M.A. Khan, C. James, and S.J. Milne, Mater. *Chem. Phys.*,2009. 113(1) 135-139.
[67] Boyle, T.J., T.M. Alam, D. Dimos, G.J. Moore, C.D. Buchheit, H.N. Al-Shareef, E.R. Mechenbier, B.R. Bear, and J.W. Ziller, *Chem. Mater.,* 1997. 9(12) 3187-3198.

In: Ferroelectrics: New Research
Editors: Xui Li Huang and Su Lyn Ma

ISBN: 978-1-61942-282-7

Chapter 9

LOW TEMPERATURE FABRICATION OF HIGH-*K* AND FERROELECTRIC THIN FILMS FOR CAPACITIVE APPLICATIONS

Xiaohong Zhu[1] and Emmanuel Defaÿ[2]
[1] Department of Materials Science, Sichuan University, Chengdu, China
[2] CEA, LETI, MINATEC, Grenoble, France

ABSTRACT

A low temperature process can ensure the applicability of high-*k* (high dielectric permittivity) and ferroelectric thin films in above-IC technologies and at the same time benefit from a low thermal budget. Here we report on the growth mechanism and dielectric properties of several important high-*k* and ferroelectric thin films, such as $Bi_{24}Fe_2O_{39}$ (BFO), $Ba_{0.7}Sr_{0.3}TiO_3$ (BST), etc, all of which were successfully prepared by a low temperature process composed of sputtering (ion beam sputtering or rf magnetron sputtering) and postdeposition annealing. The effect of annealing temperature on the structural and dielectric properties of BST thin films was systematically investigated. A sharp transition in their tunable dielectric behaviours was observed in good agreement with the evolution of crystal structure from amorphous to crystalline phase. It was demonstrated that the perovskite phase could crystallize in BST films at very low temperatures (~450 °C), and the low temperature processed BST films showed very good dielectric properties with a high dielectric tunability, typically 66.2% at E (electric field intensity) = 800 kV/cm, and a low dielectric loss, roughly 0.01 at zero field. Similarly, high-*k* BFO thin films were prepared on technologically desirable platinized silicon substrates by the aforementioned low temperature process (≤450 °C). The BFO film not only exhibited high dielectric permittivity (ε_r=113) and relatively low loss tangent ($\tan\delta$=0.012), but also showed a fairly small quadratic voltage coefficient of capacitance (α~800 ppm/V^2) and a small temperature coefficient of capacitance (α_T~790 ppm/°C). The improvement of low temperature crystallization ability in these high-*k* and ferroelectric films is believed to stem from an enhancement of adatom activity by an intensified bombardment of energetic species in the sputtering process. The attractive dielectric and electrical properties obtained make the low temperature processed high-*k* and ferroelectric thin films promising candidates for various capacitive applications in silicon-based integrated circuits and devices.

I. INTRODUCTION

With the rapid development in microelectronic market, there is a significant increase in the need for, and applications of, high integrated, powerful, miniaturized, and low cost capacitor systems. Capacitors are key passive components in most of the integrated circuits which are used in analog filtering, dc decoupling, analog-to-digital conversion, and data storage. The miniaturization of capacitors with high capacitance values requires use of dielectrics possessing a high capacitance density per unit area, namely, high-*k* dielectrics. Accordingly, the replacement of traditional dielectrics, SiO_2 and Si_3N_4, by oxide materials with higher permittivity (high *k*) is an inevitable trend of development because the aforementioned traditional dielectrics have too low dielectric constant ($\varepsilon_r<10$). Nowadays, there is a lot of ongoing effort to develop high-*k* materials such as HfO_2, Ta_2O_5, ZrO_2, and Al_2O_3, etc [1-10]. These high-*k* materials, however, usually show either a still relatively low dielectric permittivity (ε_r~20) or a large quadratic voltage coefficient of capacitance ($\alpha\geq1000$ ppm/V^2) [9,11]. Jorel et al. [11] have suggested that it is rather difficult to obtain both a high capacitance density and a low α value in a single high-*k* material, unless a bilayer such as $SrTiO_3/ZrO_2$ is used. Therefore, exploitation of high-*k* materials exhibiting simultaneously high ε_r, low α, good thermal stability, and minimized leakage current density is quite challenging but very important for practical applications. Recently, high permittivity (ε_r=62.8) with good stability under external electric fields has been reported in epitaxial $SrHf_{0.5}Ti_{0.5}O_3$ films; however, a high substrate temperature of 750 °C was necessary for their crystallization [12]. As is well known, besides the integration and component size reduction requirements, capacitors are increasingly required to be integrated on-chip, i.e., compatible with modern multilevel interconnect and metallization technologies. Hence, a low processing temperature is also very important for ensuring the applicability of high-*k* materials in modern above-IC technologies.

On the other hand, barium strontium titanate ($Ba_{1-x}Sr_xTiO_3$, hereafter represented as BST) thin films have been intensively investigated for applications in tunable microwave/rf components, such as frequency-agile filters, voltage-controlled oscillators, phase shifters, and antennas, because of their large electric-field-dependent tunability and relatively low dielectric loss [13-16]. BST is a continuous solid solution between barium titanate ($BaTiO_3$) and strontium titanate ($SrTiO_3$) over the entire range of concentration, and its Curie temperature decreases linearly with increasing strontium titanate concentration. As a result, the transition temperature and, hence, the electrical and optical properties of BST can be tailored over a broad range for various electronic applications. Compared to mechanically and magnetically tuning techniques, the tunable devices based on ferroelectric films are fast, small and lightweight and, because they work using an electric field, have low power consumption. Since silicon wafer plays a dominant role in the integrated electronics industry, the integration of BST thin films with conventional silicon microelectronic technology is of great significance. However, a rather high processing temperature, around 600-800 °C [13,16,17,18], is generally required to achieve sufficient crystallinity and, thus, good physical properties of BST thin films, which limits inevitably the applicability of these thin films in integrated circuits (IC), especially in the issues of above-IC technologies.

In this chapter, we report on the growth of high permittivity (ε_r=113) $Bi_{24}Fe_2O_{39}$ (BFO) [19] and ferroelectric $Ba_{0.7}Sr_{0.3}TiO_3$ [20,21] thin films on technologically desirable platinized silicon substrates by a low temperature process (~450 °C). The excellent dielectric responses and low leakage currents in these favorably low temperature processed films confirm high-*k* $Bi_{24}Fe_2O_{39}$ and ferroelectric $Ba_{0.7}Sr_{0.3}TiO_3$ thin films as promising candidates for various capacitive applications in silicon-based integrated circuits and devices.

II. EXPERIMENTAL

BFO thin films were deposited at room temperature on Pt(111)/TiO_2/SiO_2/Si(100) substrates by rf magnetron sputtering. Magnetron sputtering is a powerful and flexible technique which can be used to coat virtually any workpiece with a wide range of materials - any solid metal or alloy and a variety of compounds. Sputtering is the removal of atomised material from a solid due to energetic bombardment of its surface layers by ions or neutral particles. The substrate is placed in a vacuum chamber with a target of the material to be deposited. Plasma is generated in a passive source gas (e.g. argon) in the chamber, and the ion bombardment is directed towards the target, causing material to be sputtered off the target and condense on the chamber walls and the substrate. Figure 1(a) illustrates the working principle of sputter deposition. A strong magnetic field (magnetron) can be used to concentrate the plasma near the target to increase the deposition rate. Rf power supply is used for deposition of insulating materials. Two identical commercial powder targets made from a mixture of binary oxides Bi_2O_3 and Fe_2O_3 (Super Conductor Materials, Inc., USA), having a nominal composition of $Bi_{1.15}FeO_3$, were employed simultaneously in the film deposition. A high-purity sputtering gas composed of 90% Ar+10% O_2 was used for deposition under a working pressure of 5×10^{-3} mbar, and the sputtering power was maintained at 200 W. The as-deposited thin film samples were then postannealed at 450 °C in pure Ar atmosphere for 1 h. The BFO film thickness was examined to be about 140 nm by using a Dektak 3030 profilometer.

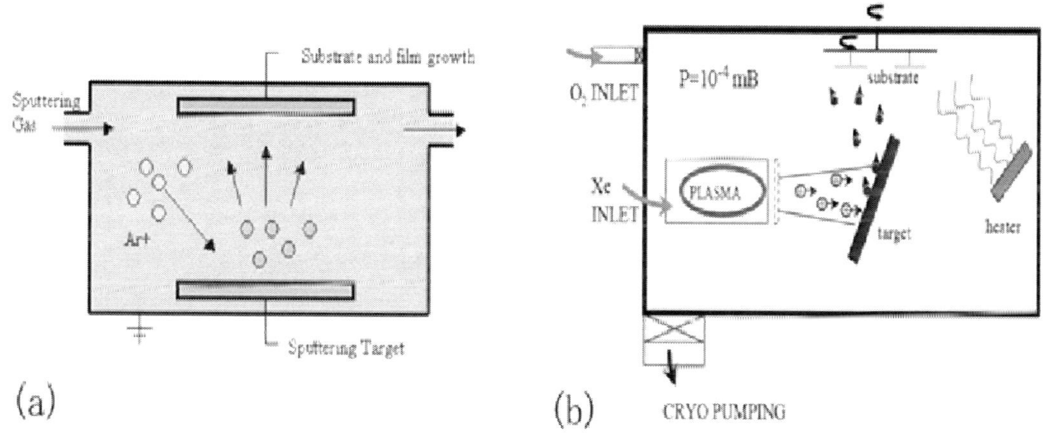

Figure 1. Schematic illustration of the working principle of sputter deposition (a) and ion beam sputtering deposition system (b).

BST thin films with thickness of 500 nm were deposited on $Pt/TiO_2/SiO_2/Si(100)$ substrates by ion beam sputtering (IBS) from a single stoichiometric $Ba_{0.7}Sr_{0.3}TiO_3$ target in an Oxford 500 apparatus. When compared to conventional sputtering methods, the IBS technique exhibits several very attractive advantages: (i) The plasma is isolated from the substrate, thereby avoiding plasma damage to the deposited films. (ii) Very high kinetic energy ion beam leads to excellent adhesion and high density for film deposition. The sputtered material impacts on the substrate with a kinetic energy which may be several orders of magnitude higher than in conventional sputter deposition techniques. (iii) The IBS process can be precisely controlled and easily automated so that films can be deposited with extremely high accuracy and repeatability. Figure 1(b) presents a schematic illustration of our IBS system. Xe ions rather than Ar ions are used in order to achieve a better stoichiometry of the BST films, particularly for precise Ba and Sr composition in the films, because the atomic mass of Xe is comparatively closer to those of Ba and Sr elements. In this system, substrates can be heated up to 150 °C by IR lamps, and they are continuously rotated during the deposition for good film thickness uniformity. In the present work, the ion beam voltage and ion beam current were set at 500 V and 30 mA, respectively. The substrates were not heated during the deposition and their temperature was only dependent on ion bombardment. After deposition, the films were postannealed at a wide range of temperatures, ranging from 300 to 700 °C, for 30 min in a conventional furnace in air ambient.

The crystallographic structures of the BFO and BST thin films were analyzed by x-ray diffraction (XRD) with a Siemens D5000 diffractometer using Cu Kα radiation. The film surface morphology and microstructure was characterized by atomic force microscopy (AFM) using a tapping mode (Digital Instruments/Veeco Dimension 3000). In order to measure the dielectric and electrical properties of the films, square platinum top electrodes of $90\times90\ \mu m^2$ in area and 100 nm in thickness were fabricated by dc sputtering and then a lift-off lithography process. The dielectric property measurements were performed by using an HP4275A multifrequency *LCR* meter and an Agilent 4294A precision *LCR* meter. The leakage current characteristics of the films were measured with an HP4155A semiconductor parameter analyzer and an HP4140B pA meter/dc voltage source.

III. Results and Discussion

(1) High-*K* BFO Thin Films

Figure 2 shows typical θ-2θ XRD scans of the BFO thin film with intensity shown in the log scale. Except for a small amount of $BiFeO_3$ impurity phase, the diffraction peaks can be mainly indexed to a tetragonal lattice of $Bi_{24}Fe_2O_{39}$ [22,23], displaying highly (201)-preferred orientation. The Bi-rich and Fe-deficient stoichiometry in the $Bi_{24}Fe_2O_{39}$ film is probably ascribed to the careful tailoring of sputtering parameters, thereby obtaining significantly dissimilar sputtering yields of Bi and Fe. Consequently, such a specific Bi-enriched film composition is kinetically helpful in forming the majority $Bi_{24}Fe_2O_{39}$ phase. It has been verified that the variation in annealing parameters, such as a longer annealing duration, a relatively higher annealing temperature, and an oxygen-containing annealing atmosphere, could not change remarkably the phase formation in the film, revealing a stable characteristic of $Bi_{24}Fe_2O_{39}$ phase. The good reproducibility of dielectric properties over randomly different sample regions, which will be presented in the following context, confirms the homogenous distribution of local composition in the $Bi_{24}Fe_2O_{39}$ film. It should be pointed out that crystalline $Bi_{24}Fe_2O_{39}$ phase has even been observed in the 400 °C annealed film with just a little worse crystallinity. Such a low temperature process (≤450 °C) can ensure the applicability of these $Bi_{24}Fe_2O_{39}$ films in above-IC technologies, and at the same time, benefit form a low thermal budget. Similar to their counterpart of $BiFeO_3$ thin films from the same bismuth iron oxide material system [24], the improvement of low temperature crystallization ability in the $Bi_{24}Fe_2O_{39}$ films is believed to stem from an enhancement of adatom activity by an intensified bombardment of energetic species in the sputtering process.

Figure 3 presents the relative dielectric constant (ε_r) and loss tangent (tanδ) of the BFO thin film as a function of the applied dc bias voltage, measured at a frequency of 100 kHz and at room temperature (300 K), with dc bias voltage swept cyclically from negative bias (-3 V) to positive bias (+3 V) and vice versa. It can be seen from the figure that there is no hysteresis and the measured curves are repeatable for the two different bias-sweeping directions, indicating an almost negligible effect of space charge defects and thus good quality of the BFO film. Very interestingly, the ε_r of the BFO film, reaching as high as 113 at zero bias, is much higher than those obtained in the aforementioned common high-*k* dielectric thin films (ε_r~20); meanwhile, such a high ε_r is accompanied by a relatively low loss tangent (tanδ=0.012) in the BFO film. Furthermore, it is important to note that the dielectric permittivity and loss tangent of the BFO film show excellent stability under external dc bias field. With the increase in applied dc bias, the dielectric permittivity increases very slowly, displaying a tiny upward variation trend. This is in marked contrast to the large downward variation of ε_r with dc bias field that was observed in $BiFeO_3$ thin films [25]. It is known that an important issue regarding capacitor performances is the voltage linearity, which shows the dependence of capacitance (C) on the applied bias (V). For high-*k* oxides, a quadratic voltage dependence of capacitance is generally observed and can be expressed as

$$[C(V)-C(0)]/C(0)=\alpha V^2+\beta V, \quad (1)$$

where $C(0)$ is the capacitance at zero bias, and α and β are the quadratic and linear coefficients, respectively. For usual working voltages (1-3 V), the quadratic contribution prevails ($\alpha V^2>\beta V$) and the most important parameter that must be controlled is the quadratic coefficient α [11]. In the case of our BFO film, the quadratic voltage dependence of capacitance is fitted and the coefficient α is derived to be about 800 ppm/V^2, as presented in the inset of Figure 3. This α value obtained in the BFO film is comparable to or even lower than those obtained in most of the high-*k* dielectrics (α: 100-5000 ppm/V^2) [1-10]. The capacitance density ($C_S=\varepsilon_0\varepsilon_r/d$), another important parameter for practical applications, is calculated to be 7.15 fF/μm^2 for the 140 nm thick BFO film, which could be further enhanced by reasonably decreasing the film thickness (d) in order to meet the long term ITRS requirements [11].

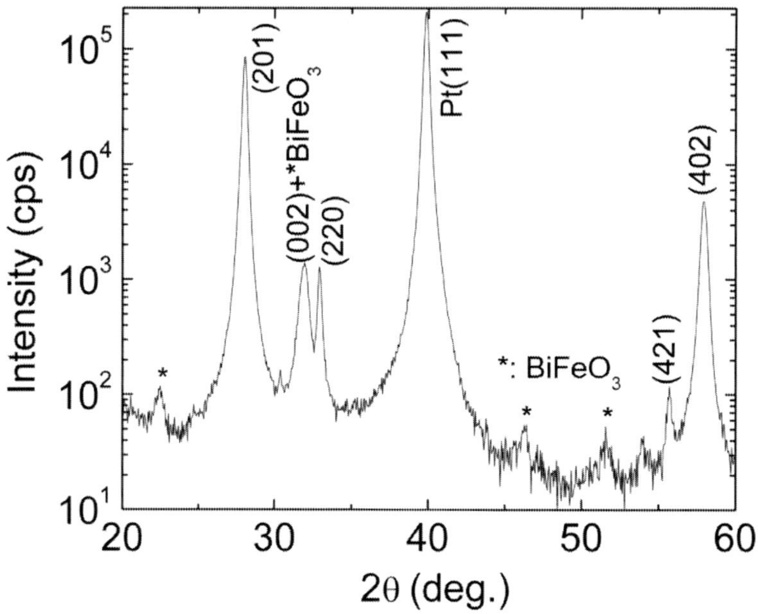

Figure 2. Typical θ-2θ XRD scans of the $Bi_{24}Fe_2O_{39}$ thin film grown on Pt(111)/TiO_2/SiO_2/Si(100) substrate.

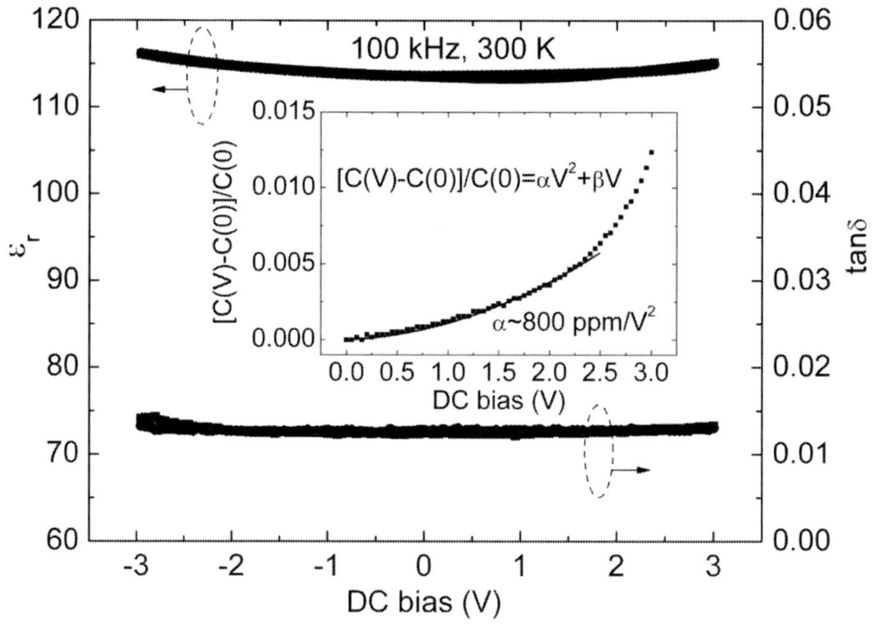

Figure 3. Relative dielectric constant (ε_r) and loss tangent (tanδ) of the $Bi_{24}Fe_2O_{39}$ thin film as a function of applied dc bias voltage, measured at 100 kHz and 300 K. Note that the measured curves are perfectly repeatable for the two different bias-sweeping directions. The inset shows a fitted curve of the quadratic voltage dependence of capacitance.

Figure 4(a) shows the frequency dependence of the dielectric constant and loss tangent between 100 Hz and 1 MHz, measured at 300 K but with different ac signal amplitudes (30,

50, 100, and 200 mV). It is clear that both the measured dielectric constant and loss tangent are independent of the ac signal amplitude, and the dielectric constant exhibits a slight dispersion with frequency, indicating that the BFO film has good film homogeneity and only a slight concentration of space charge defects [26], which is consistent with the above ε_r vs. V measurement. The temperature dependence of the dielectric constant and loss tangent at 100 kHz over 20 to 120 °C is depicted in Figure 4(b). Clearly, the dielectric constant and loss tangent of the BFO film show a similar linear variation trend with temperature. The temperature coefficient of capacitance (α_T) is defined as

$$\alpha_T = (C' - C)/(C \times \Delta T), \qquad (2)$$

where C and C' are the capacitance at 20 and 120 °C, respectively. As shown in Figure 4(b), the α_T is calculated to be as low as 790 ppm/°C, which is of the same order of magnitude (10^{-4} °C^{-1}) as those obtained in most of high-k oxides [9], revealing a good temperature stability of dielectric properties for the BFO film. Combining the aforementioned C-V variation and the temperature dependence of capacitance (C-T) presented here, one can find that the $Bi_{24}Fe_2O_{39}$ film exhibits positive variations both with dc bias field and temperature (i.e., both α and α_T positive). There are various explanations reported in literature for the observed C-V and C-T variations, such as enhanced electron injection [27], logarithmic temperature-dependent mobility [28], electrode coupling and "hopping" conduction between vacancy sites [29], thermal expansion leading to an entropy increase [9], ionic motion [9,30], and so on. In particular, the model based on ionic motion of Al^{3+} between two different sites has been used for the interpretation of both positive α and α_T coefficients observed in Al_2O_3 [30]. Similarly, it is proposed that the positive C-V and C-T variations observed in the BFO film could be attributed to the metallic cation displacement in the tetragonal lattice.

The variations of leakage current density (J) with the applied dc bias field (E) are plotted in Figure 5. As can be seen in the figure, the leakage current density remains at a relatively low level with the increase in dc bias field, roughly J~10^{-6}-10^{-4} A/cm^2 under E=400 kV/cm. It is noted that the asymmetric J-E curve may arise from the different interfacial qualities between the film and top/bottom electrodes. The leakage behavior for both high positive and negative biases, however, can be similarly explained in terms of the Fowler–Nordheim (FN) tunneling mechanism [31], for which

$$J = BE^2 \exp\left(-\frac{C\Phi_i^{3/2}}{E}\right), \qquad (3)$$

where B and C are constants and Φ_i is the potential barrier height. It is shown in the insets of Figure 5 that a negative slope is obtained in the ln(J/E^2) versus $1/E$ plot for both positive and negative biases, coherent with a contribution from FN tunneling in the leakage current, i.e., an injection of charge carriers into the BFO film from electrodes by tunneling through the interfacial energy barrier.

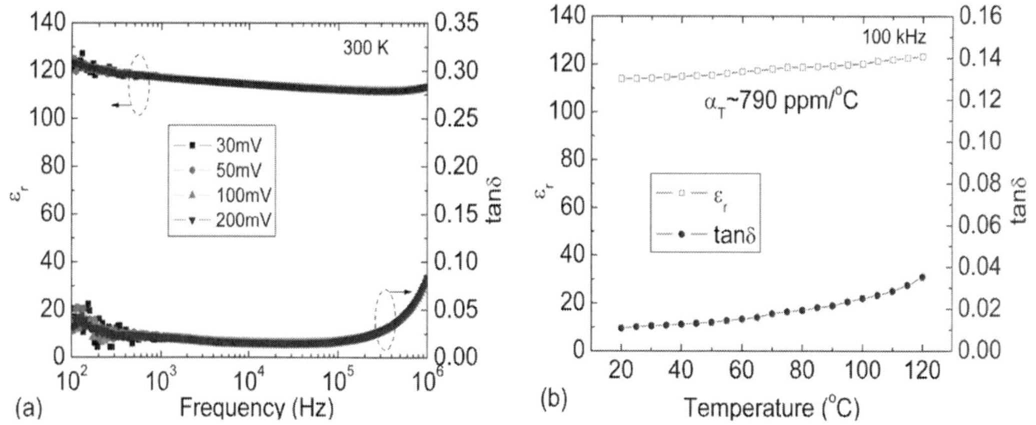

Figure 4. (a) Frequency dependence, measured at room temperature (300 K) but with different ac signal amplitudes and (b) temperature dependence, measured at 100 kHz, of the dielectric constant and loss tangent for the $Bi_{24}Fe_2O_{39}$ thin film.

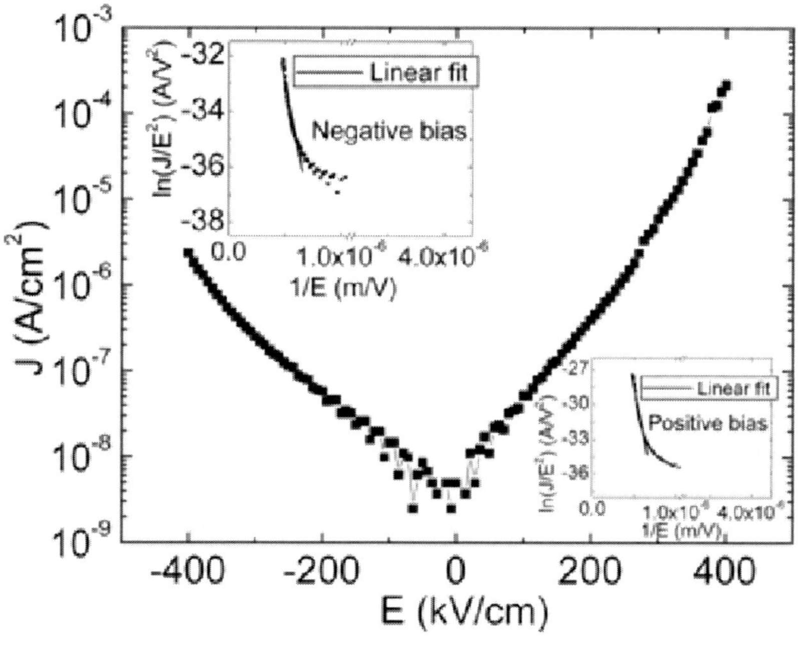

Figure 5. Variations of leakage current density with applied dc electric field for the $Bi_{24}Fe_2O_{39}$ thin film, measured at room temperature. The two insets show the leakage current behavior plotted with $\ln(J/E^2)$ vs $1/E$, in which the lines are fits for the FN tunneling mechanism.

(2) Ferroelectric BST Thin Films

Figure 6 shows θ-2θ XRD scans of the BST thin films that were annealed at different temperatures in the range of 435-685 °C. Judging from the major diffraction peaks (100), (110) and (200), especially from peak (110), one can see a gradual evolution in the perovskite crystallization with increasing the annealing temperature. For the 435 °C annealed film, it is still amorphous. However, the BST film starts to nucleate and grow in the perovskite phase when the annealing temperature is raised to 440 °C. Accordingly, a weak (110) diffraction peak can be visible. Then, a further increase in the annealing temperature results in an enhancement of peak intensity and, thus, an improvement in the BST film crystallinity. As a result, the perovskite phase can be clearly evidenced in the 450 °C annealed film. Such a low crystallization temperature is promising for the integration of these thin-film materials into silicon integrated circuits and devices. The dramatic lowering of the crystallization temperature in our case is believed to originate from the high energetic process nature of IBS technique, which is helpful in enhancing the reactive activation of sputtered atoms and effectively overcomes the energy barrier for perovskite crystallization in films.

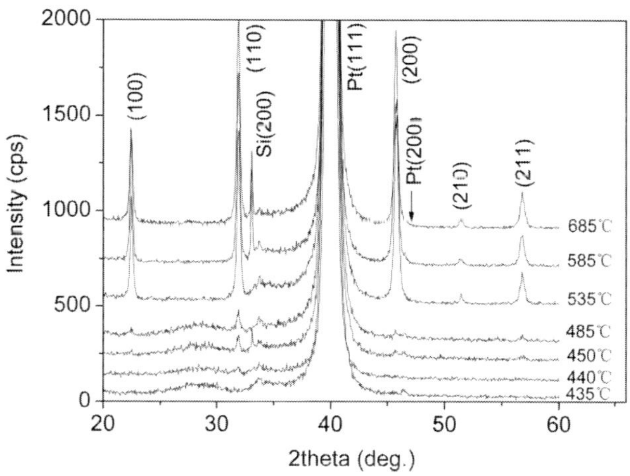

Figure 6. XRD θ-2θ scans of the BST thin films 500 nm in thickness that were deposited on Pt(111)/TiO_2/SiO_2/Si substrates and postannealed at different temperatures ranging from 435 to 685 °C.

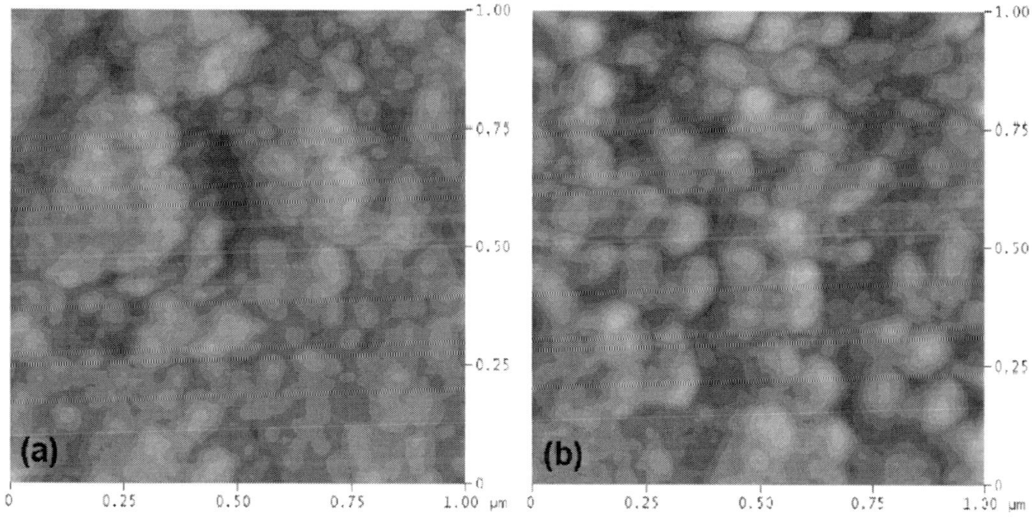

Figure 7. AFM micrographs of the surface morphology for the 535 °C annealed BST films with thickness of (a) 0.5 μm and (b) 1 μm. The scan area is 1×1 μm^2 for both.

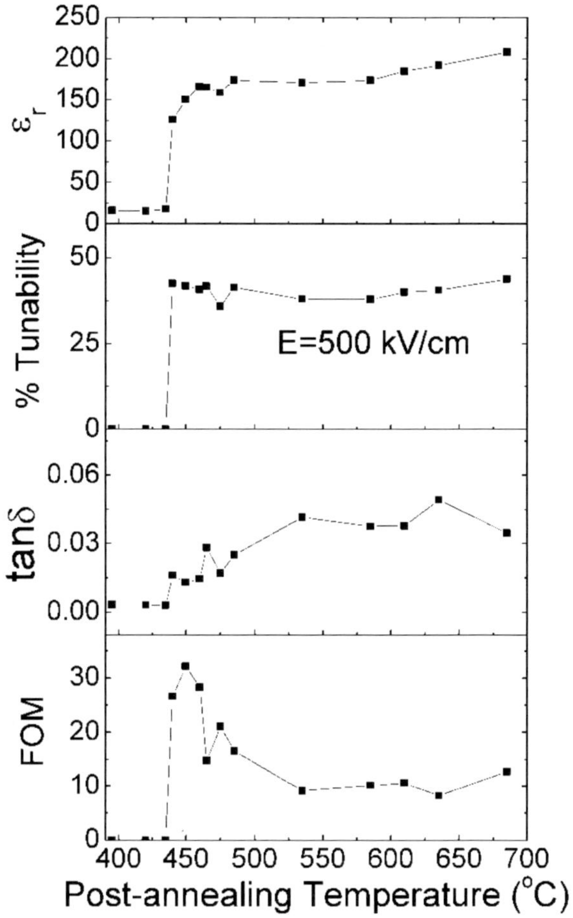

Figure 8. Postannealing temperature dependence of the relative dielectric constant at zero bias, percent tunability with a dc bias field of 500 kV/cm, loss tangent at zero bias, and figure of merit, measured at 300 K and 100 kHz.

It is noted that the crystallization temperature increases with the increase in film thickness; for example, the postannealing temperature is required to be above 500 °C for obtaining noticeable crystallization in the 1 μm thick BST film [20]. Figure 7 shows two-dimensional AFM images over a scan area 1×1 μm^2 for the 0.5 and 1 μm thick BST films, both of which were annealed at 535 °C. The AFM micrographs indicate that the BST films are well crystallized, very dense and relatively smooth. The root-mean-square (RMS) roughness over the scanned areas is about 3 nm for the two films, while the average grain size of the 1 μm thick film is slightly larger than that of the 0.5 μm thick one.

The dielectric properties of the BST films were investigated systematically as a function of the annealing temperature. Figure 8 shows the annealing temperature dependence of the relative dielectric constant (ε_r) at zero bias, percent tunability with a dc bias field of 500 kV/cm, loss tangent ($\tan\delta$) at zero bias, and figure of merit (FOM), all of which were measured at room temperature and at a frequency of 100 kHz. The dielectric tunability is defined as:

$$\text{tunability} = \frac{\varepsilon(0) - \varepsilon(E)}{\varepsilon(0)} \times 100\% \tag{4}$$

where $\varepsilon(0)$ and $\varepsilon(E)$ are the dielectric constant at the zero electric field and applied electric field E, respectively. The figure of merit, defined as tunability/$(\tan\delta)_{0\ kV/cm}$, is commonly used to evaluate the overall dielectric performance of the materials for tunable microwave devices [32]. As shown in the figure, the dielectric constant and tunability display a sharp transition to high values at 435-440 °C where the BST films undergo a structural transition from amorphous to crystalline phase. Also interesting to note is that the tunability for the BST films annealed at low temperatures (440 °C and 450 °C), roughly 42%, is comparable to those results obtained in high temperature annealed films, although the crystallinity is much inferior in the films annealed at low temperatures. This unexpected but exciting result could be understood as follows. The incompletely crystallized films, which were annealed at low temperatures, can be considered as crystalline/amorphous composite [33,34]. The big contrast in the values of dielectric permittivity of the amorphous and crystalline regions affects the redistribution of the electric field around the amorphous regions. Therefore, the dielectric permittivity of the crystallites under applied electric field becomes inhomogeneously distributed over the volume of the crystallites. The redistribution of the electric field surrounding the amorphous regions can enhance the local tuning ability of the crystallites, which can effectively compensate for the suppression of the tunability caused by a reduced total volume of crystallites [35].

It is worth mentioning that the BST films annealed at low temperatures not only show high tunability but also relatively lower dielectric loss. As shown in Figure 8, a fairly low $\tan\delta$, 0.013 in value, is achieved in the 450 °C annealed film, leading to the highest figure-of-merit factor in this film. High-performance BST thin films by such a low temperature process (450 °C) can ensure their applicability in above-IC technologies, and at the same time, benefit from a low thermal budget.

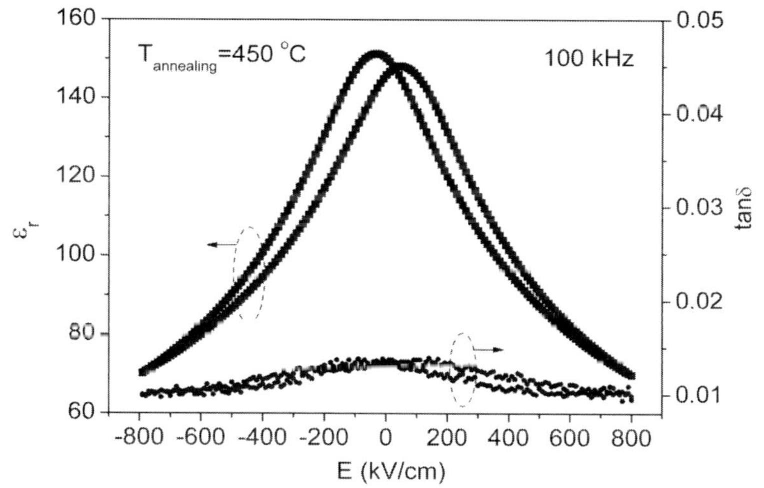

Figure 9. Dielectric constant and loss tangent as a function of the applied dc electric field for the 450 °C annealed film, measured at 300 K and 100 kHz.

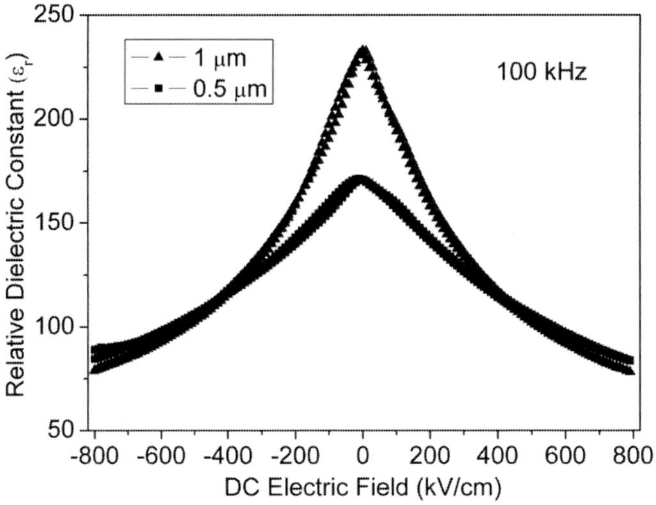

Figure 10. Dielectric constant as a function of the applied dc electric field for the 535 °C annealed BST films with two different thicknesses, measured at a frequency of 100 kHz and at 300 K.

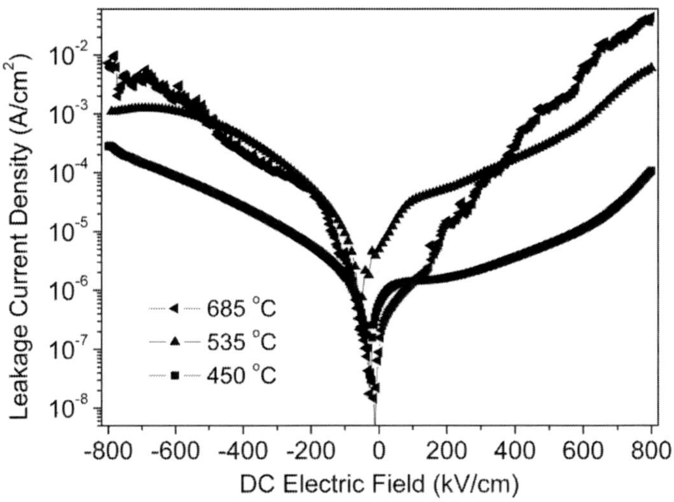

Figure 11. Variations of leakage current density with the applied dc electric field up to 800 kV/cm for the BST films annealed at different temperatures. All the leakage current measurements were carried out at room temperature.

The dielectric constant and loss tangent for the 450 °C annealed film are shown as a function of the applied dc bias field in Figure 9. Compared to the aforementioned tunability of 42% at an electric field intensity of 500 kV/cm, a higher tunability of 53% is obtained at 800 kV/cm, indicative of a great potential for the improvement in dielectric tunability upon increasing the applied dc field. Moreover, the loss tangent displays similarly a downward trend with the applied dc electric field, making the dielectric loss fairly low over the entire range of operating voltages. It should be pointed out that the observed butterfly-shaped hysteresis effect might be attributed to the presence of space charges, such as oxygen vacancies, trapped at the grain boundaries and/or at the film/electrode interfaces, which give rise to local polar regions in the film [36,37].

As the film thickness is increased, the dielectric constant and tunability of BST thin films can be enhanced. Figure 10 shows the room-temperature dielectric constant as a function of the applied dc electric field for the 535 °C annealed BST films with two different thicknesses. From the figure it can be seen that the dielectric tunability of the 0.5 μm thick film is comparatively lower, especially in the high field region. At the electric field intensity of 800 kV/cm, the tunability is about 51.0% for the 0.5 μm thick film, whereas it reaches 66.2% for the 1 μm thick film. Moreover, we noticed that the dielectric constant is also relatively lower in the 0.5 μm thick film with ε_r at zero bias around 170, whereas a higher value of about 230 is obtained in the 1 μm thick film. The degradation of dielectric properties in the thinner film is thought of as associated with a relatively larger thermal misfit strain and more severe influence from the interfacial low-dielectric layer between the BST film and the metal electrode [20,38].

Figure 11 shows the variations of leakage current density with the applied dc electric field up to 800 kV/cm for the BST films annealed at different temperatures. As can be seen in the figure, the leakage current density for the 450 °C annealed film remains at a quite low level (10^{-4} A/cm^2) when the dc electric field is increased to 800 kV/cm, which, especially on the positively biased side, is remarkably lower than those obtained in the films annealed at higher temperatures. This is probably because the high temperature thermal treatment promotes the possibilities of interdiffusion between the film and the electrodes. It is noted that the slight asymmetry in the leakage behaviour stems from the different upper electrode/film and film/bottom electrode interfacial qualities.

CONCLUSION

High permittivity $Bi_{24}Fe_2O_{39}$ and ferroelectric $Ba_{0.7}Sr_{0.3}TiO_3$ thin films have successfully been deposited on technologically desirable Pt(111)/TiO_2/SiO_2/Si(100) substrates by a low temperature process combining sputtering (ion beam sputtering or rf magnetron sputtering) at room temperature and postdeposition annealing at temperatures as low as 450 °C. The improvement of low temperature crystallization ability in these high-*k* and ferroelectric films is probably attributed to an enhancement of adatom activity by an intensified bombardment of energetic species in the sputtering process.

The 450 °C annealed BFO film has a nearly pure tetragonal crystal structure with highly (201)-preferred orientation. The BFO thin film exhibits high dielectric permittivity (ε_r=113) with relatively low loss tangent (tanδ=0.012), as well as good stability both with dc bias field and temperature. The slight positive *C-V* and *C-T* variations observed in the BFO film are probably attributed to the metallic cation displacement in the tetragonal lattice. The leakage current density of the BFO film remains at a relatively low level with the increase in applied electric field, obeying the Fowler–Nordheim tunneling mechanism.

It is also demonstrated that the perovskite phase could crystallize in BST films at a very low temperature, around 450 °C, and a sharp transition in the tunable dielectric behaviours of BST thin films has been achieved, in good agreement with the evolution of crystal structure from amorphous to crystalline phase. Very importantly, both a high dielectric tunability of

42% at *E*=500 kV/cm and a low loss tangent of 0.013 at zero bias have been obtained in the 450 °C annealed film, thereby resulting in the highest figure-of-merit factor among all the different temperature annealed films. Note that the dielectric tunability of the BST films can be further enhanced to 66.2% at *E*=800 kV/cm when the film thickness increases to 1 μm. Moreover, the 450 °C annealed film possesses superior leakage current characteristics with a low leakage current density of about 10^{-4} A/cm^2 at high *E*=800 kV/cm.

The attractive dielectric and electrical properties obtained make the low temperature processed high-*k* BFO and ferroelectric BST thin films promising candidates for various capacitive applications in silicon-based integrated circuits and devices.

ACKNOWLEDGMENTS

The work presented here was supported by EU project "CAMELIA" under Contract No. NMP3-CT-2006-033103, and partially supported by the National Natural Science Foundation of China (Grant No. 61071017), the Program for New Century Excellent Talents in University of China (Grant No. NCET-10-0582), and the Doctoral Fund of the Ministry of Education of China (Grant No. 20100181120021).

REFERENCES

[1] Lee, B. H.; Kang, L. G.; Nieh, R.; Qi, W. J.; Lee, J. C. *Appl. Phys. Lett.* 2000, 76, 1926-1928.

[2] Gilmer, D. C.; Hegde, R.; Cotton, R.; Garcia, R.; Dhandapani, V.; Triyoso, D.; Roan, D.; Franke, A.; Rai, R.; Prabhu, L.; Hobbs, C.; Grant, J. M.; La, L.; Samavedam, S.; Taylor, B.; Tseng, H.; Tobin, P. *Appl. Phys. Lett.* 2002, 81, 1288-1290.

[3] Kim, S. J.; Cho, B. J.; Li, M. F.; Yu, X.; Zhu, C.; Chin, A.; Kwong, D. L. *IEEE Electron Device Lett.* 2003, 24, 387-389.

[4] Govindarajan, S.; Boescke, T. S.; Sivasubramani, P.; Kirsch, P. D.; Lee, B. H.; Tseng, H. H.; Jammy, R.; Schroeder, U.; Ramanathan, S.; Gnade, B. E. *Appl. Phys. Lett.* 2007, 91, 062906.1-062906.3.

[5] Sun, J.; Larsson, M.; Maximov, I.; Hardtdegen, H.; Xu, H. Q. *Appl. Phys. Lett.* 2009, 94, 042114.1-042114.3.

[6] Deloffre, E.; Montès, L.; Ghibaudo, G.; Bruyère, S.; Blonkowski, S.; Bécu, S.; Gros-Jean, M.; Crémer, S. *Microelectron. Reliab.* 2005, 45, 925-928.

[7] Cheng, C. H.; Lee, J. Y. M. *Appl. Phys. Lett.* 2007, 91, 192903.1-192903.3.

[8] Qi, W. J.; Nieh, R.; Lee, B. H.; Kang, L.; Jeon, Y.; Lee, J. C. *Appl. Phys. Lett.* 2000, 77, 3269-3271.

[9] Blonkowski, S. *Appl. Phys. Lett.* 2007, 91, 172903.1-172903.3.

[10] Robertson, J. *J. Appl. Phys.* 2008, 104, 124111.1-124111.7.

[11] Jorel, C.; Vallée, C.; Gonon, P.; Gourvest, E.; Dubarry, C.; Defay, E.; *Appl. Phys. Lett.* 2009, 94, 253502.1-253502.3.

[12] Yan, L.; Suchomcl, M. R.; Grygiel, C.; Niu, H. J.; McMitchell, S. R. C.; Bacsa, J.; Clark, J. H.; Werner, M.; Chalker, P. R.; Rosseinsky, M. J. *Appl. Phys. Lett.* 2009, 94, 232903.1-232903.3.
[13] Chen, C. L.; Shen, J.; Chen, S. Y.; Luo, G. P.; Chu, C. W.; Miranda, F. A.; Van Keuls, F. W.; Jiang, J. C.; Meletis, E. I.; Chang, H. Y. *Appl. Phys. Lett.* 2001, 78, 652-654.
[14] Acikel, B.; Taylor, T. R.; Hansen, P. J.; Speck, J. S; York, R. A. IEEE Microw. *Wirel. Compon. Lett.* 2002, 12, 237-239.
[15] Tagantsev, A. K.; Sherman, V. O.; Astafiev, K. F.; Venkatesh, J; Setter, N. *J. Electroceram.* 2003, 11, 5-66.
[16] Xu, J.; Menesklou, W; Ivers-Tiffée, E. *J. Eur. Ceram. Soc.* 2004, 24, 1735-1739.
[17] Zhu, X. H.; Zheng, D. N.; Zeng, H.; Peng, W.; Zhu, J. G.; Yuan, X. W.; Yong, L. P.; Miao, J.; Li, J.; Tian, H. Y.; Xu, X. P. *Thin Solid Films* 2006, 496, 376-382.
[18] Kim, K. B.; Yun, T. S.; Lee, J. C.; Chaker, M.; Park, C. S.; Wu, K. *Electron. Lett.* 2007, 43, 757-759.
[19] Zhu, X. H.; Defaÿ, E.; Lee, Y.; Andre, B.; Aïd, M.; Zhu, J. L.; Xiao, D. Q.; Zhu, J. G. *Appl. Phys. Lett.* 2010, 97, 232903.1-232903.3.
[20] Zhu, X. H.; Guigues, B.; Defaÿ, E.; Dubarry, C.; Aïd, M. *J. Appl. Phys.* 2009, 105, 044108.1-044108.6.
[21] Zhu, X. H.; Defaÿ, E.; Guigues, B.; Le Rhun, G.; Dubarry, C.; Aïd, M. *J. Eur. Ceram. Soc.* 2010, 30, 471-474.
[22] JCPDS-ICDD 1997, card No. 42-0201.
[23] Ternon, C.; Thery, J.; Baron, T.; Ducros, C.; Sanchette, F.; Kreisel, *J. Thin Solid Films* 2006, 515, 481-484.
[24] Lee, Y. H.; Wu, J. M.; Chueh, Y. L.; Chou, L. *J. Appl. Phys. Lett.* 2005, 87, 172901.1-172901.3.
[25] Habouti, S.; Shiva, R. K.; Solterbeck, C. H. ; Es-Souni, M.; Zaporojtchenko, V. *J. Appl. Phys.* 2007, 102, 044113.1-044113.7.
[26] Zhu, X. H.; Zheng, D. N.; Peng, W.; Zhu, J. G.; Yuan, X. W.; Li, J.; Zhang, M. J.; Chen, Y. F.; Tian, H. Y.; Xu, X. P. *Mater. Lett.* 2004, 58, 3591-3596.
[27] Blonkowski, S.; Regache, M.; Halimaoui, A. *J. Appl. Phys.* 2001, 90, 1501-1508.
[28] Zhu, C.; Hu, H.; Yu, X.; Kim, S. J.; Chin, A.; Li, M. F.; Cho, B. J.; Kwong, D. L. Tech. Dig. - Int. Electron Devices Meet. 2003, 879-882.
[29] Gonon P.; Vallée, C. *Appl. Phys. Lett.* 2007, 90, 142906.1-142906.3.
[30] Bécu, S.; Crémer, S.; Autran, J. L. *Appl. Phys. Lett.* 2006, 88, 052902.1-052902.3.
[31] Zhu, X. H.; Béa, H.; Bibes, M.; Fusil, S.; Bouzehouane, K.; Jacquet, E.; Barthélémy, A.; Lebeugle, D.; Viret, M.; Colson, D. *Appl. Phys. Lett.* 2008, 93, 082902.1-082902.3.
[32] Treece, R. E.; Thompson, J. B., Mueller, J. B.; Mueller, C. H.; Rivkin, T; Cromar, M. W. *IEEE Trans. Appl. Supercond.* 1997, 7, 2363 2366.
[33] Zhu, X. H.; Guigues, B.; Defaÿ, E; Aïd, M. *J. Appl. Phys.* 2008, 104, 074118.1-074118.5.
[34] Guigues, B.; Zhu, X. H.; Sobotka, J.; Guiblin, N.; Defaÿ, E.; Dkhil, B. *Appl. Phys. Lett.* 2009, 95, 032902.1-032902.3.
[35] Sherman, V. O.; Tagantsev, A. K.; Setter, N.; Iddles, D; Price, T. *J. Appl. Phys.* 2006, 99, 074104.1-074104.10.
[36] Pontes, F. M.; Leite, E. R.; Longo, E.; Varela, J. A.; Araujo, E. B.; Eiras, *J. A. Appl. Phys. Lett.* 2000, 76, 2433-2435.

[37] Zhu, X. H.; Yong, L. P.; Tian, H. F.; Peng, W.; Li, J. Q; Zheng, D. N. *J. Phys.: Condens. Matter* 2006, 18, 4709-4718.
[38] Zhu, X. H.; Guigues, B.; Defaÿ, E.; Dubarry, C.; Aïd, M. *J. Appl. Phys.* 2009, 106, 024109.1-024109.5.

Reviewed by Prof. Dr. Jiliang Zhu
Department of Materials Science, Sichuan University, Chengdu 610064, China

In: Ferroelectrics: New Research
Editors: Xui Li Huang and Su Lyn Ma

ISBN: 978-1-61942-282-7

Chapter 10

ELECTRONIC FERROELECTRICITY AND MAGNETOELECTRIC EFFECTS IN CHARGE ORDERED OXIDES

Young Sun and Li-Qin Yan
Beijing National Laboratory for Condensed Matter Physics, Institute of Physics, Chinese Academy of Sciences, Beijing, China

ABSTRACT

In conventional ferroelectric materials the covalent bond between the anions and cations or the orbital hybridization of electrons on both ions plays a key role in establishing the dipolar arrangement. In contrast, electronic ferroelectricity in which the electric dipole depends on electron correlations rather than the covalency, has been recently proposed and experimentally verified. In this chapter, we report on recent progress in the study of electronic ferroelectricity in a series of charge ordered oxides. In addition to a general description, we focus on the electronic ferroelectricity induced by different forms of charge ordering in two specific systems: (i) strongly charge frustrated system RFe_2O_4 (R=Lu,Yb), and (ii) charge ordered perovskite manganites of the type $(R_{1-x}Ca_x)MnO_3$ (R=La,Pr). The coexistence of electronic ferroelectricity and magnetism in these systems leads to a peculiar multiferroics. Furthermore, the charge ordering can be tuned by applying electric field, laser irradiation, etc., which results in a variety of interesting effects. Especially, the magnetoelectric effects and giant dielectric tunability accompanied with electronic ferroelectricity could be promising for many applications.

INTRODUCTION

Ferroelectric materials have been widely used in modern electric devices such as memory elements, filtering devices, and high-performance insulators. In conventional ferroelectric materials the covalent bond between the anions and cations or the orbital hybridization of electrons on both ions plays a key role in establishing the dipolar arrangement [1-3]. The polarization is due to the relative displacement of positive and negative ions and a lattice distortion is generally accompanied with the ferroelectric transition. In strong contrast, a new class of ferroelectricity termed as "electronic ferroelectricity" in which the electric dipole depends on electron correlations rather than the displacement of cation and anion pairs, has been theoretically proposed and experimental discovered in the last decade. The progress is mainly driven by the intensive search for novel multiferroic materials in which ferroelectricity and magnetism coexist [4-7]. The cross coupling between ferroelectricity and magnetism in multiferroics provides additional ways for magnetization and polarization manipulation, *i.e.*, the control of polarization by a magnetic field and of magnetization by an electric field [8-11]. The first theory of electronic ferroelectricity was proposed by Portengen *et al.* for mixed-valent compounds in 1996 [12]. Now the concept has been developed to refer to the phenomena where spin and charge degrees of freedom of electrons are directly responsible for ferroelectricity [13]. In general, there are two types of electronic ferroelectricity: (i) charge driven ferroelectricity where the polarization is induced by different forms of charge order; (ii) spin driven ferroelectricity where magnetic interactions and specific spin ordering (for example, spiral spin ordering) can cause a net polarization. The spin driven ferroelectricty and multiferroicity have been intensitively studied in the past sevral years, and a number of review articles have been published [14-16]. In this paper, we focus on the charge driven electronic ferroelectricty in charge ordered compounds. Charge order (CO) in solids is often observed in systems with ions that formally have a mixed valence [17]. So far two types of charge order have been proposed: site-centered CO and bond-centered CO in which mirror planes of the CO structure can be marked at ions or bond centers (Figure 1). Because both types of CO structure do not break spatial inversion symmetry, a net polarization is not induced. In order to realize ferroelectricity induced by the CO, it is necessary to combine the site-centered and bond-centered CO, *i.e.*, the simultaneous presence of inequivalent *sites* and *bonds* in system. In some materials bonds are inequivalent because of the crystallographic structure, and a spontaneous CO that occurs below a certain ordering temperature drives the inequivalence of the sites. Or vice versa, the material can contain ions with different valences, which after a structural dimerization transition induces ferroelectricity. These two effects may also occur simultaneously, in which case there is one common phase transition.

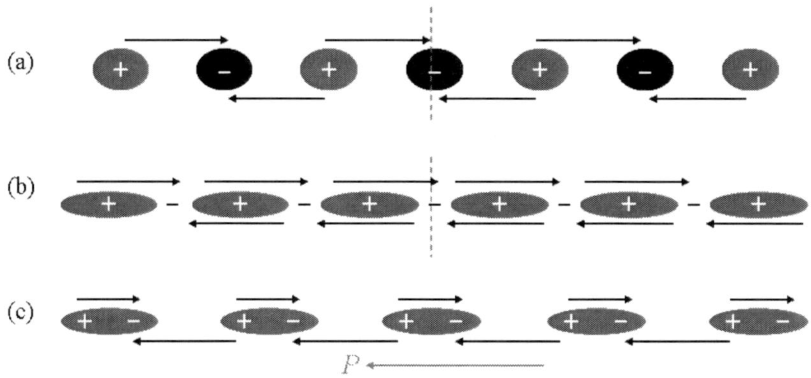

Figure 1. (a) Scheme of one-dimensional site-centered charge order; (b) bond-centered charge order; and (c) a linear combination of these two which induces a net polarization.

In this chapter, we review recent progress on the electronic ferroelectricity induced by the CO in two remarkable systems: (i) strongly charge frustrated system RFe_2O_4 (R=Lu and Yb), and (ii) charge ordered perovskite manganites of the type $R_{1-x}Ca_xMnO_3$ (R=La, Pr). In addition to the ferroelectric and dielectric properties, the manipulation of the CO by external fields (electric, light, etc.) and induced magnetoelectric effects are also described.

2. ELECTRONIC FERROELECTRICITY IN RFe_2O_4

2.1. Charge Order and Ferroelectricity

The RFe_2O_4 system where R stands for rare-earth elements from Dy to Lu and Y [18], has a rhombohedral space group ($R\overline{3}m$) with a crystal structure consists of the alternate stacking of triangular lattices of rare earth elements, iron and oxygen (Figure 2). An equal amount of Fe2+ and Fe3+ coexists at the same site in the triangular lattice. Compared to the average Fe valence of 2.5+, Fe^{2+} and Fe^{3+} ions are considered as having an excess and a deficiency of half an electron, respectively. Thus the interaction between Fe^{2+} and Fe^{3+} is accompanied by a frustration on the triangular lattice. The competing interactions between frustrated charges are settled by a superstructure of Fe^{2+} and Fe^{3+} arrangements on the bilayer structure. As shown in Figure 3, within each bilayer of the iron triangular lattice, the lower layer presents the ratio 2:1 (Fe^{2+}:Fe^{3+}) but the upper layer shows the ratio 1:2 (Fe^{2+}:Fe^{3+}). This site centered charge redistribution is closely associated with the bilayer crystallographic structure and frustrated triangular geometry of RFe_2O_4. As a result, an electric dipole oriented from the upper layer to the lower layer within each iron bilayer is formed [19,20].

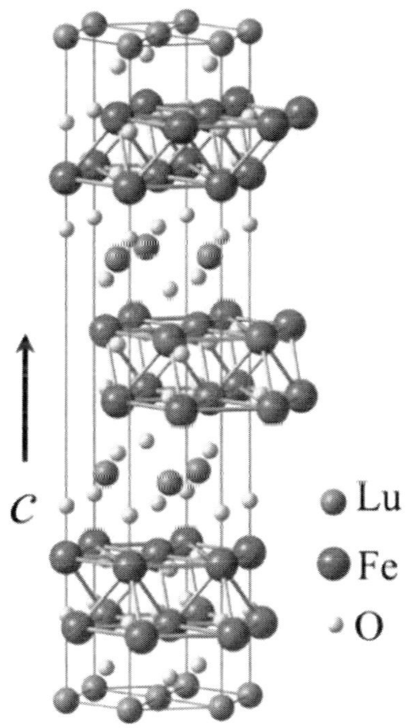

Figure 2. Crystal structure of RFe_2O_4.

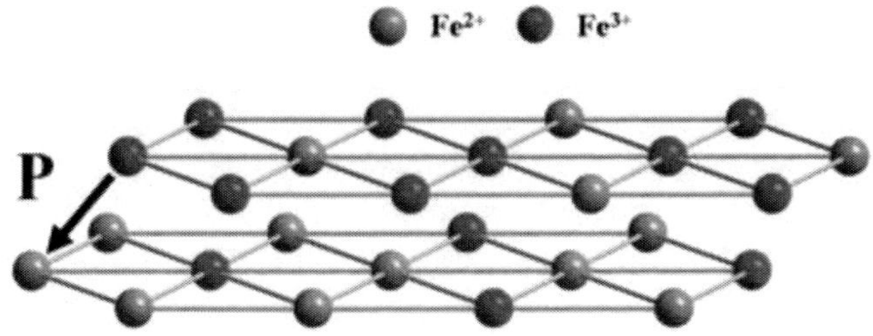

Figure 3. Charge order configuration in RFe_2O_4.

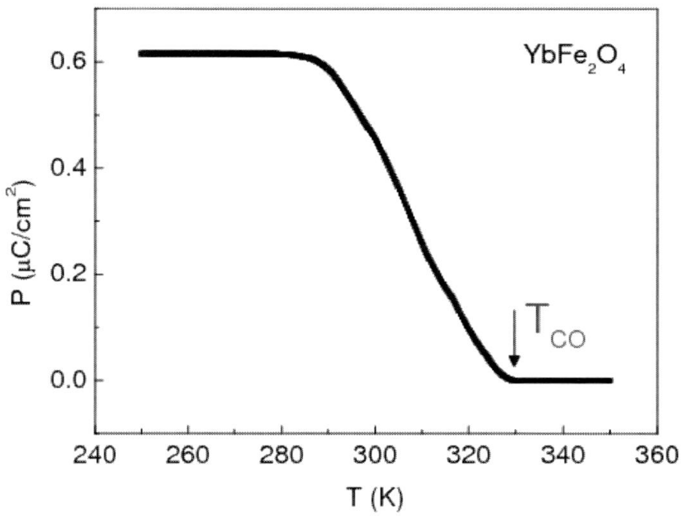

Figure 4. Temperature dependence of electric polarization of $YbFe_2O_4$. A ferroelectric transition is observed at the charge order transition temperature T_{CO}.

The CO in RFe_2O_4 has been confirmed by many experiments such as neutron scattering, electron diffraction, x-ray scattering, *etc.* [21-25]. For example, $LuFe_2O_4$ exhibits a three-dimensional (3D) charge ordered state below 330 K. Above 330 K, the 3D charge ordering transforms into a 2D ordering state. In 2005, Ikeda *et al.* first reported the discovery of ferroelectricity in $LuFe_2O_4$ and concluded that the electric polarization arises from the ordered arrangement of electrons on a triangular lattice [19]. Figure 4. presents a similar result in $YbFe_2O_4$. The plot is the integration of a pyroelectric current measurement. The current flow from the sample was recorded on heating after electric field cooling along the *c* axis. A spontaneous electric polarization appears just below the CO transition temperature T_{CO} ~ 330 K, which indicates the close correlation between the CO and ferroelectricity. The dielectric constant shows a strong dispersion consistent with order-disorder type ferroelectric materials [19].

2.2. Giant Dielectric Tunability

Ferroelectric materials with a high dielectric tunability, *i.e.*, a strong dependence of their dielectric constant on the applied dc bias electric field, have potential usefulness for many devices [26–29]. A high tunability in ferroelectrics such as $Sr_{1-x}Ba_xTiO_3$ usually requires a large dc electric field in the order of tens of kV/cm. Besides, conventional ferroelectric materials usually exhibit a high tunability in a narrow temperature range around the ferroelectric transition. These drawbacks have strongly limited the applicability of dielectric tunability in tunable devices. Nevertheless, it was found that the electronic ferroelectric RFe_2O_4 system exhibits a low-field giant dielectric tunability over a broad temperature interval around room temperature [30-32].

Figure 5. shows the dielectric constant of $LuFe_2O_4$ as a function of temperature under zero and a 50 V/cm dc bias field. A large difference between two curves appears above 250 K as the dielectric constant is greatly suppressed in a dc bias field, indicating a high dielectric tunability.

The dielectric tunability (n_r) is defined by

$$n_r = \frac{\varepsilon(0) - \varepsilon(E_0)}{\varepsilon_0} \qquad (1)$$

Figure 6. shows the tunability n_r under a 50 V/cm bias field as a function of temperature at different frequencies. The tunability in $LuFe_2O_4$ exhibits a broad peak so that a high tunability retains in a broad temperature interval which presents the merits to the conventional ferroelectric materials where the tunability is strongly temperature dependent and shows a sharp peak in a narrow temperature range. Such a broad operating temperature interval around room temperature is very promising for applications. With increasing frequency, the maximum value of n_r decays, while the broad peak shifts to higher temperatures. Interestingly, although n_r has a strong frequency dependence at low temperature, it is nearly the same around room temperature.

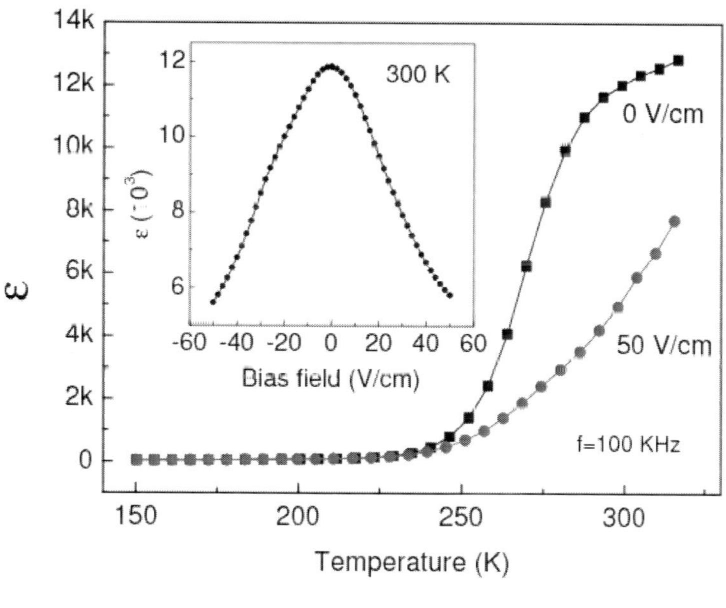

Figure 5. Influence of dc bias electric field on the dielectric constant of $LuFe_2O_4$.

The physical origin of the giant dielectric tunability effect in bulk $LuFe_2O_4$ could be correlated closely with the spectacular electronic ferroelectricity. The nature of a polar arrangement of the ordered Fe-3*d* charges in $LuFe_2O_4$ causes the energy required for charge motion of 3*d* electrons could be much lower than the case of ion displacement. Hence, a small electric field could sharply reduce the charge fluctuation on Fe ions, leading to a large reduction in permittivity. The dielectric tunability can be modified by the substitution of Fe^{2+} with other transition metal elements.

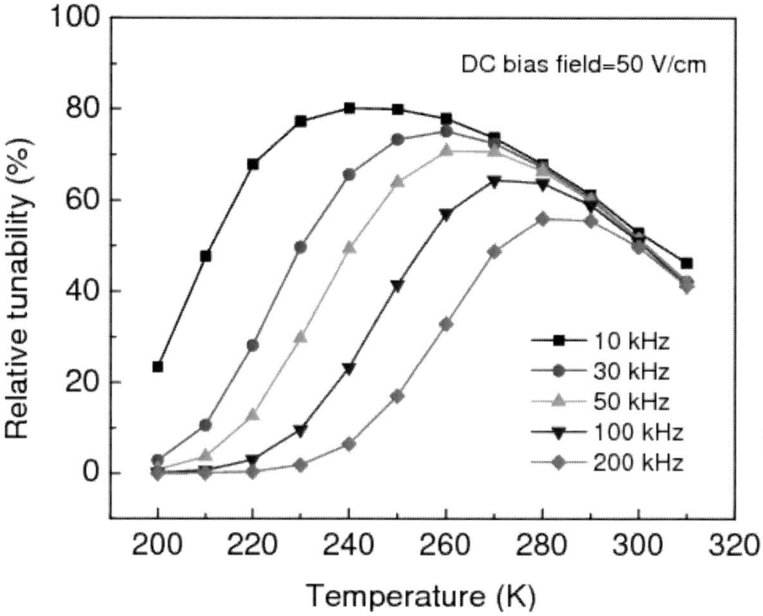

Figure 6. Temperature depencence of dielectric tunability in $LuFe_2O_4$.

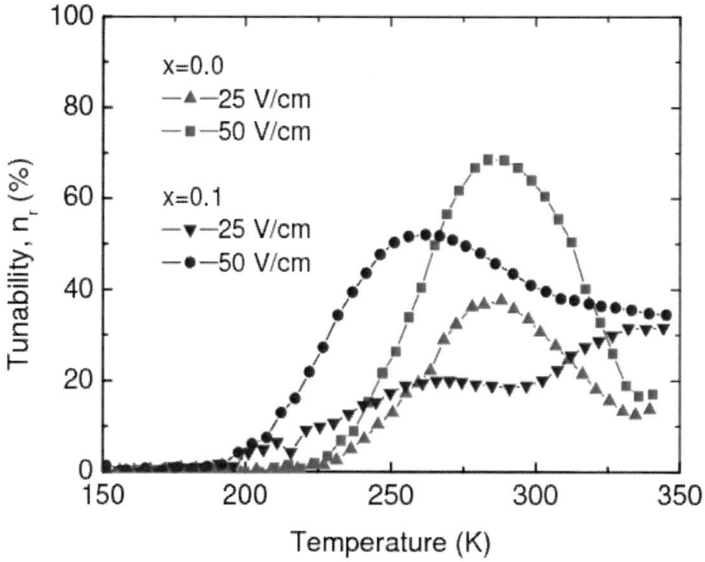

Figure 7. Giant dielectric tunability in $LuFe_{1-x}Mg_xO_4$.

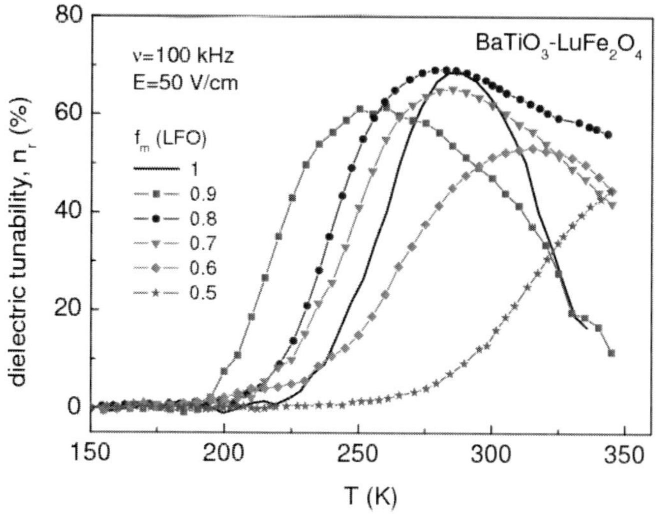

Figure 8. Dielectric tunability in $BaTiO_3$-$LuFe_2O4$ composites.

As shown in Figure 7, the replacement of 10% Mg to Fe ions induces a broader temperature peak while a high tunability remains. Such a broad temperature interval across room temperature is beneficial for practical applications. At the same time, the reduction of the maximum of tunability indicates that the giant dielectric tunability is strongly correlated with charge ordering of Fe^{2+} and Fe^{3+} ions. The substitution of Mg ions for ferrous ions directly alters the ratio of Fe^{2+} and Fe^{3+} ions and, consequently, results in the impairment of charge ordering and the disturbance of polarization fluctuation [31].

The ceramic composites consisting of conventional ferroelectric $BaTiO_3$ and electronic ferroelectric $LuFe_2O_4$ also exhibit exotic dielectric tunability response [32]. As the $LuFe_2O_4$ volume fraction approached a percolation threshold, the composites exhibited a novel percolative effect with a colossal dielectric constant and a giant dielectric tunability. Figure 8. plots the temperature dependence of relative dielectric tunability measured at 50 V/cm for the $BaTiO_3$–$LuFe_2O_4$ composites with $f_m \geq 0.5$. All the composites present a high dielectric tunability ($n_r > 40\%$) at a small bias field of 50 V/cm. The maximum peak and temperature dependence can be finely tuned by adjusting $LuFe_2O_4$ fraction. Especially, the composite with $LuFe_2O_4$ fraction $f_m = 0.8$ exhibits a high tunability covering a broad temperature interval across room temperature in strong contrast to that of traditional ferroelectrics.

2.3. Modify Charge Order by External Fields

The CO state in many systems could be manipulated by external forces such as magnetic field, electric field, laser irradiation, and pressure [33-38]. For example, an electrically driven phase transition is realized in the CO perovskite manganites [38] and magnetite (Fe_3O_4) [34] below the T_{CO}. Similarly, the CO state in RFe_2O_4 can be modified by external fields [39-42].

Figure 9. demonstrates the electric field modification on the CO in a $YbFe_2O_4$ single crystal. When a small electric current is applied in the *ab* plane of the crystal, the resistivity is high and remains constant because the CO is unaffected. However, once a high electric current is applied, which generates a high electric field in the order of several kV/cm, the measured voltage on the sample shows an abrupt drop and then stabilizes at a small value. This sharp transition is a sign of CO breakdown induced by a high electric field. The resistivity drops by more than three orders, indicating an electrically driven insulator-metal transition. In the CO state, the charge carriers are initially localized at each atomic site due to the repulsive electron-electron interaction. A high enough electric field may cause the breakdown of the CO state, which immediately releases a large number of mobile charge carriers, and consequently, a metallic state appears. This electrically driven phase transition leads to a giant electroresistance effect which could be used in merrory devices [39]. As observed in a variety of perovskite manganites, the CO state can be modified by illumination with x-ray or visible radiation [38]. A similar effect in RFe_2O_4 would be expected because its CO state is more sensitive to external stimulus. Figutr 10. shows the photocurrent effect in a $LuFe_2O_4$ single crystal. When a laser beam with the wavelength λ=532 nm is shed on the *ab* plane of the crystal, a dramatic increase of electric current is observed. The abrupt rise of electric current in response to laser illumination indicates the breakdown of CO by laser irradiation which releases the localized electrons.

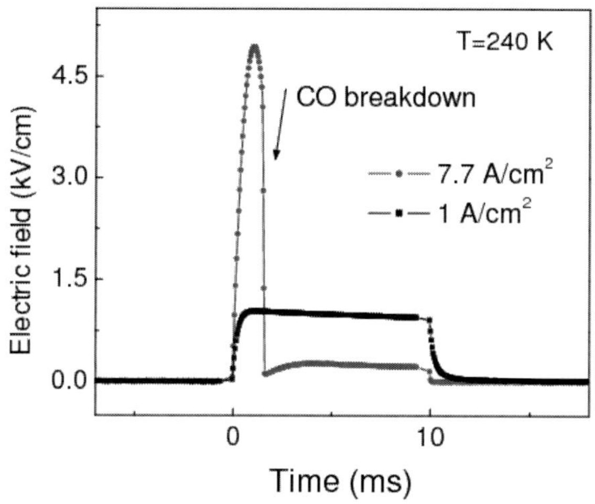

Figure 9. Electric field modification of charge order in $LuFe_2O_4$.

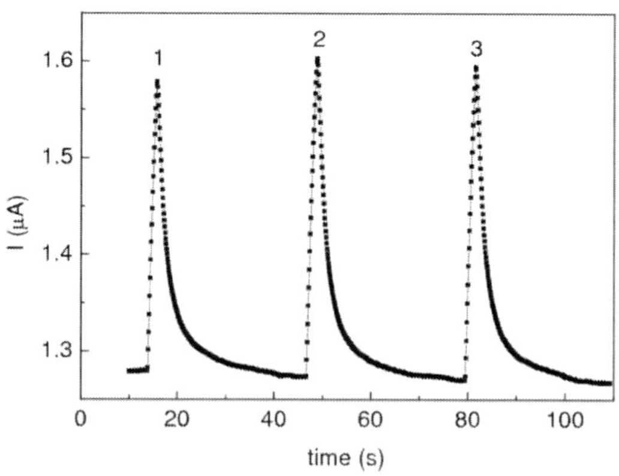

Figure 10. Modification of charge order by laser illumination in $LuFe_2O_4$.

2.4. Multiferroics and Magnetoelectric Effects

Multiferroic materials in which magnetic and ferroelectric orders coexist have attracted great interest because of their potential for technological applications as well as the intriguing fundamental physics. The RFe_2O_4 family represents a good example of charge order induced multiferroics. On one hand, the ordered arrangement of charges leads to an electronic ferroelectricity below 330 K. On another hand, the strong magnetic interactions between localized Fe moments develop as a ferrimagnetic ordering below 240 K [22]. Therefore, the coexistence of ferrimagnetic order and electronic ferroelectricity makes RFe_2O_4 a new class of multiferroics. Since both the ferroelectricity and magnetism in the system originate from the same Fe ions, it holds the promise of strong magnetoelectric coupling [43].

In order to testify the magnetoelectric effects, we investigated the magnetization of a $LuFe_2O_4$ single crystal as a function of applied voltage (Figure 11). The magnetic field is applied along the *c* axis (the easy magnetization direction) while the voltage pulse is applied in the *ab* plane where the CO is easier to be broken down than the *c*axis direction. With the increase of applied voltage, the magnetization changes smoothly when the voltage is low. This initial change could be the normal magnetic relaxation with time. When the applied voltage is above a critical voltage, the magnetization exhibits a giant jump. In zero magnetic field, the magnetization jumps down.

In contrast, the magnetization jumps up in a constant 0.1 T magnetic field. These results suggest that the magnetization of multiferroic RFe_2O_4 can be effectively controlled by electric fields. By simultaneously monitoring the current with applied voltage pulse, we find that this dramatic magnetization change is related to the breakdown of the CO [44].

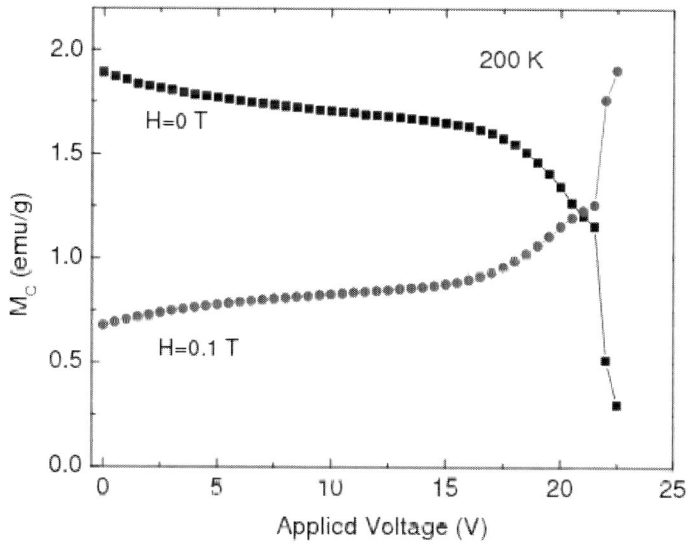

Figure 11. Magnetization as a function of applied voltage pulse in $LuFe_2O_4$. The curve with *H*=0 T was measured after filed cooling in a 5 T to 200K and removing the field. The curve with *H*=0.1 T was measured in a constant 0.1 T field after zero field cooling to 200 K.

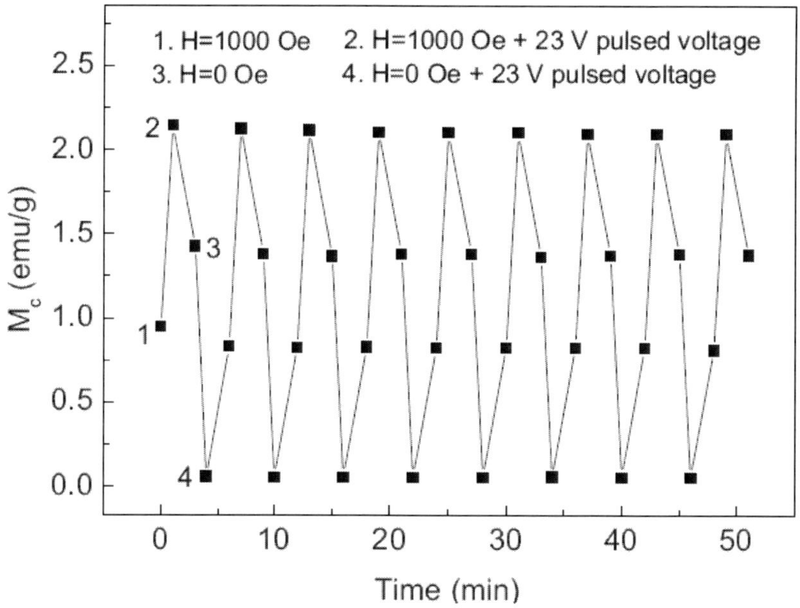

Figure 12. Reversible switch of magnetization by electric pulses in $LuFe_2O_4$.

Figure 12. shows the reversible switch of magnetic states at 200 K by employing the electrical control of magnetization in single crystal $LuFe_2O_4$. In state 1, the magnetization is set by applying a 0.1 T field; then the magnetization is switched up to a high level (state 2) by applying a voltage pulse; once the magnetic field is cutoff the magnetization drops to a remanent level (state 3); and finally, the magnetization is switched down to a very low level (state 4) by applying another voltage pulse.

Here we can see an effective electrical control of magnetization can be realized combining with a small magnetic field. The magnetization switch in $LuFe_2O_4$ only requires a low current density of 10^0 A/cm^2 or a small electric field of 10^2–10^3 V/cm, which is quite smaller than using spin-polarized/unpolarized current or electric fields, where a high current density [45] of 10^5–10^8 A/cm^2 or a high electric field [46] of 10^5–10^6 V/cm is usually required.

The electrically induced magnetization change is related to the breakdown of CO and an intrinsic spin-charge coupling in $LuFe_2O_4$. We interpreted the physical mechanism by a three-temperature model in which the temperatures of the electron, spin, and lattice are considered separately [47].

The CO breakdown triggered by external stimulus leads to an immediate rise in the temperature of the electrons, and consequently drives a temperature rise in the spin system via a strong electron–spin coupling. This temporary temperature change in the spin system during the applied current pulse would accelerate the spin relaxation, and result in a macroscopic change in magnetization. This kind of electrically assisted magnetic relaxation may be expected in other CO systems with a strong electron–spin coupling [47].

Another strong evidence for the magnetoelectric coupling in RFe_2O_4 comes from the photomagnetic effect. As shown in Figure 13, the sample was cooled in zero field from 300 to 220 K, then a 0.1 T magnetic field was applied along the *c* axis and the magnetization was measured as a function of time.

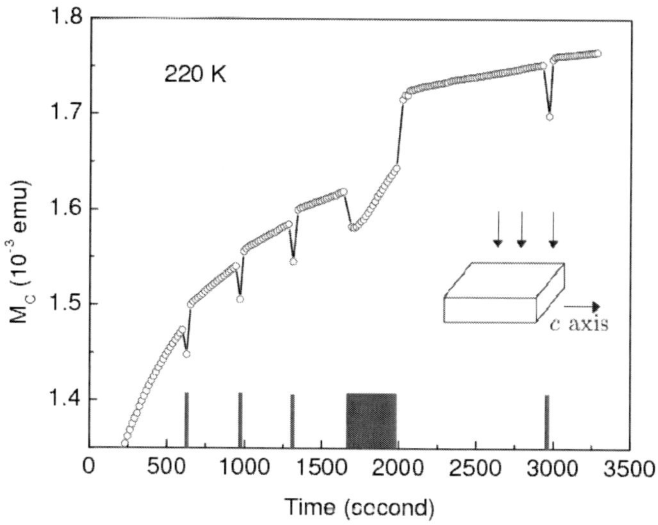

Figure 13. The photomagnetic effect in $YbFe_2O_4$.

After a time, a laser illumination was shed on the sample in the *ab* plane and the magnetization was measured simultaneously. Once a single data point was recorded, the illumination was immediately blocked and the magnetization was measured for another period.

The magnetization drops when the illumination is on, and returns when the illumination is off. We have excluded laser heating as the origin of magnetization drop [42]. The photo-induced magnetization change is an intrinsic magnetoelectric effect. Once the CO is broken by laser illumination, the magnetic state is also modified due to the spin-charge coupling.

3. Electronic Ferroelectricity and Magnetoelectric Effects in Manganites

3.1. $Pr_{1-x}Ca_xMnO_3$

Among the various mixed-valence manganites, the $R_{1-x}Ca_xMnO_3$ system is of great interest for the study of the CO and strongly correlated nature among spin, charge and lattice [48-53]. In these manganites CO typically extends over a large part of the doping phase diagram. For example, in $Pr_{1-x}Ca_xMnO_3$, CO exists for a wide doping range ($0.3<x<0.85$). For a long time, the CO in $Pr_{1-x}Ca_xMnO_3$ was thought to be the classical picture with a site-centered checkerboard pattern. However, Daoud-Aladine *et al.* [54] proposed an electronic ground state where one e_g electron is shared by two Mn ions, the so-called bond-centered Zener polaron picture which has been confirmed by recent experiments [55]. Either the site-centered or bond-centered CO does not break the inversion symmetry, and thus would not induce a spontaneous polarization. Efremov *et al.* [56] proposed that the CO configuration in $Pr_{1-x}Ca_xMnO_3$ depends on the doping level. A pure site-centered ground state with antiferromagnetic charge exchange (CE) double zigzag state sets at x=0.5 and a pure bond-centered one appears at x=0.4.

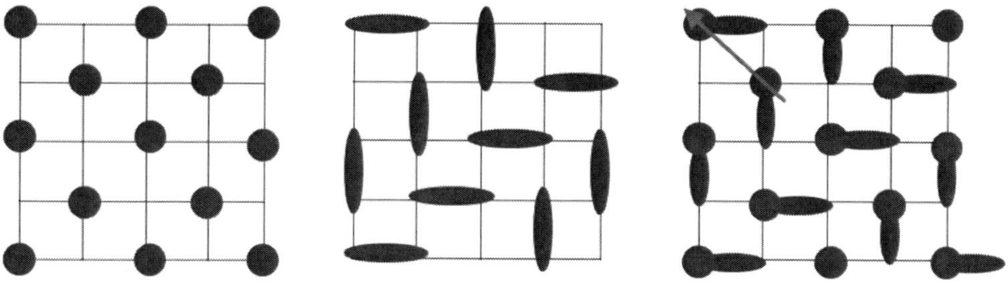

Figure 14. The charge ordering in $Pr_{1-x}Ca_xMnO_3$. Site-centered (left), bond-centered (middle), and intermediate (right). A polarization is induced in the intermediate CO configuration.

In the range of $0.4<x<0.5$, the bond-centered and the site-centered CO pictures coexist and the result breaks the inversion symmetry, leading to the appearance of a spontaneous electric polarization (Figure 14). It had been difficult to prove the predicted polarization in $Pr_{1-x}Ca_xMnO_3$ because of the relatively high conductivity of these compounds. Recently, some progress in experiments has been achieved. Jooss et al. [57] provides, by refinements of electron diffraction microscopy data, indirect evidence for canted antiferroelectricity in $Pr_{0.68}Ca_{0.32}MnO_3$.

More recently, the measurement of the electric field gradient tensor provided clues to the predicted paraelectric to ferroelectric phase transition in the charge ordered $Pr_{1-x}Ca_xMnO_3$ manganites [58].

3.2. $La_{0.5}Ca_{0.5}MnO_3$

The half-doped manganite $La_{0.5}Ca_{0.5}MnO_3$ is another prototype of CO manganites. The ground state of $La_{0.5}Ca_{0.5}MnO_3$ has been proved to be inhomogeneous, consisting typically of metallic clusters trapped in an insulating charge ordered antiferromagnetic matrix [59]. The CO configuration in $La_{0.5}Ca_{0.5}MnO_3$ is still controversial even after extensive investigations [60]. On one hand, a traditional checkerboard site-centered CO state has been proposed. On the other hand, a bond-centered CO state based on the so-called Zener polaron has been suggested. In either case, the CO configuration does not break the spacial inversion symmetry and should not induce a ferroelectricity. Recently, a theoretical work by Giovannetti *et al.* illustrated that the combination of a peculiar charge-orbital ordering and a tendency to form spin dimmers in $La_{0.5}Ca_{0.5}MnO_3$ breaks the inversion symmetry and result in polar states with relatively strong ferroelectricity [61]. Based on this theory, the CO induced ferroelectricity along with the antiferromagnetic ordering could make $La_{0.5}Ca_{0.5}MnO_3$ a possible candidate for new multiferroics with strong ME effects. In the following, we present some experimental evidences which support this theoretical prediction [62].

The dielectric constant of $La_{0.5}Ca_{0.5}MnO_3$ as a function of temperature is shown in Figure 15. A pronounced peak appears just at 200 K where the CO transition occurs. Below the peak temperature T_P, the dielectric constant shows certain dispersion behavior (frequency dependance). The dielectric constant measured with different electrodes always shows a maximum at 200 K, suggesting that the dielectric peak is an intrinsic characteristic of $La_{0.5}Ca_{0.5}MnO_3$.

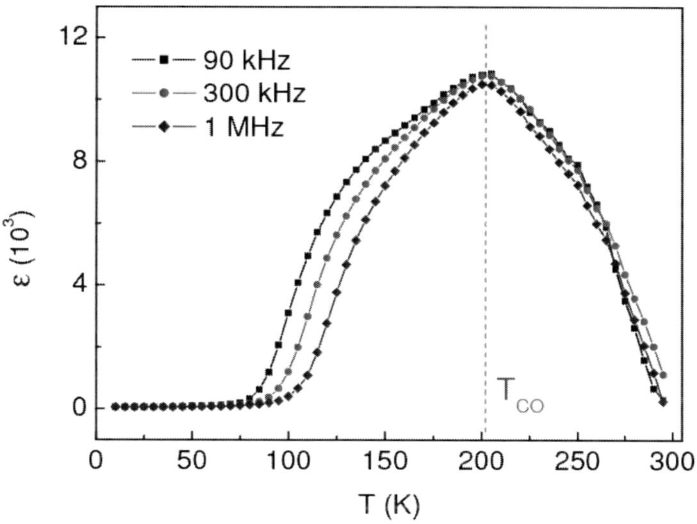

Figure 15. Temperature dependence of dielectric constant of $La_{0.5}Ca_{0.5}MnO_3$. A pronounced peak is observed at the CO temperature T_{CO}.

Since a peak in the dielectric constant was commonly observed at the ferroelectric transition, the dielectric peak at T_{CO} indicates that a polarization could be induced by the CO transition. The relaxorlike broad peak and the dispersion below T_P imply that the polar entities could be in the form of small polar clusters/domains. In addition to the dielectric peak at T_{CO}, a strong magnetodielectric effect was observed in $La_{0.5}Ca_{0.5}MnO_3$. The dielectric constant measured in a 5 T magnetic field is much higher than that in zero field (Figure 16). This giant magnetodielectric effect below T_{CO} seems to indicate a strong ME coupling in the antiferromagnetic CO state.

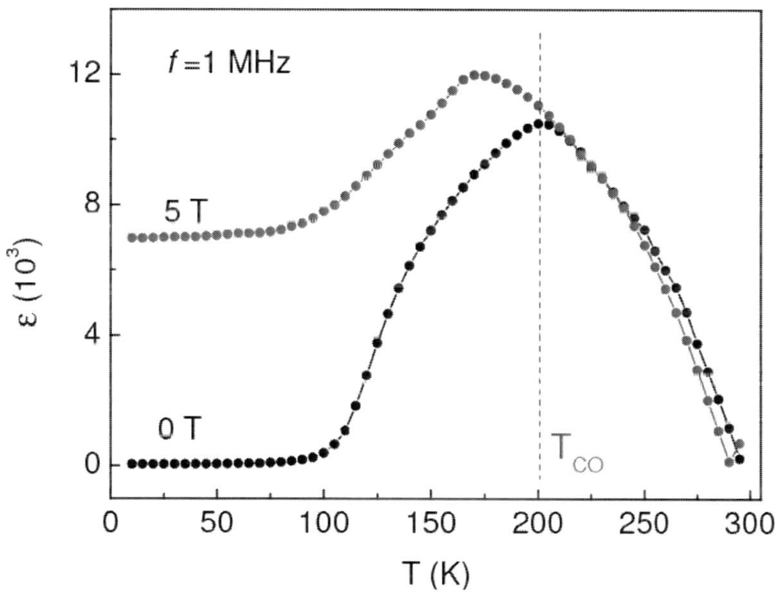

Figure 16. The magnetodielectric effect in $La_{0.5}Ca_{0.5}MnO_3$.

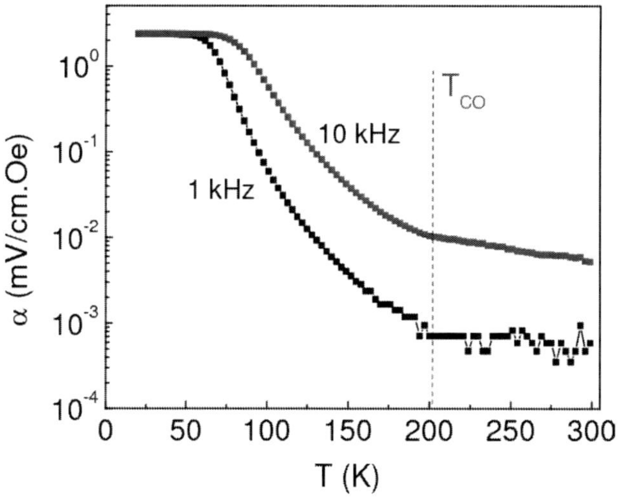

Figure 17. Temperature dependence of magnetoelectric coefficient in $La_{0.5}Ca_{0.5}MnO_3$. The coefficient increases rapidly below the CO transition temperature T_{CO}.

The measurements of ME coupling coefficient by a dynamic technique [63] provide further evidence for the CO induced multiferroics and ME effects. Figure 17. shows the temperature dependence of the longitudinal ME coefficient α in $La_{0.5}Ca_{0.5}MnO_3$. The ME coefficient is very small at high temperature, indicating a weak ME coupling in the paramagnetic state. Just below the CO transition at 200 K, the ME coefficient increases rapidly and reaches 2.38 (mV/cm.Oe) at low temperature. These results directly evidence that the ME coupling is greatly enhanced in the CO state.

CONCLUSION

We have introduced the concept of electronic ferroelectricity and reviewed recent progress in charge-order driven ferroelectricity and multiferroics in two systems: (1) strongly charge frustrated system RFe_2O_4 (R=Lu,Yb), and (2) the $(R_{1-x}Ca_x)MnO_3$ (R=La,Pr) perovskite manganites. The ferroelectricity in RFe_2O_4 has been well verified by both experiments and theory. It originates from the peculiar CO on a frustrated triangular bilayer lattice. The CO state in RFe_2O_4 can be manipulated by external stimulus, which results in a variety of interesting effects. The giant dielectric tunability and strong magnetoelectric effects (electrical control of magnetization) are promising for applications. For the CO manganites, theoretical work goes much ahead of experiments. Although some signs have been recently observed, more solid experimental evidences for electronic ferroelectricity in $(R_{1-x}Ca_x)MnO_3$ manganites are required.

Compared with conventional displacive ferroelectricity, this electronic ferroelectricity is fragile and can be easily modified by external stimulus because the charge carriers are easier to fluctuate around localized sites than anion/ion displacement. This feature makes it difficult to realize a stable high polarization at room temperature. In addition to the two systems described in this paper, other materials or possible candidates for the CO induced ferroelectricity include magnetite Fe_3O_4 [64], layered perovskite manganites $Pr(Sr_xCa_{1-}$

$_x)_2Mn_2O_7$ [65], quasi-two dimentional organic salt [66], and so on. However, the ferroelectricity in these materials is either weak or well below room temperature. In the future, more efforts are deserved in this field to find new materials with strong electronic ferroelectricity around room temperature.

REFERENCES

[1] R. E. Cohen, *Nature* 358, 136 (1992).

[2] R. D. King-Smith and D. Vanderbilt, *Phys. Rev. B* 47, 1651 (1993).

[3] R. Resta, *Rev. Mod. Phys.* 66, 899 (1994).

[4] M. Fiebig, *J. Phys. D* 38, R123 (2005).

[5] W. Eerenstein, N. D. Mathur, and J. F. Scott, *Nature* 442, 759 (2006).

[6] R. Ramesh and N. A. Spaldin, *Nature Mater.* 6, 21 (2007).

[7] J. van den Brink and D. I. Khomskii, *J. Phys.: Condens. Matter* 20, 43 (2008).

[8] T. Kimura, T. Goto, H. Shintani, K. Ishizaka, T. Arima, and Y. Tokura, *Nature* 426, 55 (2003).

[9] Y. Kitagawa, Y. Hiraoka, T. Honda, T. Ishikura, H. Nakamura, and T. Kimura, *Nature Mater.* 9, 797 (2010).

[10] N. Hur, S. Park, P. A. Sharma, J. S. Ahn, S. Guha, and S. W. Cheong, *Nature* 429, 392 (2004).

[11] B. G. Ueland, J. W. Lynn, M. Laver, Y. J. Choi, and S. W. Cheong, *Phys. Rev. Lett.* 104, 147204 (2010).

[12] T. Portengen, Th. Östreich, and L. J. Sham, *Phys. Rev. B* 54, 17452 (1996).

[13] S. Ishihara, *J. Phys. Soc. Japan* 79, 011010 (2010).

[14] T. Kimura, *Annu. Rev. Mater. Resear.* 37, 387 (2007).

[15] Y. Tokura, *J. Mag. Magn. Mater.* 310, 1145 (2007).

[16] S.-W. Cheong and M. Mostovoy, *Nature Materials* 6, 13 (2007).

[17] M. Coey, *Nature*, 430, 155 (2004).

[18] Y. Yamada, K. Kitsuda, S. Nohdo, and N. Ikeda, *Phys. Rev. B* 62, 12167 (2000).

[19] N. Ikeda, H. Ohsumi, K. Ohwada, K.Ishii, T. Inami, K. Kakurai, Y. Murakami, K. Yoshii, S. Mori, Y. Horibe, and H. Kito, *Nature* 436, 1136 (2005).

[20] A. Nagano, M. Naka, J. Nasu, and S. Ishihara, *Phys. Rev. Lett.* 99, 217202 (2007).

[21] Y. Zhang, H. X. Yang, C. Ma, H. F. Tian, and J. Q. Li, *Phys. Rev. Lett.* 98, 247602 (2007).

[22] M. Angst, R. P. Hermann, A. D. Christianson, M. D. Lumsden, C. Lee, M. H. Whangbo, J. W. Kim, P. J. Ryan, S. E. Nagler, W. Tian, R. Jin, B. C. Sales, and D. Mandrus, *Phys. Rev. Lett.* 101, 227601 (2008).

[23] A. D. Christianson, M. D. Lumsden, M. Angst, Z. Yamani, W. Tian, R. Jin, E. A. Payzant, S. E. Nagler, B. C. Sales, and D. Mandrus, *Phys. Rev. Lett.* 100, 107601 (2008).

[24] A. M. Mulders, S. M. Lawrence, U. Staub, M. Garcia-Fernandez, V. Scagnoli, C. Mazzoli, E. Pomjakushina, K. Conder, and Y. Wang, *Phys. Rev. Lett.* 103, 077602 (2009).

[25] X. S. Xu, M. Angst, T. V. Brinzari, R. P. Hermann, J. L. Musfeldt, A. D. Christianson, D. Mandrus, B. C. Sales, S. McGill, J.-W. Kim, and Z. Islam, *Phys. Rev. Lett.* 101, 227602 (2008).
[26] A. K. Tagantsev, V. O. Sherman, K. F. Astafiev, J. Venkatesh, and N.Setter, *J. Electroceram.* 11, 5 (2003).
[27] O. G. Vendik, E. K. Hollmann, A. B. Kozyrev, and A. M. Prudan, *J. Supercond.* 12, 325 (1999).
[28] K. M. Johnson, *J. Appl. Phys.* 33, 2826 (1962).
[29] A. Feteira, D. C. Sinclair, I. M. Reaney, Y. Somiya, and M. T. Lanagan, *J. Am. Ceram. Soc.* 87, 1082 (2004).
[30] C.-H. Li, X.-Q. Zhang, Z.-H. Cheng, and Y. Sun, *Appl. Phys. Lett.* 92, 182903 (2008).
[31] Y. Liu, C.-H. Li, X.-Q.Zhang, Z.-H. Cheng, and Y. Sun, *J. Appl.Phys.* 104, 104112 (2008).
[32] Y. Liu, T. Zou, F. Wang, C.-H. Li, L. Zheng, and Y. Sun, *Physica B* 406, 1263 (2011).
[33] A. Asamitsu, Y. Tomioka, H. Kuwahara, and Y. Tokura, *Nature* (London) 388, 50 (1997).
[34] S. Lee, A. Fursina, J. T. Mayo, C. T. Yavuz, V. L. Colvin, R. G. S. Sofin,I. V. Shvests, and D. Natelson, *Nature Mater.* 7, 130 (2008).
[35] V. Kiriukhin, D. Casa, J. P. Hill, B. Keimer, A. Vigliante, Y. Tomioka, and Y. Tokura, *Nature* 386, 813 (1997).
[36] K. Miyano, T. Tanaka, Y. Tomioka, and Y. Tokura, *Phys. Rev. Lett.* 78, 4257 (1997).
[37] M. Matsubara, Y. Okimoto, T. Ogasawara, S. Iwai, Y. Tomioka, H. Okamoto, and Y. Tokura, *Phys. Rev. B* 77, 09441013 (2008).
[38] E. Dagotto, T. Hotta, and A. Moreo, *Phys. Rep.* 344, 1 (2001).
[39] C. Li, X. Zhang, Z. Cheng, and Y. Sun, *Appl. Phys. Lett.*93, 152103 (2008).
[40] L. Zeng, H. Yang, Y. Zhang, H. Tian, C. Ma, Y. Qin, Y. Zhao, and J. Li, *Eur. Phys. Lett.* 84, 57011 (2008).
[41] X. Shen, C. H. Xu, C. H. Li, Y. Zhang, Q. Zhao, H. X. Yang, Y. Sun, J. Q. Li, C. Q. Jin, and R. C. Yu, *Appl. Phys. Lett.* 96, 102909 (2010).
[42] C.-H. Li, Y. Liu , F. Wang, X. Luo, Y.-P. Sun, X.-Q. Zhang, Z.-H. Cheng, and Y. Sun, *Chin. Phys. Lett.* 26, 127501 (2009).
[43] M. Naka, A. Nagano, and S. Ishihara, Phys. Rev. B 77, 224441 (2008).
[44] C.-H. Li, F. Wang, Y. Liu, X.-Q. Zhang, Z.-H. Cheng, and Y. Sun, *Phys. Rev B* 79, 172412 (2009).
[45] M. Yamanouchi, D. Chiba, F. Matsukura, and H. Ohno, *Nature* 428, 539 (2004).
[46] T. Lottermoser, T. Lonkai, U. Amann, D. Hohlwein, J. Ihringer, and M. Fiebig, *Nature* 430, 541 (2004).
[47] F. Wang, C.-H. Li, T. Zou, Y. Liu, and Y. Sun, *J. Phys.: Condens. Matter* 22,496001 (2010).
[48] P. G. Radaelli, D. E. Cox, M. Marezio, and S-W. Cheong, *Phys.Rev.B* 55, 3015 (1997).
[49] S. Moris, C. H. Chen, and S.-W. Cheong, *Nature* 392, 473 (1998).
[50] J. M. D. Coey, M. Viret, and S. von Molnar, *Advan. in Phys.* 48, 167 (1999).
[51] E. L. Nagaev, *Phys. Rep.* 346, 387 (2001).
[52] James C. Loudon, Neil D. Mathur, and Paul A. Midgley, *Nature* 420, 797 (2002).
[53] G. C. Milward, M. J. Calderon, and P. B. Littlewood, *Nature* 433, 607 (2005).

[54] A. Daoud-Aladine, J. Rodríguez-Carvajal, L. Pinsard-Gaudart, M. T. Fernández-Díaz, and A. Revcolevschi, *Phys. Rev. Lett.* 89, 097205 (2002).
[55] L. Wu, R. F. Klie, Y. Zhu, and Ch. Jooss, *Phys. Rev. B* 76, 174210 (2007).
[56] D. V. Efremov, J. van den Brink, and D. I. Khomskii, *Nature Mater.* 3, 853 (2004).
[57] Ch. Jooss, L.Wu, T. Beetz, R. F. Klie, M. Beleggia, M. A. Schofield, S. Schramm, J. Hoffmann, and Y. Zhu, *PNAS* 104, 13597 (2007).
[58] A. M. L. Lopes, J. P. Araujo, V. S. Amaral, J. G. Correia, Y. Tomioka, and Y. Tokura, *Phys. Rev. Lett.* 100, 155702 (2008).
[59] M. Pissas and G. Kallias, *Phys. Rev. B* 68, 134414 (2003).
[60] C. H. Patterson, *Phys. Rev. B* 72, 085125 (2005).
[61] G. Giovannetti, S. Kumar, J. van den Brink, and S. Picozzi, *Phys. Rev. Lett.* 103, 037601 (2009).
[62] Tao Zou, Fen Wang, Yi Liu, Li-Qin Yan, and Young Sun, *Appl. Phys. Lett.* 97, 092501 (2010).
[63] M. Mahesh, Kumar, A. Srinivas, S. V. Suryanarayana, G. S. Kumar, and T. Bhimasankaram, *Bull. Mater. Sci.* 21, 251 (1998).
[64] K. Yamauchi, T. Fukushima, and S. Picozzi, *Phys. Rev. B* 79, 212404 (2009).
[65] Y. Tokunaga, T. Lottermoser, Y. Lee, R. Kumai, M. Uchida, T. Arima, and Y. Tokura, *Nature Mater.* 5, 937 (2006).
[66] S. Iwai, K. Yamamoto, A. Kashiwazaki, F. Hiramatsu, H. Nakaya, Y. Kawakami, K. Yakushi, H. Okamoto, H. Mori, and Y. Nishio, *Phys. Rev. Lett.* 98, 097402 (2007).

In: Ferroelectrics: New Research
Editors: Xui Li Huang and Su Lyn Ma

ISBN: 978-1-61942-282-7

Chapter 11

COPLANAR WAVEGUIDE FERROELECTRIC THIN FILM MICROWAVE PHASE SHIFTERS

Su Sheng* and C. K. Ong
Centre for Superconducting and Magnetic Materials,
Department of Physics, National University of Singapore,
Singapore

ABSTRACT

Ferroelectric thin film, such as barium strontium titanate ($Ba_xSr_{1-x}TiO_3$ or BST), has shown great promise for the fabrication of tunable RF and microwave components. An important tunable microwave device using ferroelectric film is a wideband phase shifter which is the most important component of the phased array antenna. Phase shifters with low insertion loss, low drive power, proper power handling capability, continuous tunability, good impedance match, fast response speed, and low production cost are the key to the development of lightweight phased array antennas. This chapter looks at microwave phase shifter applications of ferroelectric thin films. It starts with the introduction of microwave phase shifter and coplanar waveguide transmission lines, with emphasis laid on the significance and applications of ferroelectric materials. The basic theory of distributed transmission line phase shifters is then discussed. The physical layout design, device fabrication, and microwave characterization of distributed coplanar waveguide transmission line phase shifter using ferroelectric thin film varactors with parallel-plate electrodes are presented in the final two sections.

1. INTRODUCTION

Modern radio frequency (RF) and microwave engineering is an exciting and dynamic field, due in large part to the explosive growth of commercial wireless markets and the symbiosis between recent advances in modern electronic device technology. In recent years, there is a rapidly growing demand for electrically tunable RF and microwave devices in

---

* E-mail: ssheng@nus.edu.sg.

advanced radar and mobile communication systems. Ferroelectric thin film has shown great promise for the fabrication of tunable RF and microwave components, such as tunable microwave resonators, filters, and phase shifters [1-3]. The $Ba_xSr_{1-x}TiO_3$ ($0 < x < 1$) (BST) thin films are currently considered to be important materials for tunable microwave devices because BST thin films have high dielectric constants, high tunability, low dielectric loss tangent and low leakage current; most importantly, it can be integrated on a traditional Si substrate to replace the current silicon oxide and nitride dielectrics, which would have considerable commercial impact. Electronically tunable microwave devices that use integrated monolithic phase shifters for digital–analog beam-forming are required in several telecommunications and defense applications. For example, they play a very important role in phased-array antennas. Changing the phase of each array element allows the main antenna beam to be steered. Critical phase-shifter parameters such as phase agility, insertion loss, switching time, power-handling capacity, and temperature stability seriously depend on its material technology and the corresponding fabrication techniques. In most cases, phase shifters require inductors consuming lots of circuit area. However, small circuits are favorable concerning costs. The control complexity should be as low as possible. Moreover, the amplitude variations versus phase control should be small. All those constraints make the design of integrated phase shifters to a challenging task. In modern information and communication techniques, planar integrated microwave circuits play an important role. The planarization of RF interconnections and wires using microstrip lines, strip lines and coplanar waveguide (CPW) provides compatibility with ferroelectric thin films and with semiconductors with design flexibility, compactness and reduction of weight with superior performance. A CPW type phase shifter based on ferroelectric varactors [4] has advantages over other type phase shifters because of both simple structure and low production cost due to its simple device fabrication process.

In the following, we start with the introduction of microwave phase shifter and coplanar waveguide transmission lines, with emphasis laid on the significance and applications of ferroelectric materials. The basic theory of distributed transmission line phase shifters is then discussed. Finally, the physical layout design, device fabrication, and microwave characterization of distributed coplanar waveguide transmission line phase shifter using ferroelectric thin film varactors with parallel-plate electrodes are presented.

2. Microwave Phase Shifters

A microwave phase shifter is a two-port device capable of producing a true delay of a microwave signal flowing through it. The produced time delay can be a constant or a variable for a device. And the phase shifter is accordingly called fixed phase shifter or tunable phase shifter. Electronically tunable phase shifters are key elements in phased array antennas and microwave communication systems for improving wireless communications, surveillance, sensing and tracking. Current wireless and information technology require continual upgrade on the phase shifters which have compact size, low fabrication cost, as well as large phase shifts, low insertion loss with broadband operating frequencies, and the ability to handle wide ranges of power levels depending on their applications. Especially, demand of phase shifters which operate above 2 GHz up to 12 GHz is high for wireless communication systems.

The insertion phase of a phase shifter is the phase delay experienced by a radiofrequency (RF) signal propagating between port 1 and port 2 and is the angle of S_{21}, the transmission coefficient from port 1 to port 2. If the angle of S_{21} equals the angle of S_{12}, the transmission coefficient from port 2 to port 1, the phase shifter is reciprocal, while if these two angles are not equal the device is non-reciprocal. Phase shifters can be analog or digital. Analog phase shifters are devices whose phase shift changes continuously with the control input and therefore offer almost unlimited resolution with monotonic performance. There are basically two general methods of producing phase shifters: mechanical shifting and electronic shifting. Larger mechanical devices are expensive and can't take advantage of the economies of scale needed to make affordable phase shifters; therefore, the focus of this discussion will be on electronically phase shifters. Electronically phase shifters have been fabricated using various materials of ferrite [5-7], semiconductor [8, 9], and ferroelectric [10-15].

2.1. Ferrite Phase Shifters

Ferrite phase shifters have been widely used for phased array antennas due to their lightweight and smaller size. The ferrite phase shifter relies on the interaction between the electromagnetic waves and the spinning electrons in a magnetized ferrite. The spinning procession varies with the applied magnetic field, which changes the permeability of the ferrite. Therefore, by electronically controlling the applied magnetic field, it is possible to change the propagation constant of the electromagnetic waves, resulting in a change in their phase. These ferrite phase shifters can be realized in various designs such as waveguides, coaxial lines, striplines and microstriplines operating either in analog or digital mode. Traditional ferrite phase shifters of millimeter and sub-millimeter wave bands are based on Faraday rotation for electromagnetic waves in axially magnetized ferrite rods in waveguides. A nonreciprocal phase shifter that has different phase shifts in opposite directions of propagation in a waveguide may be obtained with one ferrite slab or with two slabs symmetrically placed, with respect to the center of the guide, but oppositely magnetized. An adaptation of the phase shifter that provides reciprocal phase shift is referred to as the "dual mode" device and has been successfully employed in several antenna designs. These ferrite phase shifters provide a variable insertion phase by varying the magnitude of the bias magnetic field, resulting in a variation in the equivalent inductance of the waveguide and yielding a variable propagation delay through the device. Another type of ferrite phase shifter does not use this phenomenon but rather makes use of the variation in the direction of the bias magnetic field to effect change in insertion phase with no change in the propagation delay. Because of the similarity of this phase shifter to the rotary-vane mechanical phase shifter, it has been called the rotary field phase shifter.

Phase shifters using ferrite materials have demonstrated large phase shifts, low insertion loss, and high power. However, ferrite phase shifter devices require an electromagnet or solenoidal magnetic field for EM control since a bias magnetic field H is necessary for the operation and frequency tuning of these devices. This requires high electrical power for operation and involves large power dissipation, and it is bulky. Moreover, such a magnetic tuning is inherently slow, noisy and cannot be used in applications where rapid beam scanning is required. Just as critical, it should be noted this makes the device miniaturization difficult and incompatible with integrated circuit technology.

2.2. Semiconductor Phase Shifters

Semiconductor phase shifters, based primarily on GaAs, but also on SiGe and InP, have enjoyed steady progress for the past three decades. Their small size and relatively low power consumption compared to ferrite devices has created new insertion opportunities. The most commonly used semiconductor control elements in analog phase shifters are varactor diodes. Varactor diodes operating in a reverse-biased condition provide a junction capacitance that varies with applied voltage and can be used as an "electrically variable" capacitor in tuned circuits. The variable capacitance of a varactor diode is a nonlinear function of voltage; therefore, the drive voltage is too distorted for a linear relationship between dc voltage and phase. The varactor capacitance changes rapidly (in less than a nanosecond) with voltage around zero volt bias; therefore, the continuous phase shifters generate harmonics and increase losses for higher-incident power levels at a low bias voltages. The dc voltage has to be larger than the RF voltage to control the phase shift. Semiconductor phase shifters can usually be categorized as PIN diode phase shifters and field effect transistor (FET) phase shifters depending on which electronic control element is used as an electronic switch. Since semiconductor phase shifters with PIN diodes or FET circuits are very expensive and have significant loss at microwave frequencies, there is a need for low loss, cost-effective components for phase array antennas. The semiconductor devices and their circuitry are generally faster switching and inherently reciprocal, a useful antenna property. Semiconductor phase shifters are smaller than ferrites but their application is limited because of the high insertion loss at high frequencies and poor power-handling capability.

2.3. Ferroelectric Phase Shifters

Phase shifters using voltage controllable ferroelectric thin films can provide a continuous phase shift due to change in phase velocity of the RF signal by applying a bias voltage. Compared with semiconductor and ferrite phase shifters, the ferroelectric phase shifters are characterized by high speed, high reliability, radiation resistance, completely integrated technology, and a high level of permissible microwave power. The high tunability of BST, desirable to achieve a large phase shift with a physically relatively small device coupled with relatively low loss at microwave frequencies, fuels the further development of this promising ferroelectric material. Phase shifters using BST thin films have been developed by several groups. Some groups have investigated a transmission-type phase shifter making use of ferroelectric material, which forms either an entire or a fraction of a substrate so that the phase velocity can be controlled by changing capacitance [12, 16-19]. Also, a reflection-type phase shifter, which is composed of a Lange coupler and BST varactors, has been reported [20, 21].

Phase shifters using voltage controllable thin films can be fabricated into microstrip or coplanar waveguide (CPW) configurations. Phase shifters using microstrip transmission lines can be fabricated on bulk ceramic or ferroelectric substrates. Monolithic integration of these phase shifters are desirable because of reliability and cost issues. However, microstrip phase shifters using non-semiconducting substrates have limitations in monolithic integration with another passive circuit and active circuits. Hence, coplanar waveguide phase shifters on semiconductor substrate have been investigated intensively, based on Schottky contact,

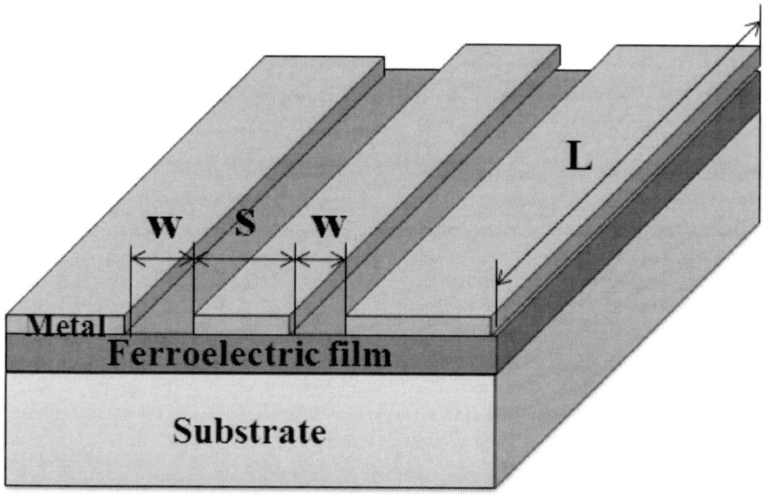

Figure 1. Schematic of a coplanar waveguide (CPW) on ferroelectric thin film.

ferroelectric material or varactor. Among them, a coplanar waveguide (CPW) transmission line phase shifter based on ferroelectric varactors [4] has advantages over other type phase shifters because of both simple structure and low production cost due to its simple device fabrication process. In recent years, a lot of interests were focused on distributed analog phase shifter circuits [15, 22-23], where the wave propagates through periodic structures comprising of lumped capacitive elements. Although these capacitances can in fact be with either parallel-plate or planar-type capacitors, in comparison with the coplanar design [24-26], the devices with parallel-plate electrodes [11, 27] offer lower control voltage and higher tunability.

However, because of the high dielectric permittivity and the small thickness, the parallel-plate varactors have large capacitance which limits their high frequency applications; also in some ferroelectric CPW phase shifters, parallel plate varactors involved an elaborate process and required bottom electrode patterning. Patterning bottom electrode typically limits the minimum bottom electrode dimension and requires stringent lithography facility.

Moreover, most of the distributed circuit designs need very small capacitance in order to attempt an impedance match. So, in this chapter, a distributed phase shifter circuit with two series connection parallel plate varactors, which can increase the dimension of varactor and simultaneously maintain the small capacitance, will be presented.

3. Coplanar Waveguide Transmission Lines

A coplanar waveguide (CPW) fabricated on a dielectric substrate was first demonstrated by C. P. Wen in 1969 as a novel surface strip transmission line [28]. Since that time, tremendous progress has been made in CPW based microwave integrated circuits (MICs) as well as monolithic microwave integrated circuits (MMICs). A conventional CPW transmission line on a dielectric substrate consists of a center strip conductor with semi-infinite ground planes on two sides. The CPW structure here consists of three layers from bottom to top: the substrate, the ferroelectric thin film and the conductors formed a center strip separated by a narrow gap from a pair of ground planes on either side, as shown in

Figure 1. As the strip conductor is deposited on the same surface as the two ground planes in close proximity, the CPW transmission line has demonstrated several advantageous features over the conventional microstrip line, for instance, easy surface mounting of external devices, easy fabrication of both shunt and series passive elements, low-frequency dispersion, and easy adjustment of desirable characteristic impedance. Since its proposal, the CPW technology has been progressively gaining a tremendous application in exploration of advanced RF and microwave integrated circuits, modules, and subsystems. The dimensions of the center strip, the gap, the thickness and permittivity of the dielectric substrate determined the effective dielectric constant (ε_{eff}), characteristic impedance (Z_0), and the attenuation (α) of the transmission line. So far, extensive work in theory has been done to characterize a variety of CPW transmission lines in terms of per unit length characteristic impedance and propagation constants by developing the quasi-static and full-wave analysis approaches [29, 30]. The characteristic impedance limits are obtained by combining maximum strip width with minimum slot space for the lower limit, and minimum strip width with maximum slot space for the upper limit. CPW is not very sensitive to substrate thickness and allows a wide range of impedance values (20–250 Ω) on relatively thick substrates.

The coplanar waveguide, in contrast to the microstrip and the stripline, has shielding between adjacent lines, which creates a better isolation between them.

The CPW transmission line has basically two dominant modes of propagation: quasi-TEM (called the even mode or coplanar waveguide mode) and non-TEM mode (called the odd mode or slotline mode). In CPW microwave circuits, the quasi-TEM mode is the most desired due to its low radiation properties. In this way, the excitation of the unwanted slotline mode can be avoided by either maintaining the symmetry of the structure or using air-bridges as bond wires to connect the ground planes of the CPW.

Also, the cross-sectional dimensions, such as slot and strip widths, are readily selected electrically short in the frequency region of operation, so as to minimize the longitudinal magnetic field component.

4. Theory of Distributed Transmission Line Phase Shifters

The equivalent circuit of distributed ferroelectric phase shifter is shown in Figure 2. The phase can be controlled by tuning of the transmission line. According to ref. 4, one can define a unit cell for this periodic structure, which consists of a section of transmission line of length (L_{sect}) and a shunt varactor (C_{var}) to ground. The transmission line section can be approximated as a lumped inductance (L_t) and capacitance (C_t).

The shunt varactor is added periodically as discrete elements to the transmission line. This varactor loading makes the distributed phase shifter a periodic structure. A result of creating a periodic structure is the existence of a cutoff frequency or Bragg frequency near the point where the guided wavelength approaches the periodic spacing of the discrete components. The Bragg frequency is given by [31]:

$$f_{Bragg} = \frac{1}{\pi\sqrt{L_t(C_t+C_{var})}} \tag{1}$$

$$L_t = \frac{L_{sect}}{v_i} Z_i \ , C_t = \frac{L_{sect}}{Z_i v_i} \qquad (2)$$

where L_t and C_t are the inductance and capacitance per unit cell and Z_i and v_i are the impedance and velocity on the high impedance line, respectively.

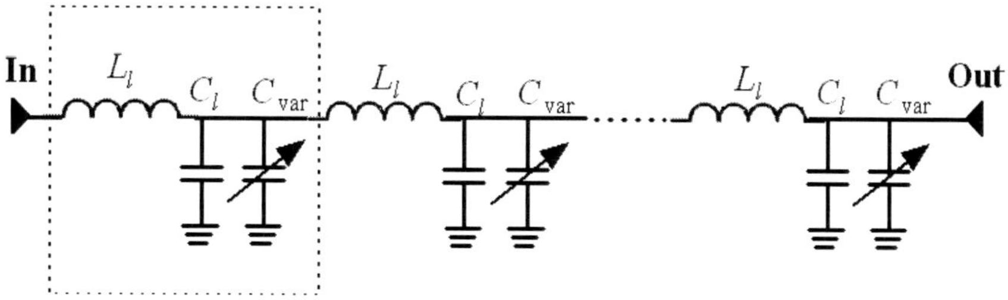

Figure 2. Equivalent circuit of distributed ferroelectric phase shifter.

When frequencies are well below the Bragg frequency, the distributed transmission line phase shifter may be treated as a synthetic transmission line. The capacitance per unit length for the synthetic transmission line has been increased due to the periodic loading.

The inductance per unit length of the synthetic transmission line is unchanged from that of the original unloaded line. Since the loading varactors are voltage dependent, the properties of this synthetic transmission line, such as its characteristic impedance and phase velocity, are voltage dependent. Therefore, it is possible to control the phase shift for a given length of transmission line by changing the bias voltage. The phase velocity, the characteristic impedance (Z_l), and the insertion phase (φ) of this transmission line are given by

$$v_{phase} = 1/\sqrt{L_l(C_l + C_{var}/L_{sect})} \qquad (3)$$

$$Z_L = \sqrt{L_l/(C_l + C_{var}/L_{sect})} \qquad (4)$$

$$\varphi = 2n\pi f L_{sect}\sqrt{L_l(C_l + C_{var}/L_{sect})} \qquad (5)$$

where $L_l - Z_i/v_i$ and $C_l - 1/(Z_i v_i)$ are the inductance and capacitance per unit length of the CPW transmission line, C'_{var} is the capacitance of the loaded varactor, L_{sect} is length of a periodical cell, and n is number of sections. The phase shift of each section of the distributed phase shifter varies as capacitance of varactor is tuned by dc bias. At any given frequency f, the maximum differential phase shift of each section is given by

$$\Delta\varphi = 2n\pi f L_{sect}\sqrt{L_l}\left(\sqrt{C_l + C_{var}^{max}/L_{sect}} - \sqrt{C_l + C_{var}^{min}/L_{sect}}\right) \qquad (6)$$

The terms C_{var}^{max} and C_{var}^{min} denote the extremes of the values C_{var} can assume with tuning. The equation (6) indicates that changing the capacitance ration ($C_{var}^{max}/C_{var}^{min}$) results in more differential phase shift.

5. Ferroelectric Thin Film Microwave Phase Shifters

5.1. Design and Implementation

The layout of distributed transmission line phase shifter of coplanar waveguide using ferroelectric thin film varactors with parallel plate electrodes based on 6 unit cells is shown in Figure 3 (a). Coplanar line is deposited on top of ferroelectric film. The shunt capacitor is added periodically as discrete elements to the transmission line. Tuning the periodic loading reactance alters the phase velocity of the transmission line, resulting in a phase shift.

For the optimization of the distributed transmission line, such as for better impedance match, the structure of phase shifter was designed by a high-performance full-wave electromagnetic field simulator (HFSS). The primary parameters used to simulate phase shifter were adopted shown in Table 1. The capacitance of shunt varactor used in simulation was variable from 0.07 pF to 0.4 pF.

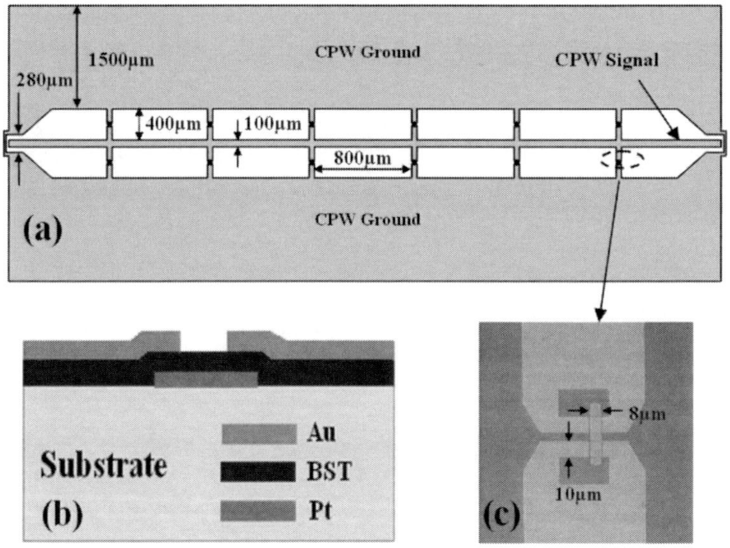

Figure 3. (a) Layout of the distributed transmission line phase shifter of coplanar waveguide. (b) The side view of a parallel plate varactor. (c) The photograph of top view of a varactor.

Table 1. The primary parameters used to simulate phase shifter

Parameter	Value
The dimension of Si substrate	$5 \times 10 \times 0.5$ mm^3
The number of unit cells	6
The dielectric constant of substrate	12
The width of center conductor	100 μm
The spacing between two varactors	800 μm
The gap dimension of transmission line	400 μm

For different capacitance of shunt varactor used in simulation, different characteristic impedance of the phase shifter circuit was obtained by HFSS. Figure 4 shows the matching impedance of phase shifter circuit as a function of capacitance of the shunt varactor. The capacitance of the shunt varactors used in the phase shifter circuit was 0.24 pF for obtaining characteristic impedance closed to 50 ohm.

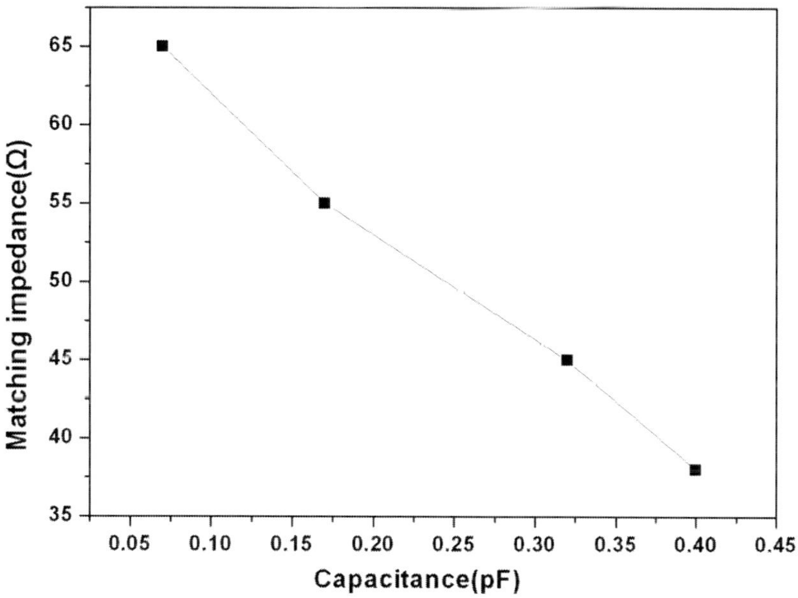

Figure 4. The matching impedance of phase shifter circuit as a function of capacitance of shunt varactor.

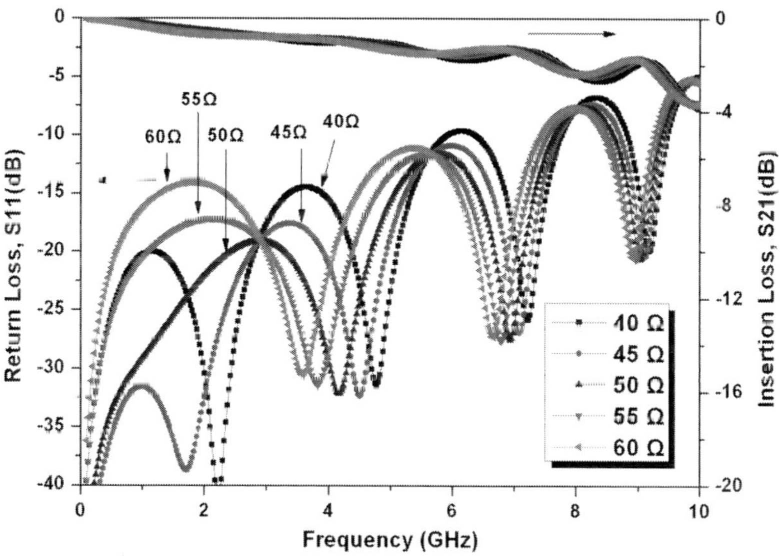

Figure 5. The simulation results of the phase shifter under different full port impedance.

For the capacitance of shunt varactor of 0.24 pF, the scattering matrixes of the simulation of distributed phase shifter were shown in Figure 5 under different full port impedance. When the full port impedance is 50 ohm, the return loss is smallest, which indicates that the characteristic impedance of the phase shifter circuit with the capacitance of shunt varactor of 0.24 pF is closer to 50 ohm. Because of two series connection parallel plate varactors used in the circuit, the capacitance of ferroelectric thin film parallel-plate varactor designed to be about 0.48 pF. $Ba_{0.25}Sr_{0.75}TiO_3$ (BST) thin film of thickness ~300 nm was grown using pulsed laser deposition (PLD) method on pre-patterned high-resistivity silicon substrate (5mm×10mm) that had ion beam etched Pt metal as the bottom electrode of parallel-plate varactor. The BST thin film was grown using a KrF excimer laser operating at 3 Hz and 250 mJ. The substrate temperature and oxygen pressure were maintained at 700 ˚C and 0.2 mbar, respectively. After deposition, the sample was annealed for half an hour at about 650 ˚C in 1×10^3 mbar oxygen pressure. In order to decrease the dielectric loss at high frequency, a layer of 2-μm-thick Au was electroplated on the prepared seed layers of Au/Cr by RF magnetron sputtering.

The schematic side view of a parallel-plate varactor is shown in Figure 3 (b) and the photograph of top view of a varactor is also shown in Figure 3 (c).

For the highest operating frequency of the phase shifter being well below cutoff frequency and also impedance match of the circuit, the dimensions of the conductor layer were selected appropriately. Because larger capacitance of the varactors is favorable for both the purpose of high tunability and easy fabrication, the impedance of unloaded CPW should be kept as high as possible to achieve impedance match while varactors are loaded.

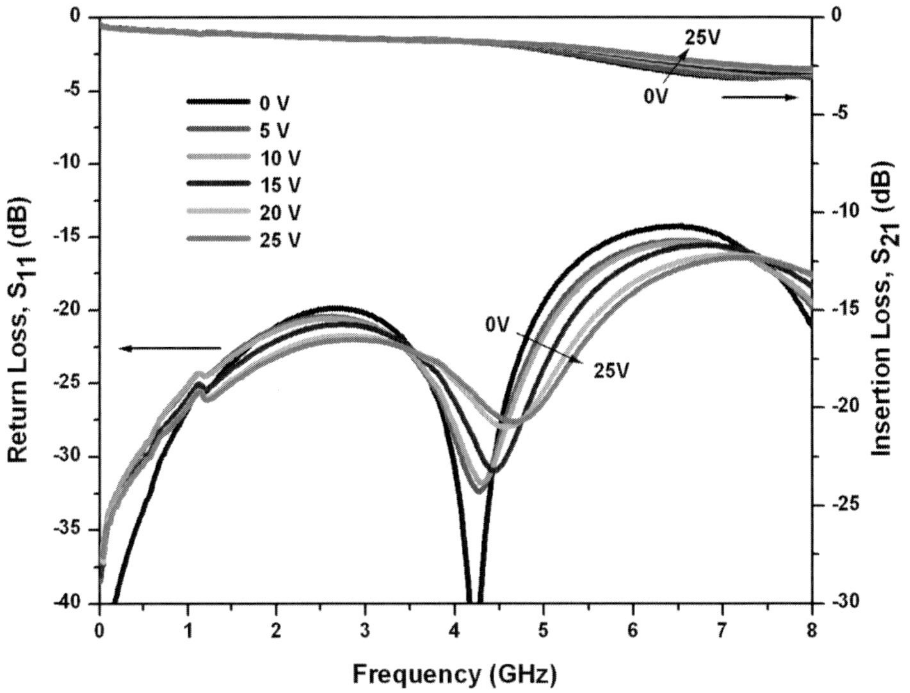

Figure 6. Frequency dependent insertion loss and return loss of the distributed ferroelectric phase shifter based on 6 unit cells at different bias voltages.

Based on the simulation results, the CPW transmission line with gap dimension of 400 μm and center conductor width of 100μm periodically loaded with ferroelectric thin film parallel-plate varactors (8μm × 10μm) with spacing $L_{sect} = 800$ μm. The phase shifter characteristics are controlled mainly by the properties of the varactor according to the equation (6).

5.2. Experimental Results and Discussion

The scattering matrix of the CPW phase shifter shown in Figure 6 was measured by PNA Series Network Analyzer (Agilent N5230A) with external bias tees. Off wafer short-open-load-thru (SLOT) calibration was performed setting the reference planes at the probe tips. All measurements were taken from 10 MHz to 8 GHz at steps of 10 MHz. Pasternack PE1611 bias tee and Keithley 237 voltage source were used to supply dc bias voltage through signal probe and the copper pad.

A periodic resonance could be observed in the frequency dependent return loss and is due to the coupling of the inductance from the center signal transmission line and capacitance of ferroelectric thin film varactor. When the dc bias is applied, the dielectric constant of the ferroelectric film is decreased due to the soft mode hardening, and therefore the resonance moves to a higher frequency. The insertion loss is decreased as the applied dc bias is increased because the dielectric loss in the ferroelectric thin film is reduced. The return loss was better than 16 dB over all phase states. The maximum insertion loss was 3.05 dB at 6.5 GHz with no dc bias. The differential phase shift with respect to the zero bias is plotted for 5 different bias voltages in Figure 7.

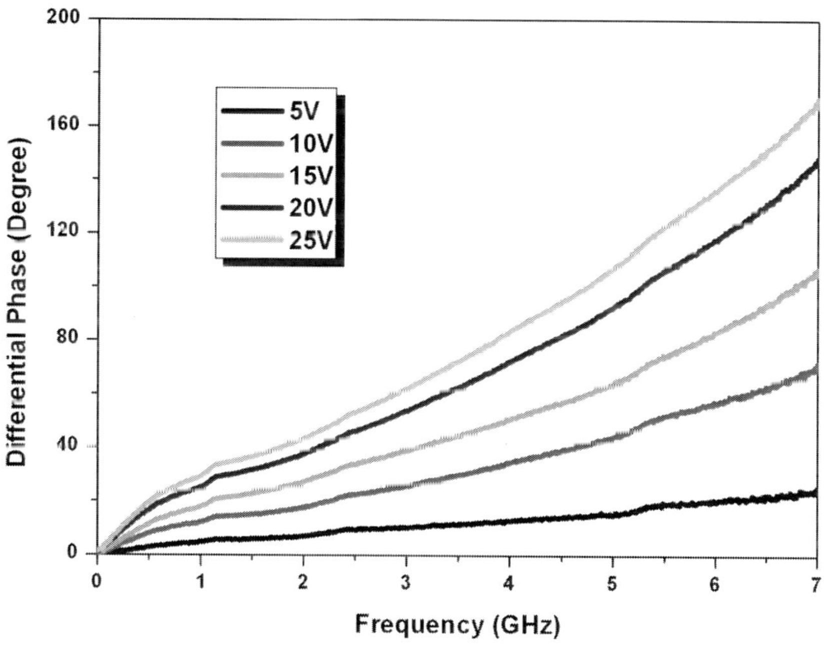

Figure 7. Differential phase shift of the phase shifter as a function of frequency at different bias voltages.

The circuit of phase shifter was capable of a continuous 0-170° phase shift with an insertion loss of only 2.3 dB at 7 GHz. The figure of merit F of the phase shifter, defined as the maximum differential phase shift divided by the insertion loss averaged across the full voltage range, is rather high, $F \approx 78\,°/dB$. These values are significant since large differential phase shift are achieved with such a low bias voltage of 25 V.

A high performance microwave phase shifter with a 360° phase shift at a low bias voltage can be produced with more unit cells. The insertion loss for these circuits would be decreased by using thicker bottom electrodes.

CONCLUSION

In this chapter, the physical layout design, device fabrication, and microwave characterization of distributed CPW transmission line phase shifter using ferroelectric thin film varactors with parallel-plate electrodes are presented. The equivalent circuit model was simulated by the EM simulation software in an effort to optimize the design. The fabricated phase shifter exhibited differential phase shift of 170° at 7 GHz with a low bias voltage of 25 V. The maximum insertion loss was 3.05 dB at 6.5 GHz with no dc bias and the return loss was better than 16 dB over all phase states. These results provide a simple method for obtaining excellent microwave properties of CPW phase shifter using ferroelectric thin film varactors. The high performance microwave phase shifters using ferroelectric films with a 360° phase shift are the most critical component of modern electronically phased array antenna systems for mobile communications services.

REFERENCES

[1] G. Subruamanyam, F. V. Keuls, and F. A. Miranda, "A K-band tunable microstrip bandpass filter using a thin film conductor/ferroelectric/dielectric multilayer configurations", *IEEE Microw. Guided Wave Lett.* 8, 78 (1998).

[2] B. Acikel, T. R. Taylor, P. J. Hansen, J. S. Speck, and R. A. York, "A new high performance phase shifter using BaSrTiO thin films", *IEEE Microw. Wireless Compon. Lett.* 12, 237 (2002).

[3] G. Velu, K. Blary, L. Burgnies, A. Marteau, G. Houzet, D. Lippens, and J. C. Carru, "A 360° BST phase shifter with moderate bias voltage at 30 GHz", *IEEE Trans. Microw. Theory Tech.* 55, 438 (2007).

[4] A. S. Nagra and R. A. York, "Distributed analog phase shifters with low insertion loss", *IEEE Trans. Microwave Theory Tech.* 47, 1705 (1999).

[5] H. How, H. Hu, C. Vittoria, L. C. Kempel, and K. D. Trot, "Single crystal yttrium iron garnet phase shifter at X band", *J. Appl. Phys.*, 85, 4853 (1999).

[6] I. Viswanathan, S. D. Yoon, T. Sakai, A. L. Geiler, J. W. Wang, C. N. Chinnasamy, C. Vittoria, and V. G. Harris, "High Performance Compact Microstripline Phase Shifter at C-Band Using Yttrium Iron Garnet", *IEEE Transactions on Magnetics*, 45 (10), 4176 (2009).

[7] J. D. Adam, L. E. Davis, G. F. Dionne, E. F. Schloemann, and S. N. Stitlzer, "Ferrite devices and materials", *IEEE Trans. Microwave Theory Tech.*, 50(3), 721 (2002).

[8] C. Tsironis and P. Harrop, "Dual gate GaAs MESFET phase shifter with gain at 12 GHz", *Electron. Lett.*, 16, 553(1980).

[9] M. Kumar, R. J. Menna, and H. Huang, "Broadband active phase shifter using dual-gate MESFET", *IEEE Trans. Microw. Theory Tech.*, MTT-29, 1098 (1981).

[10] S. Sheng, Peng Wang, Xin Chen, Xiao-Yu Zhang, and C. K. Ong, "Two paralleled $Ba_{0.25}Sr_{0.75}TiO_3$ ferroelectric varactors series connected coplanar waveguide microwave phase shifter", *J. Appl. Phys.*, 105, 114509 (2009).

[11] L. Y. Chen, R. Forse, A. H. Cardona, T.C. Watson, and R. A. York, "Compact Analog Phase Shifters using Thin-Film (Ba,Sr)TiO_3 Varactors", *IEEE/MTT-S International Microwave Symposium*, v1-6, pp.666, Honolulu, HI, (2007).

[12] P. Wang, C. Y. Tan, Y. G. Ma, W. N. Cheng, C. K. Ong, "Planar tunable high-temperature superconductor microwave broadband phase shifter with patterned ferroelectric thin film", *Supercond. Sci. Technol.*, 20, 77 (2007).

[13] D. Kim, Y. Choi, M. Ahn, M.G. Allen, J.S. Kenney, and P. Marry, "*2.4 GHz Continuously Variable Ferroelectric Phase Shifters Using All-Pass Networks*", *IEEE Microwave Wireless Comp. Lett.*, 13, 434 (2003).

[14] P. M. Suherman, T. J. Jackson, Y. Y. Tse, I. P. Jones, R. I. Chakalova, M. J. Lancaster, and A. Porch, "Microwave properties of $Ba_{0.5}Sr_{0.5}TiO_3$ thin film coplanar phase shifters", *J. Appl. Phys.*, 99, 104101 (2006).

[15] G. Vélu, K. Blary, L. Burgnies, J. C. Carru, E. Delos, A. Marteau, D. Lippens, "A 310°/3.6-dB K-band phase shifter using paraelectric BST thin films", *IEEE Microwave Wireless Comp Lett.*, 16, 87 (2006).

[16] F. D. Flaviis, N. G. Alexopoulos and O. M. Stafsudd, "Planar microwave integrated phase shifter design with high purity ferroelectric material", *IEEE Trans. on Microwave Theory and Tech.*, 45(6), 963 (1997).

[17] A. Kozyrev, V. Osadchy, A. Pavlov, and L. Sengupta, "Application of Ferroelectrics in Phase Shifter Design", *IEEE MTT-S Int. Microwave Symp. Dig.*, 3, 1355 (2000).

[18] Y. Liu, A. S. Nagra, E. G. Erker, P. Periaswamy, T. R. Taylor, J. Speck, and R. A. York, "$BaSrTiO_3$ Interdigitated Capacitors for Distributed Phase Shifter Applications," *IEEE Microwave and Guided Wave Lett.*, 10(11), 448 (2000).

[19] S. J. Lee, S. E. Moon, M. H. Kwak, H. C. Ryu, Y. T. Kim, and K. Y. Kang, "High Dielectric Tunability of (Ba,Sr)TiO_3 Thin Films and Their Coplanar Waveguide Phase Shifter Application," *Jpn. J. Appl. Phys.*, 43, 6750 (2004).

[20] D. S. Kim, Y. S. Choi, M. G. Allen, J. S. Kenny, and D. Kiesling, "A Wide Bandwith Monolithic BST Reflection Type Phase Shifter Using a Coplanar Waveguide Lange Coupler," *IEEE MTT-S International Microwave Symposium Digest*, 3, 1471 (2002).

[21] O. G. Vendik, M. A. Nilol'ski, and S. P. Zubho, "Optimization of Reflection Type Ferroelectric Phase Shifter with Respect to Sensitivity to Statistical Dispersion of Circuit Characteristics at Microwaves," *Integ. Ferroelectrics*, 49, 141 (2002).

[22] V. O. Sherman, T. Yamada, A. Noeth, N. Setter, M. Mandeljc, B. Malic, M. Kosec, and M. Vukadinovic, "Microwave Phase Shifters Based on Sol-Gel Derived $Ba_{0.3}Sr_{0.7}TiO_3$ Ferroelectric Thin Films", *37th European Microwave Conference*, Munich, p1295 (2007).

[23] T. Ji, H. Yoon, J.K. Abraham, and V.K. Varadan, "Ku-band antenna array feed distribution network with ferroelectric phase shifters on silicon", *IEEE Trans. Microwave Theory Tech.*, 54, 1131 (2006).

[24] P. M. Suherman, T. J. Jackson, Y. Y. Tse, I. P. Jones, R. I. Chakalova, M. J. Lancaster, and A. Porch, "Microwave properties of $Ba_{0.5}Sr_{0.5}TiO_3$ thin film coplanar phase shifters", *J. Appl. Phys.*, 99,104101 (2006).

[25] A. Yu, C. Yang, H. Chen and J. Zhang, "Distributed ferroelectric phase shifter with slots periodically inserted into ground", *Electronics Letters*, 44(3), 207 (2008).

[26] I. Paul, K. J. Vinoy, "Design optimization of micromachined ferroelectric based phase shifter using modified ground coplanar waveguide", *Microwave and Optical Technology Letters*, 46(2), 185 (2005).

[27] W. F. Hu, D. Zhang, M. J. Lancaster, T. W. Button, and B. Su, "Investigation of Ferroelectric Thick-Film Varactors for Microwave Phase Shifters", *IEEE Trans. Microw. Theory Tech.*, 55, 418 (2007).

[28] C. P. Wen, "Coplanar Waveguide: A Surface Strip Transmission Line Suitable for Nonreciprocal Gyromagnetic Device Applications", *IEEE Trans. Microwave Theory Tech.*, 17, 1087 (1969).

[29] R. N. Simons, "Coplanar Waveguide Circuits, Components, and Systems", *Wiley*, New York (2001).

[30] I. Wolff, "Coplanar Microwave Integrated Circuits", *Wiley-Interscience*, New Jersey (2006).

[31] M. J. W. Rodwell, S. T. Allen, R. Y. Yu, M. G. Case, U. Bhattacharya, M. Reddy, E. Carman, M. Kamegawa, Y. Konishi, J. Pusl, R. Pullela, and J. Esch, "Active and nonlinear wave propagation devices in ultrafast electronics and optoelectronics", *Proc. IEEE*, 82, 1035 (1994).

In: Ferroelectrics: New Research
Editors: Xui Li Huang and Su Lyn Ma
ISBN: 978-1-61942-282-7

Chapter 12

NOVEL FERROELECTRIC SOLAR CELLS BASED ON BIFEO

Jingzhong Xiao[1, 2, *], José. A. Paixiao[1], Maria M. R. Costa[1], Lan Xu[1] and Dunlu Sun[3]

[1]Department of Physics, University of Coimbra, Rua Larga, Coimbra, Portugal

[2]International Centre for Materials Physics, Chinese Academy of Sciences, Shenyang, China

[3]Anhui Institude of Optics and Fine Mechanics, Chinese Academy of Sciences, Hefei, China

ABSTRACT

The study in the ferroelectric photovoltaic (PV) effect has been greatly achieved and becomes a key direction in fields of condensed physics and materials sciences. The interest in developing ferroelectric PV devices with specific properties has drastically increased due to the rapidly growing needs in solar energy. In this field, advances in ferroelectric PV cells based on $BiFeO_3$ (BFO) play an important role, and the extended research performed around the world in order to optimize the properties of the ferroelectrics is required. This paper focuses on common and important $BiFeO_3$ ferroelectric materials used for PV effect, and presents a review of the latest developments.

Keywords: ferroelectrics, photovoltaics, solar cells.

Nowadays the global energy problem is rapidly intensifying, especially, the recent nuclear-energy safety crisis in Japanese nuclear power facilities induced by an earthquake, have led to urgent demands for renewable, clean and safe energy, such as solar energy that obtained via solar cells. The finite nature of fossil fuels also have led to increased interest in

---

* E-email: xiao@pollux.fis.uc.pt.

the direct conversion of sunlight into electric power by solar cells, which has emerged as one of the most sustainable and environmentally friendly approaches for solving the energy source issues. [1]

Solar cells, based on the photovoltaic effect (photo-for light, and voltaic-for voltage) shown in Figure 1, are materials that absorb sunlight and convert it into electricity, in which the amount of electricity produced varies with the amount of light and intensity. In conventional Si semiconductor solar cells or photovoltaic (PV) cells, photons with energy higher than the bandgap are absorbed to produce electron-hole pairs, which are separated in the p–n junction. [2] The photogenerated electron–hole pair is separated by the electric field in the depletion region; the electrons and holes move out of this region under the acceleration induced by the electric field. The open-circuit voltage (*V*oc) in this case is limited by the bandgap (*E*g) of the semiconductor.

Howver, the conventional silicon-based solar cells exhibit disadvantages of high-cost, complicated processing requirements, property instability. To solve this problem, there appear currently many research activities, which can be divided into three areas: making current solar cells cheaper and/or more efficient to effectively compete with other energy sources; developing new technologies based on new solar cell architectural designs; and developing new materials to serve as light absorbers and charge carriers.

Since several decades ago, a unique photovoltaic effect had been discoveried in some ferroelectrics.[3] In oxide perovskite ferroelectrics possessing spontaneous polarization, a steady photovoltaic voltage and current can exist in a homogeneous medium under uniform illumination, to produce such a ferroelectric PV effect, making ferroelectrics to be a new materials to develop solar cells.[3-8] This PV effect being fundamentally different from that based on Si exhibits fascinating advantages such as extremely large photovoltage (several orders of magnitude greater than the bandgap and can reach 1000-100000 V), a photocurrent proportional to the polarization magnitude, and charge-carrier separation in domain walls, etc.

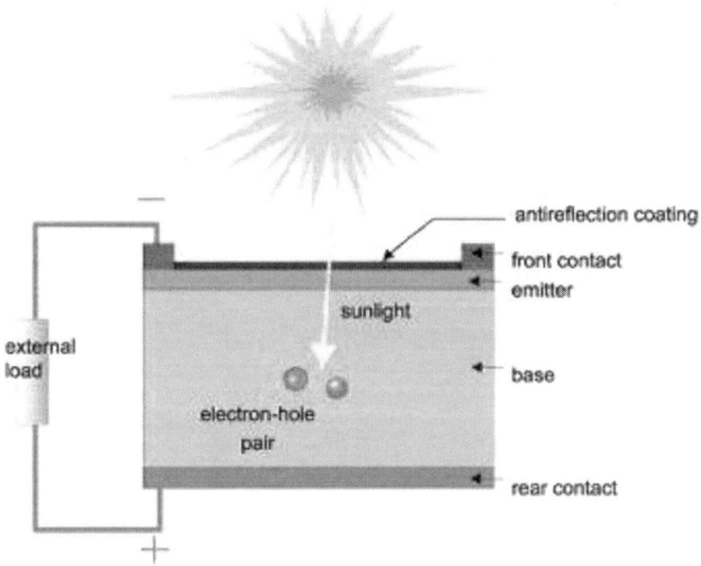

Figure 1 the basic work principle of Si-based solar cells.

The anomalous photovoltaic effect in ferroelectrics can be defined as both the steady-state shortcircuit photovoltaic current and the anomalous opencircuit photovoltage (PV) V>>Eg. The observed transient photocurrents in a ferroelectric may be related to the influence of nonequilibrium photoexcited carriers (such as impurity centers) on the spontaneous polarization P0. Previously, the ferroelectric PV effect is proposed to be associated with the photorefractive effect and a change in the spontaneous polarization due to illumination. However, the nature of the photovoltaic effect in ferroelectrics is far from fully understood and only the first steps have been taken in developing the theory of the PV effect. A detailed and systematic account of the experimental investigations and a comparison with the proposed possible mechanisms on the assumption had been described. [3]

Considering measurements of the photovoltages in ferroelectrics under open-circuit conditions (PV effect), photovoltages of V >> Eg magnitude had been previously observed in an investigation of photoelectric phenomena in ferroelectric SbSIBr and BaTiO3 single crystals, [9, 10] as well as the PV effect was discovered in lithium niobate and several other ferroelectric niobates. [11 25] As sumarized in Ref.3, there was no controversy in the early experimental works on the PV effect. A steady-state photovoltaic current generated by uniform illumination of ferroelectrics without an external field was reliably. This photovoltaic current indicated that the the charging of a crystal were associated with the anomalous photovoltages in ferroelectrics, which are several orders of magnitude greater than the band gap. New carrier transport mechanisms allowing for the explanation of the photovoltaic current should be hypothesised based on the characteristics of the excitation, recombination, and scattering in a pyroelectric crystal. There is no doubt that studies of the PV effect, will make an important contribution to the physics of ferroelectricity, which is of intrinsic interest.

However, the ferroelectric photovoltaics had been negnected for a long time since most oxide ferroelectrics have a bandgap of at least 3 eV, absorbing primarily in the ultraviolet (UV) region, which comprises only 8% of the solar spectrum.[4] Very recently, a big step forward in this research had been achieved in BiFeO3 ferroelectrics with a small bandgap (2.67 eV) within the visible range of 2.5-2.9 eV, in which a substantial visible-light photovoltaic effect was observed.[5-8] BiFeO3 exhibits a large remnant ferroelectric polarization and a high light absorption coefficient. The photovoltage observed in BiFeO3 nanoscale thin films was significantly higher than its electronic bandgap. [5] Despite nowadays the development of PV cells based on silicon and other functional materials has been greatly achieved, the use of ABO3 ferroelectric PV cells seems to provide another potential pathway for further boosting conversion efficiency and lowing cost, and may lead to a new future trende for solar-energy research.[6]

Ferroelectric BiFeO3 (BFO) exhibits a rhombohedral R3c structure with a perovskite pseudocubic unit cell ($a \approx 3.96$ Å, $\alpha \approx 89.4^{\circ}$) and the polarization elongate along the [111] direction below critical temperature (T_c) ≈ 1100 K, and contains transition metal ions with unpaired d electrons. Their d electrons can result in a relatively small optical gap and give rise to a high concentration of charged impurities and defects. The electronic transport properties and a diode effect associated with the direction of bulk electric polarization had been recently discovered in BiFeO3 crystals with a single ferroelectric domain.[8] For typical p-n junction diodes, the current density follows an exponential relationship of $J \propto \exp(qV/\alpha kBT)$, where q is the electron charge, V is voltage, α is a constant called the ideality factor, and kB is the Boltzmann constant. There existed substantial currents that were nonlinear with applied electric field E and also depended strongly on the direction of E in in BFO crystal specimens. The magnitude of E here is much less than ferroelectric coercivity, so polarization switching does not occur during the E sweep for curves of current density (J) versus E. But at 300 and 350 K, BFO shows linear-scale J(E) curves, which exhibit diodelike behavior, from which the diode directions switch whenever ferroelectric polarization is reversed by external pulses, and the diode forward direction is always along the ferroelectric polarization direction. The observed diodelike behavior of dc conduction implies a possibility of zero-bias PV effect, and in BFO, a substantial PV effect with semitransparent Au electrodes illuminated by visible light with a total power density being low than $20mW/cm^2$ also had been found, showing that visible light induces substantial photocurrent.

When a ferroelectric is under intense light illumination, the second-order optical responses combined with a linear term induce an asymmetric polarization, which can result in an optical rectification effect, such as a PV effect. This effect is supposed to be maximal when the polarized-light electric field is along the ferroelectric polarization and follows a sinusoidal angular dependence.

This second-order optical response is an intrinsic bulk effect. On the otherhand, a noticeable variation in the magnitude of the rectification and PV effects in different samples, suggesting the importance of impurities and defects with respect to the transport mechanism.

This investigation of switchable diode and photovoltaic effects in BFO reveals unusual and intriguing charge conduction behavior in leaky ferroelectrics, and should open advance studies of BFO-based multifunctional devices.[8] Although, the bulk photovoltaic effect in ferroelectric BFO materials have been mainly studied in research activities, the underlying mechanism for such an effect was not very clear. It was generally believed that the effect was associated with the loss of inversion symmetry in the distribution of defects, impurities, space charges and polarizations in ferroelectrics.

Recently, a very high photovoltaic voltage originated from the domain walls of the ferroelectric material was observed in BFO thin films by Yang et al.[8] BFO films with thicknesses between 100 and 500 nm were grown on single-crystalline (110) DyScO3 (DSO) substrates by metal–organic chemical vapour deposition (MOCVD).

Through careful design of the experiments by using BiFeO3 thin films of varying thicknesses and domain types, they convinced unambiguously that the photovoltaic effect in BiFeO3 was originated from the 71°and/or 109° domain walls, owing to the asymmetry in polarization, instead of bulk photovoltaic effect. Figure 2 shows the band diagram in a ferroelectric solar cell, which is different from that in a conventional semiconductor solar cell. [6]

It is well known the biggest difference between the photovoltaic effect in a ferroelectric material and that in a conventional semiconductor p-n junction is the magnitude of the electric field that separates the photogenerated electron-hole pair; the effective electric field in a ferroelectric material is around one order of magnitude higher than in a p-n junction. The domain wall in BFO acts as the functions as p-n junction in a classical p-n junction. Since the polarization divergence at the domain walls is non-zero (that is, $P \neq 0$) and the imbalance charges acumulates at the domain walls, inducing an electrostatic potential offset varing from 0.02 eV to 0.20 eV in various materials, which causes a band bending across the domain wall, similar to the band bending in the depletion layer of a typical p-n junction. Therefore, the stronger electric field and efficient charge separation occurs at domain wall (as thin as 2 nm) in films. While in a domain in bulk $BiFeO_3$, the photoinduced charge carriers could be

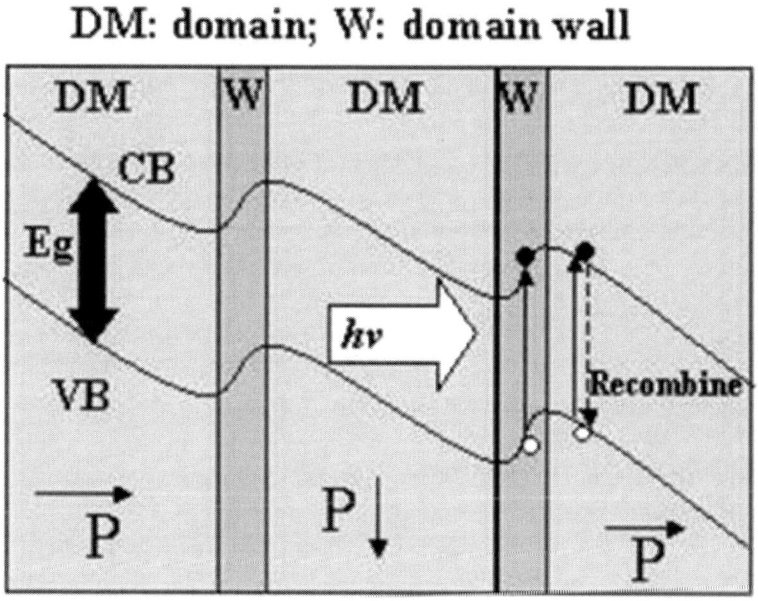

Figure 2 the principles of operation and band diagram for a ferroelectric solar cell, the band bending occurs across the domain walls (W)[6].

localized and tightly bound, allowing them to quickly recombine before their diffusing to either side of a domain wall. This may explain why the bulk photovoltaic effect in $BiFeO_3$ is negligibly small.

On the otherhand, the domain wall photovoltaic effect can be deduced form the fact that the open-circuit voltage increases linearly with the number of domain walls, which implies a high voltage output in $BiFeO_3$ can be obtained from well arranged multiple domain walls. In contrast, a single-domain sample cannot generate appreciable change in open-circuit voltage with increasing sample thickness.

However, the power conversion efficiency of ferroelectric solar cell based on BFO thin film is still poor and much lower than that of commercially solar cells. The main reason is that the leaky ferroelectric domains in BFO cannot withstand strong electric polarization and it is quite difficult to increase the bulk conductivity of ferroelectric domains, then difficult to achieve a high conversion efficiency. Furthermore, the polarization-associated charge density across a domain wall in leaky ferroelectrics will be reduced for lowering the open-circuit voltage.

Although $BiFeO_3$ (BFO), as a multiferroic material, in which the ferroelectric, ferromagnetic and ferroelastic orderings interact with each other, can provide novel opportunity to tune the output signal of a solar cell made of these ferroelectric materials by various mechanical, electrical and magnetic means. Yao *et al* had already carried out some preliminary attractive experiments in applying an electric field higher than the coercive field in the manipulation of ferroelectric domain structures by an external field, to turn on and off the photovoltaic voltage. [7] They prepared BFO thin films by RF magnetron sputtering and deposited the films on STO (001) substrates, in which appreciable photovoltaic response in the visible waves could be observed. The as-deposited BFO films are self-polarized and produce large open-circuit photovoltage up to 0.3V (18kVcm/1). The large portion of the photovoltage and photocurrent (about two thirds) is switchable in response to the switching of the ferroelectric polarization, with the direction of the photocurrent opposite to that of the polarization vector, which is the main driving force of the observed photovoltaic effect. Due to their low cost and simple fabrication process, polycrystalline BFO (poly-BFO) films seem more suitable for practical applications rather than other BFO epitaxial films, although the photovoltaic efficiency of the polycrystalline films may be smaller than that of epitaxial films. Thus It is necessary to study and improve the photovoltaic effect in polycrystalline ferroelectric films. Very recently, poly-BFO thin films on Pt/Ti/SiO_2/Si substrates by the sol–gel method had been prepared, the effect of the top electrodes on the photovoltaic response also were studied. [26] It is found that the oxide electrodes could provide a larger photovoltaic response, and a Sn doped In_2O_3 (ITO)/poly-BFO/Pt capacitor has almost 25 times larger photovoltaic response than that of a Au/Poly-BFO/Pt capacitor. Moreover, an almost 1000-fold photo-conductivity enhancement is obtained in a ITO/poly-BFO/Pt capacitor. These experimental results reveal that a combined system of narrow-band-gap ferroelectric BFO and oxide ITO electrodes can give a considerably high photovoltaic response. In conclusion, the photovoltaic effect in ferroelectrics is deeply correlated to the physics of ferroelectricity and has potential application in solar cells or photovoltaic (PV) cells for the exploiture of solar energy to satisfy the increasing demands for clean and sustainable powers. In Si photovoltaics, electron-hole pairs are created by light absorption and separated by the electric field across a depletion region, with a maximum photovoltage equal to the electronic bandgap. Comparatively, ABO_3 ferroelectrics provide an another pathway to accomplish light obsorption and charge separation for the PV generation, with many unique features such as extremely large photovoltage higher than the bandgap, and photocurrent proportional to the polarization magnitude. But most ferroelectrics have a bandgap larger than 3 eV that absorb very little of solar spectra, causing the ferroelectric photovoltaics to have been overlooked by solar energy researchers for decades. The recently development of solar cells on $BiFeO_3$, with a large polarization and a direct bandgap of 2.67 eV (465 nm) within the visible range, has prompted renaissance in research on ferroelectric photovoltaics. Under white-light illumination, $BiFeO_3$ ferroelectrics (especially thin films) exhibit a domain wall photovoltaic effect and novel photovoltage generation significantly higher than the electronic bandgap. This discovery seems to be encouraged for increasing significantly the energy conversion efficiency in ferroelectric PV cells. However, $BiFeO_3$ ferroelectrics exhibit a high coercive field and high leakage current, leading to an intrinsically low conductivity of ferroelectric domains and deteriorating the polarization performance. In addition, the physical origin and mechanism of the PV effect in $BiFeO_3$ ferroelectrics remains unclear and controversial. These could be the major obstacles in achieving high power conversion efficiency in $BiFeO_3$ ferroelectric PV cells, and the recently observed efficiency in such ferroelectric PV cells is still poor. To solve these peoblem is one of the most significance tasks in the future development of ferroelectric solar cells.

REFERENCES

[1] Ginley, D., Green, M. A., Collins, R. (2008) Solar energy conversion toward 1 terawatt. *Mater. Res. Soc. Bull.* 2008; 33: 355-364.

[2] Gur, I., Fromer, N. A., Geier, M. L., Alivisatos, A. P. (2005) Air-stable all-inorganic nanocrystal solar cells processed from solution. *Science* 2005; 310: 462-465.
[3] Fridkin, V. M., Popov, B. N. (1978) Anomalous photovoltaic effect in ferroelectrics. Usp. *Fiz. Nauk* 1978;126**:** 657-671.
[4] Joseph, W. B., Ilya G., Andrew M. R. (2008) New Highly Polar Semiconductor Ferroelectrics through *d* Cation-O Vacancy Substitution into PbTiO3: A Theoretical Study. *J. AM. CHEM. SOC.* 2008; 130: 17409–17412
[5] Yang,S. Y., Seidel, J., Byrnes, S. J., Shafer, P.C., Yang, H., Rossell, M. D., Yu, P., Chu, Y.-H., Scott, J. F., Ager, J.W., Martin, L. W., Ramesh, R. (2010) Above-bandgap voltages from ferroelectric photovoltaic devices. *NATURE NANOTECHNOLOGY* 2010; 5: 143-147.
[6] Huang H. (2010) Ferroelectric photovoltaics.*Nature Photonics* 2010; 4:143-145.
[7] Ji, W., Yao, K., Liang, Y. C. (2010) Bulk Photovoltaic Effect at Visible Wavelength in Epitaxial Ferroelectric BiFeO3 Thin Films. *Advanced materials* 2010; 22: 1763-1766.
[8] Lee, T. C S., Choi, Y. J., Kiryukhin, V., Cheong S.-W. (2009) Switchable Ferroelectric Diode and Photovoltaic Effect in BiFeO3. *Science* 2009; 324: 63-66.
[9] Grekov A. A., Malitskaya M. A., Spitsyna V. D., and Friedkin V. M., *Kristallografiyu* 1970; 15: 500.
[10] Volk T. R., Grekov A. A., Kosonogov N. A., and Malitskaya M., *Sov. Phys. Solid State* 1972;14: 3214.
[11] Glass A. M., Von der Linde D., Auston D. H., and Negran T. J. J. *Electron. Mater.* 1975; 4: 915.
[12] Ionov P. V., *Sov. Phys. Solid State* 1973; 15: 2827.
[13] Glass A. M. and Von der Linde D., *Ferroelectrics* 1976; 10: 163.
[14] Giinter P. and Micheron F., *Ferroelectrics* 1978; 18: 27.
[15] Micheron. F, *Ferroelectrics* 1978; 18: 153.
[16] Glass M., Von der Linde D., and Negran T. J., *Appl. Lett.* 1974; 25: 233.
[17] Fridkin V. M., Popov B.N., and Ionov P. V., Ferroelectrics 1978; 18: 165.
[18] Popov B.N., Fridkin V. M., *Sov. Phys. Solid State* 1978; 20: 413.
[19] Chen F. S., *J. Appl. Phys.* 1969; 40: 3389.
[20] Ohmori Y., Yamaguchi M., Yoshino K., and Inuishl Y., Jpn. *J. Appl. Phys.* 1976; 15: 2263.
[21] Simon M, Wevering St, Buse K and Kr atzig E, J. Phys. D: *Appl. Phys.* 1997; 30: 144.
[22] Kang B., Rhee B. K.u, Joo G., Lee S., Lim K.S., *Optics Communications* 2006; 266:203.
[23] Gabrielyan V. T.; Lebedeva E. L.; PirozerskiA. L.; Normatov S. A., *Ferroelectric* 2002; 281:151.
[24] Cabafies A, G, and Cabrera J. M. J. Phys : *Condens Matter*, 1993, 5: 2267.
[25] Nakamura M., Takekawa S., Liu Y., Kumaraguruharan S., Moorthy B., Hideki H., Kenji K., *Optical Materials* 2008; 31:280.
[26] Chen B., Li M., Liu Y., Zuo Z., Fei Z., Zhan Q. and Li R., *Nanotechnology* 2011; 22:195201.

INDEX

C

D

E

F

G

H

I

J

K

L

M

N

O

P

Q

R

S

T